Federated Learning for Future Intelligent Wireless Networks

Federated Learning for Future Intelligent Wireless Networks

Edited by

Yao Sun
University of Glasgow, UK

Chaoqun You
Singapore University of Technology and Design, Singapore

Gang Feng
University of Electronic Science and Technology of China, China

Lei Zhang
University of Glasgow, UK

IEEE PRESS
WILEY

Published by John Wiley & Sons, Inc., Hoboken, New Jersey.
Published simultaneously in Canada.

For general information on our other products and services or for technical support, please contact our Customer Care Department within the United States at (800) 762-2974, outside the United States at (317) 572-3993 or fax (317) 572-4002.

Wiley also publishes its books in a variety of electronic formats. Some content that appears in print may not be available in electronic formats. For more information about Wiley products, visit our web site at www.wiley.com.

Library of Congress Cataloging-in-Publication Data Applied for:

[Hardback ISBN: 9781119913894]

Cover Design: Wiley
Cover Image: © Blue Planet Studio/Shutterstock

Set in 9.5/12.5pt STIXTwoText by Straive, Chennai, India

Contents

9   Robust Federated Learning with Real-World Noisy Data   *215*

Jingyi Xu, Zihan Chen, Tony Q. S. Quek, and Kai Fong Ernest Chong

About the Editors

Yao Sun is currently a lecturer with James Watt School of Engineering, the University of Glasgow, Glasgow, United Kingdom. He has extensive research experience and has published widely in wireless networking research. He has won the IEEE Communication Society of TAOS Best Paper Award in 2019 ICC, IEEE IoT Journal Best Paper Award in 2022, and Best Paper Award in 22nd ICCT. He has been the guest editor for special issues of several international journals. He has served as TPC Chair for UCET 2021 and TPC member for a number of international flagship conferences, including ICC 2022, VTC Spring 2022, GLOBECOM 2020, and WCNC 2019. His research interests include intelligent wireless networking, semantic communications, blockchain systems, and resource management in next-generation mobile networks. He is a senior member of IEEE.

Chaoqun You is a postdoctoral research fellow at the Singapore University of Technology and Design (SUTD). She received the BS degree in communication engineering and the PhD degree in communication and information systems from the University of Electronic Science and Technology of China (UESTC) in 2013 and 2020, respectively. She was a visiting student at the University of Toronto from 2015 to 2017. Her current research interests include mobile edge computing, network virtualization, O-RAN, federated learning, and 6G.

Gang Feng received his BEng and MEng degrees in electronic engineering from the University of Electronic Science and Technology of China (UESTC) in 1986 and 1989, respectively, and the PhD degree in Information Engineering from the Chinese University of Hong Kong in 1998. At present, he is a professor with the National Key Laboratory of Wireless Communications, UESTC of China. His research interests include resource management in wireless networks, next-generation cellular networks, etc. Dr. Feng is a senior member of IEEE.

Lei Zhang is a Professor of Trustworthy Systems at the University of Glasgow. He has combined academia and industry research experience on wireless communications and networks and distributed systems for IoT, blockchain, and autonomous systems. His 20 patents have been granted/filed in 30+ countries/regions. He published 3 books and 150+ papers in peer-reviewed journals, conferences, and edited books. He received the IEEE Internet of Things Journal Best Paper Award 2022, IEEE ComSoc TAOS Technical Committee Best Paper Award 2019, and IEEE ICEICT'21 Best Paper Award.

Preface

It has been considered one of the key missing components in the existing 5G network and is widely recognized as one of the most sought-after functions for next-generation 6G communication systems. Nowadays, there are more than 10 billion Internet-of-Things (IoT) equipment and 5 billion smartphones that are equipped with artificial intelligence (AI)-empowered computing modules such as AI chips and GPU. On the one hand, the user equipment (UE) can be potentially deployed as computing nodes to process certain emerging service tasks such as crowdsensing tasks and collaborative tasks, which paves the way for applying AI in edge networks. On the other hand, in the paradigm of machine learning (ML), the powerful computing capability on these UEs can decouple ML from acquiring, storing, and training data in data centers as conventional methods.

Federated learning (FL) has been widely acknowledged as one of the most essential enablers to bring network edge intelligence into reality, as it can enable collaborative training of ML models while enhancing individual user privacy and data security. Empowered by the growing computing capabilities of UEs, FL trains ML models locally on each device where the raw data never leaves the device. Specifically, FL uses an iterative approach that requires a number of global iterations to achieve a global model accuracy. In each global iteration, UEs take a number of local iterations up to a local model accuracy. As a result, the implementation of FL at edge networks can also decrease the costs of transmitting raw data, relieve the burden on backbone networks, reduce the latency for real-time decisions.

This book would explore recent advances in the theory and practice of FL, especially when it is applied to wireless communication systems. In detail, the book covers the following aspects:
1) principles and fundamentals of FL;
2) performance analysis of FL in wireless communication systems;

3) how future wireless networks (say 6G networks) enable FL as well as how FL frameworks/algorithms can be optimized when applying to wireless networks (6G);
4) FL applications to vertical industries and some typical communication scenarios.

Chapter 1 investigates the optimization design of FL in the edge network. First, an optimization problem is formulated to manage the trade-off between model accuracy and training cost. Second, a joint optimization algorithm is designed to optimize the model compression, sample selection, and user selection strategies, which can approach a stationary optimal solution in a computationally efficient way. Finally, the performance of the proposed optimization scheme is evaluated by numerical simulation and experiment results, which show that both the accuracy loss and the cost of FL in the edge network can be reduced significantly by employing the proposed algorithm.

Chapter 2 studies non-IID data model for FL, derives a theoretical upper bound, and redesigns the federated averaging scheme to reduce the weight difference. To further mitigate the impact of non-IID data, a data-sharing scheme is designed to jointly minimize the accuracy loss, the energy consumption, and latency with constrained resource of edge systems. Then a computation-efficient algorithm is proposed to approach the optimal solution and provide the experiment results to evaluate our proposed schemes.

Chapter 3 theoretically analyzes the performance and cost of running FL, which is imperative to deeply understand the relationship between FL performance and multiple-dimensional resources. In this chapter, we construct an analytical model to investigate the relationship between the FL model accuracy and consumed resources in FL-enabled wireless edge networks. Based on the analytical model, we explicitly quantify the model accuracy, computing resources, and communication resources. Numerical results validate the effectiveness of our theoretical modeling and analysis and demonstrate the trade-off between the communication and computing resources for achieving a certain model accuracy.

Chapter 4 proposes an efficient device association scheme for radio access network (RAN) slicing by exploiting a federated reinforcement learning framework, with the aim to improve network throughput, while guaranteeing user privacy and data security. Specially, we use deep reinforcement learning to train local models on UEs under a hybrid FL framework, where horizontally FL is employed for parameter aggregation on BS, while vertically FL is employed for access selection aggregation on the encrypted party. Numerical results show that our proposed scheme can achieve significant performance gains in terms of network throughput and communication efficiency in comparison with some known state-of-the-art solutions.

Chapter 5 proposes a deep FL algorithm that utilizes knowledge distillation and differential privacy to safeguard privacy during the data fusion process. Our approach involves adding Gaussian noise at different stages of knowledge distillation-based FL to ensure privacy protection. Our experimental results demonstrate that this strategy provides better privacy preservation while achieving high-precision IoT data fusion.

Chapter 6 presents a novel systematic beam control scheme to tackle the formulated beam management problem, which is difficult due to the nonconvex objective function. The double deep Q-network (DDQN) under a FL framework is employed to solve the above optimization problem, thereby fulfilling adaptive and intelligent beam management in mmwave networks. In the proposed beam management scheme based on federated learning (BMFL), the non-raw-data aggregation can theoretically protect user privacy while reducing handoff costs. Moreover, a data cleaning technique is used before the local model training, with the aim to further strengthen the privacy protection while improving the learning convergence speed. Simulation results demonstrate the performance gain of the proposed BMFL scheme.

Chapter 7 proposes a double-layer blockchain-based deep reinforcement federated learning (BDRFL) scheme to ensure privacy-preserved and caching-efficient D2D networks. In BDRFL, a double-layer blockchain is utilized to further enhance data security. Simulation results first verify the convergence of BDRFL-based algorithm and then demonstrate that the download latency of the BDRFL-based caching scheme can be significantly reduced under different types of attacks when compared with some existing caching policies.

Chapter 8 aims to design a dynamic scheduling policy to explore the spectrum flexibility for heterogeneous federated edge learning (FEEL) so as to facilitate the distributed intelligence in edge networks. This chapter proposes a heterogeneity-aware dynamic scheduling problem to minimize the global loss function, with consideration of straggler and limited device energy issues. By solving the formulated problem, we propose a dynamic scheduling algorithm (DISCO), to make an intelligent decision on the set and order of scheduled devices in each communication round. Theoretical analysis reveals that under certain conditions, learning performance and energy constraints can be guaranteed in the DISCO. Finally, we demonstrate the superiority of the DISCO through numerical and experimental results, respectively.

Chapter 9 discusses FedCorr, a general multistage framework to tackle heterogeneous label noise in FL, which does not make any assumptions on the noise models of local clients while still maintaining client data privacy. Both theoretical analysis and experiment results demonstrate the performance gain of this novel FL framework.

Chapter 10 provides a general overview of the analog over-the-air federated learning (AirFL) system. Specially, we illustrate the general system architecture and highlight the salient feature of AirFL that adopts analog transmissions for fast (but noisy) aggregation of intermediate parameters. Then, we establish a new convergence analysis framework that takes into account the effects of fading and interference noise. Our analysis unveils the impacts from the intrinsic properties of wireless transmissions on the convergence performance of AirFL. The theoretical findings are corroborated by extensive simulations.

Chapter 11 investigates a FEEL-based training framework to DL-based channel state information (CSI) feedback. In FEEL, each UE trains an autoencoder network locally and exchanges model parameters via the base station. Therefore, data privacy is better protected compared with centralized learning because the local CSI datasets are not required to be uploaded. Neural network parameter quantization is then introduced to the FEEL-based training framework to reduce communication overhead. The simulation results indicate that the proposed FEEL-based training framework can achieve comparable performance with centralized learning.

Chapter 12 proposes a user-centric online training strategy in which the UE can collect CSI samples in the stable area and adjust the pretrained encoder online to further improve CSI reconstruction accuracy. Moreover, the proposed online training framework is extended to the multiuser scenario to improve performance sequentially. The key idea is to adopt decentralized FL without BS participation to combine the sharing of channel knowledge among UEs, which is called crowd intelligence. Simulation results show that the decentralized FL-aided framework has higher feedback accuracy than the AE without online training.

November 2023

Yao Sun
Chaoqun You
Gang Feng
Lei Zhang

1

Federated Learning with Unreliable Transmission in Mobile Edge Computing Systems

Chenyuan Feng[1], Daquan Feng[1], Zhongyuan Zhao[2], Howard H. Yang[3], and Tony Q. S. Quek[4]

[1] Shenzhen Key Laboratory of Digital Creative Technology, The Guangdong Province Engineering Laboratory for Digital Creative Technology, The Guangdong-Hong Kong Joint Laboratory for Big Data Imaging and Communication, College of Electronics and Information Engineering, Shenzhen University, Shenzhen, Guangdong, China
[2] State Key Laboratory of Networking and Switching Technology, School of Information and Communication Engineering, Beijing University of Posts and Telecommunications, Beijing, China
[3] Zhejiang University/University of Illinois at Urbana-Champaign Institute, Zhejiang University, The College of Information Science and Electronic Engineering, Haining, Zhejiang, China
[4] Information Systems Technology and Design Pillar, Singapore University of Technology and Design, Singapore

1.1 System Model

Consider the deployment of FL in an MEC scenario, which consists of an edge access point E and multiple users $U_1, \ldots, U_M$. An edge computing server S_E is equipped with E, while a local computing unit S_m is equipped with U_m, $m = 1, \ldots, M$. As shown in Figure 1.1, the edge computing server S_E and local computing units $S_1, \ldots, S_M$ can act as the computing server and the clients, respectively, which can interact with each other via the wireless channels between E and $U_1, \ldots, U_M$.

As introduced previously, FL can be implemented by updating the local models and the global model iteratively. In particular, we focus on the tth iteration, which can be introduced as follows.

1.1.1 Local Model Training

In this phase, each user updates the local model independently based on its local collected data. Without loss of generality, we focus on a specific user U_m, the local model of U_m can be updated as follows:

$$\mathbf{w}_{t,m} = \mathbf{w}_{t-1,m} - \eta_t \nabla F(\mathbf{w}_{t-1,m}, D_{t,m}), \tag{1.1}$$

Federated Learning for Future Intelligent Wireless Networks, First Edition.
Edited by Yao Sun, Chaoqun You, Gang Feng, and Lei Zhang.
© 2024 The Institute of Electrical and Electronics Engineers, Inc. Published 2024 by John Wiley & Sons, Inc.

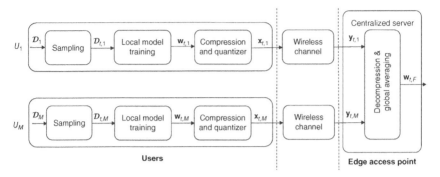

Figure 1.1 The system model of wireless FL.

where $\mathbf{w}_{t,m}$ and $\mathbf{w}_{t-1,m}$ denote the update results of U_m's local model during the tth and $(t-1)$-th iterations, respectively, $\mathcal{D}_{t,m}$ denotes the training dataset for updating $\mathbf{w}_{t,m}$, which is randomly selected from $\mathcal{D}_m$, $\mathcal{D}_{t,m} \subseteq \mathcal{D}_m$, $\mathcal{D}_m$ denotes the local dataset located at U_m, η_t is the learning rate of the tth iteration, and $\nabla F(\mathbf{w}_{t-1,m}, \mathcal{D}_{t,m})$ is the gradient of loss function $F(\mathbf{w}_{t-1,m}, \mathcal{D}_{t,m})$ with respect to $\mathbf{w}_{t-1,m}$. In this chapter, the loss function is defined as the empirical risk with respect to $\mathbf{w}_{t,m}$, which can be defined as follows:

$$F(\mathbf{w}_{t,m}, \mathcal{D}_{t,m}) = \frac{1}{N_{t,m}} \sum_{\mathbf{x} \in D_{t,m}} l(\mathbf{w}_{t,m}; \mathbf{x}), \tag{1.2}$$

where $l(\mathbf{w}_{t,m}; \mathbf{x})$ denotes the loss function of the data element $\mathbf{x}$, and $N_{t,m}$ denotes the size of $\mathcal{D}_{t,m}$.

1.1.2 Update Result Feedback via the Wireless Channels

When the local model training procedure is accomplished, U_m should transmit its update result $\mathbf{w}_{t,m}$ to E via the wireless channel. In the existing works, the server randomly selects the users since it is assumed that the communications between the computing server and the clients are ideal. However, it cannot be ensured in the MEC systems due to the unreliable wireless transmission circumstances, which will cause accuracy loss of FL models. Therefore, in this chapter, only the users with high communication reliability and low model accuracy loss are scheduled to participate in each iteration of global model averaging. In particular, the scheduling status of U_m for the tth iteration of global averaging is characterized by a Boolean variable $z_{t,m}$, i.e.,

$$z_{t,m} = \begin{cases} 1, & \text{if } U_m \text{ is scheduled} \\ 0, & \text{if } U_m \text{ is not scheduled.} \end{cases} \tag{1.3}$$

If U_m is scheduled, its update result $\mathbf{w}_{t,m}$ can be modeled as a $d \times 1$ vector, which is usually with high dimension, especially for the deep neural network models. Therefore, to improve the efficiency of update result feedback, **model sparsification** and **parameter quantization** techniques should be employed to compress $\mathbf{w}_{t,m}$. As introduced previously, $\mathbf{w}_{t,m}$ can be transformed into a sparse form via model sparsification, which can be expressed as follows:

$$\mathbf{s}_{t,m} = \mathbf{A}_{t,m}\mathbf{w}_{t,m}, \tag{1.4}$$

where $\mathbf{A}_{t,m}$ denotes a $d \times d$ sparsification matrix for $\mathbf{w}_{t,m}$.

Next, each element of $\mathbf{s}_{t,m}$ is quantized independently by employing uniform quantization. The quantization error can be approximated as an additive Gaussian noise, which is independent with $\mathbf{w}_{t,m}$. Then the quantized parameter vector can be expressed as

$$\mathbf{x}_{t,m} = \mathbf{s}_{t,m} + \mathbf{q}_{t,m} = \mathbf{A}_{t,m}\mathbf{w}_{t,m} + \mathbf{q}_{t,m}, \tag{1.5}$$

where $\mathbf{q}_{t,m}$ denotes a $d \times 1$ quantization noise vector, i.e., $\mathbf{q}_{t,m} \sim \mathcal{CN}(\mathbf{0}, \mathbf{\Omega}_{t,m})$, and $\mathbf{\Omega}_{t,m}$ denotes the covariance matrix. Due to the implementation of independent quantization, each element of $\mathbf{q}_{t,m}$ is independent with each other, i.e., $\mathbb{E}\{q_i^{t,m}(q_i^{t,m})^H\} = \omega_i^{t,m}$, $\mathbb{E}\{q_i^{t,m}(q_j^{t,m})^H\} = 0$, $i \neq j$, where $q_i^{t,m}$ and $q_j^{t,m}$ denote the ith and jth elements of $\mathbf{q}_{t,m}$, respectively. Therefore, $\mathbf{\Omega}_{t,m}$ is a diagonal matrix, which can be denoted as $\mathbf{\Omega}_{t,m} = \mathrm{diag}\{q_1^{t,m}, \ldots, q_d^{t,m}\}$.

After model sparsification and parameter quantization, $\mathbf{x}_{t,m}$ is suitable for baseband processing and wireless transmissions. In this chapter, the flat fading model is employed to characterize the wireless channels between U_m and E. Therefore, the channel fading can be assumed to be unchanged during the transmission of $\mathbf{x}_{t,m}$. Then the observation of $\mathbf{x}_{t,m}$ at E can be expressed as

$$\mathbf{y}_{t,m} = h_{t,m}\mathbf{x}_{t,m} + \mathbf{n}_{t,m}, \tag{1.6}$$

where $h_{t,m}$ captures the flat channel fading of the wireless link between U_m and E, $\mathbf{n}_{t,m}$ denotes the additive white Gaussian noise at E, i.e., $\mathbf{n}_{t,m} \sim \mathcal{CN}(\mathbf{0}, \sigma_t^2 \mathbb{I}_L)$, $\mathbb{I}_L$ denotes a $d \times d$ identity matrix, and σ_t^2 is the power of noise.

1.1.3 Global Model Averaging

To recover the update results of local models, $\mathbf{y}_{t,m}$ should be first decompressed by E. In this chapter, the minimum mean-square error (MMSE) criterion is employed, and the decompression result can be written as

$$\overline{\mathbf{w}}_{t,m} = \arg \min_{\mathbf{w}^q \in C} \| \mathbf{D}_{t,m}\mathbf{y}_{t,m} - \mathbf{w}^q \|^2, \tag{1.7}$$

where $\mathbf{D}_{t,m}$ is a $d \times d$ decompression matrix of $\mathbf{y}_{t,m}$, C denotes a set that consists of all the possible quantized parameter vectors, i.e., $\mathbf{w}^q \in C$. Since each element of the quantized model parameter vector can be detected individually, recalling the computational complexity of MMSE, the complexity of this detection is a linear function of the vector dimension.[1]

Then the global model can be updated by averaging the decompressed results of local models. As introduced in Konecný et al. [2016], the update result of global model can be expressed as

$$
\begin{aligned}
\mathbf{w}_{t,F} &= \mathbf{w}_{t-1,F} + \sum_{m=1}^{M} \frac{N_m}{N} z_{t,m} \left(\overline{\mathbf{w}}_{t,m} - \mathbf{w}_{t-1,F} \right) \\
&= \sum_{m=1}^{M} \frac{N_m}{N} \left[(1 - z_{t,m}) \mathbf{w}_{t-1,F} + z_{t,m} \overline{\mathbf{w}}_{t,m} \right],
\end{aligned} \tag{1.8}
$$

where $\mathbf{w}_{t,F}$ and $\mathbf{w}_{t-1,F}$ denote the global model for the tth and $(t-1)$-th iterations, respectively, $z_{t,m}$ is defined by (1.3), $N = \sum_{m=1}^{M} N_m$.

After global model averaging, $\mathbf{w}_{t,F}$ is sent back to the users. Since $\mathbf{w}_{t,F}$ are transmitted via downlink transmissions, which can acquire more radio resource and higher transmit power than the local model update phase. Therefore, it can be assumed that $\mathbf{w}_{t,F}$ is received successfully by all the users. Then the local model of each user can be updated as $\mathbf{w}_{t,m} = \mathbf{w}_{t,F}, m = 1, \dots, M$.

1.2 Problem Formulation

The performance of existing learning techniques is mainly determined by the accuracy of generated learning models. It is difficult to be modeled in a tractable form, and thus cannot be optimized by employing the existing resource management schemes in the MEC systems. In this section, we first derived a closed-form upper bound of model accuracy loss, which is an efficient metric to evaluate the quality of FL models. Then an optimization problem is formulated to improve the model accuracy and training efficiency of FL with limited budget of computation and communication resources.

1.2.1 Model Accuracy Loss

As introduced in Konecný et al. [2016], the objective of model training is to minimize the expected risk, which can be estimated by employing the the empirical risk given by (1.2). Therefore, the model accuracy loss, which can be defined as the

---

1 To implement (1.7) at edge server, it requires to feedback sparsification matrix $\mathbf{A}_{t,m}$, and this communication overhead can be reduced significantly when $\mathbf{w}_{t,m} \in \mathbb{R}^d$ is divided into multiple segments. Moreover, the detection can be done with channel estimation, each device can feedback $\mathbf{D}_{t,m}$ instead of channel state information.

expected mean square error (MSE) with respect to the the empirical risk and the expected risk, is an efficient metric to evaluate the accuracy performance of generated models. In this chapter, the model accuracy loss after t iterations of global averaging can be expressed as

$$\mathcal{L}_t = \mathbb{E}\left\{\| F\left(\mathbf{w}_{t,F}\right) - G(\mathbf{w}^*) \|^2\right\}, \text{ where } F(\mathbf{w}_{t,F}) = \sum_{m=1}^{M} \frac{N_m}{N} F(\mathbf{w}_{t,F}, D_{u_m})$$

(1.9)

where $F(\mathbf{w}_{t,F})$ denotes the empirical risk with respect to $\mathbf{w}_{t,F}$ for the tth iteration, $G(\mathbf{w}^*)$ denotes the minimum expected risk of our studied FL model, i.e., $G(\mathbf{w}^*) = \min_{\mathbf{w}} \mathbb{E}\{l(\mathbf{w}, \mathbf{x})\}$, and $\mathbf{w}^*$ denotes the theoretically optimal parameter vector.

As introduced in Shai et al. [2010], to guarantee the convergence of FL, the following features with respect to the empirical risk should be satisfied:

1. $F(\mathbf{w}, D)$ is L-Lipchitz, i.e., for any $\mathbf{w}_1$ and $\mathbf{w}_2$, we have

   $$\| F(\mathbf{w}_1, D) - F(\mathbf{w}_2, D) \| \leqslant L \| \mathbf{w}_1 - \mathbf{w}_2 \| .$$

2. $F(\mathbf{w}, D)$ is λ-strongly convex. i.e., for any $\mathbf{w}_1$ and $\mathbf{w}_2$, and $0 \leqslant \alpha \leqslant 1$, the following inequality can be established:

   $$F\left(\alpha\mathbf{w}_1 + (1-\alpha)\mathbf{w}_2, D\right) \leqslant \alpha F\left(\mathbf{w}_1, D\right) + (1-\alpha)F\left(\mathbf{w}_2, D\right) - \frac{\lambda}{2}\alpha(1-\alpha) \| \mathbf{w}_1 - \mathbf{w}_2 \|^2.$$

By substituting (1.8) into (1.9), an upper bound of $\mathcal{L}_t$ can be expressed as follows due to the convexity of $F(\mathbf{w}, D)$:

$$
\begin{aligned}
\mathcal{L}_t &= \mathbb{E}\left\{\left\| F\left(\sum_{m=1}^{M} \frac{N_m}{N}\left[(1 - z_{t,m})\mathbf{w}_{t-1,F} + z_{t,m}\overline{\mathbf{w}}_{t,m}\right]\right) - G(\mathbf{w}^*) \right\|^2\right\} \\
&\leqslant \mathbb{E}\left\{\left\| \sum_{m=1}^{M} \frac{N_m}{N}\left[(1 - z_{t,m})F(\mathbf{w}_{t-1,F}) + z_{t,m}F(\overline{\mathbf{w}}_{t,m}) - G(\mathbf{w}^*)\right] \right\|^2\right\} \\
&\leqslant \sum_{m=1}^{M} \frac{2N_m}{N}\left[(1 - z_{t,m})\underbrace{\mathbb{E}\left\{\| F(\mathbf{w}_{t-1,F}) - G(\mathbf{w}^*)\|^2\right\}}_{\mathcal{L}_{t-1}}\right. \\
&\qquad \left. + z_{t,m}\underbrace{\mathbb{E}\left\{\| F(\overline{\mathbf{w}}_{t,m}) - G(\mathbf{w}^*)\|^2\right\}}_{\epsilon_{t,m}}\right].
\end{aligned}
$$

(1.10)

In the paradigms of FL, each iteration can be optimized independently. Without loss of generality, we focus on the optimization of the tth iteration in this chapter, and thus $\mathcal{L}_{t-1}$ in (1.10) can be treated as a constant. Therefore, $\mathcal{L}_t$ is mainly decided by $\epsilon_{t,m}$, which can be derived as follows:

$$\epsilon_{t,m} = \mathbb{E}\left\{\|\,(F(\overline{\mathbf{w}}_{t,m}) - F(\mathbf{w}_{t,m})) + (F(\mathbf{w}_{t,m}) - G(\mathbf{w}_{t,m})) + (G(\mathbf{w}_{t,m}) - G(\mathbf{w}^*))\|^2\right\}$$

$$\leqslant 3 \left(\underbrace{\mathbb{E}\left\{\| F(\overline{\mathbf{w}}_{t,m}) - F(\mathbf{w}_{t,m})\|^2\right\}}_{\epsilon_{t,m}^C} + \underbrace{\mathbb{E}\left\{\| F(\mathbf{w}_{t,m}) - G(\mathbf{w}_{t,m})\|^2\right\}}_{\epsilon_{t,m}^S} + \underbrace{\mathbb{E}\left\{\| G(\mathbf{w}_{t,m}) - G(\mathbf{w}^*)\|^2\right\}}_{\epsilon_{t,m}^M} \right), \tag{1.11}$$

where $F(\mathbf{w}_{t,m})$ is given by (1.2), and $G(\mathbf{w}_{t,m})$ denotes the expected risk with respect to $\mathbf{w}_{t,m}$. As shown in (1.11), $\epsilon_{t,m}$ is determined by the communication loss $\epsilon_{t,m}^C$, sample selection loss $\epsilon_{t,m}^S$, and model training loss $\epsilon_{t,m}^T$, which can be derived as follows.

1.2.2 Communication Loss $\epsilon_{t,m}^C$

The communication loss is jointly decided by the model compression loss and the communication error caused by wireless transmissions. Then an upper bound of $\epsilon_{t,m}^C$ can be provided by the following lemma.

Lemma 1.1 *When $F(\mathbf{w}, \mathcal{D})$ is L-Lipchitz, an upper bound of communication loss $\epsilon_{t,m}^C$ in (1.11) can be expressed as*

$$\epsilon_{t,m}^C \leqslant \overline{\epsilon}_{t,m}^C = L^2 \, tr \left(h_{t,m}^2 \mathbf{D}_{t,m} \mathbf{A}_{t,m} \mathbf{A}_{t,m}^H \mathbf{D}_{t,m}^H + a_{t,m} \mathbf{D}_{t,m} \mathbf{D}_{t,m}^H + \mathbf{I}_d \right.$$
$$\left. - (h_{t,m} \mathbf{D}_{t,m} \mathbf{A}_{t,m} + h_{t,m}^H \mathbf{A}_{t,m}^H \mathbf{D}_{t,m}^H) \right), \tag{1.12}$$

where $a_{t,m} = \sigma_{t,m}^2 h_{t,m}^2 + \varrho_t^2$, $\sigma_{t,m}^2$ is the power of quantization noise, and σ_t^2 is the power of Gaussian noise.

1.2.3 Sample Selection Loss $\epsilon_{t,m}^S$

Although the training dataset $\mathcal{D}_{t,m}$ is generated based on the distribution of $\mathbf{x}$, the sample selection bias still exists due to the limited size of $\mathcal{D}_{t,m}$. Since $\mathbf{w}_{t,m}$ is trained based on $\mathcal{D}_{t,m}$, the sample selection bias causes error propagation during the model training procedure, which is characterized as sample selection loss in this chapter.

As shown in (1.11), $\epsilon_{t,m}^{S}$ is defined as the difference between the empirical risk with respect to $D_{t,m}$ and the expected risk with respect to the distribution of data elements. As introduced in Devroye et al. [1996], an upper bound of $\epsilon_{t,m}^{S}$ can be provided as follows.

Remark 1.1 ((Devroye et al. [1996]), Theorem 12.6) When $F(\mathbf{w}, D)$ is convex, a tractable upper bound of $\epsilon_{t,m}^{S}$ can be expressed as follows:

$$\epsilon_{t,m}^{S} \leqslant \bar{\epsilon}_{t,m}^{S} = \frac{128(d \log N_{t,m} + 4)}{N_{t,m}}. \tag{1.13}$$

1.2.4 Model Training Loss $\epsilon_{t,m}^{M}$

As introduced in Section 1.1, $\mathbf{w}_{t,m}$ is generated by employing stochastic gradient descent (SGD) method, which is optimized in an iterative way. Theoretically, it can approach the optimal training result $\mathbf{w}^{*}$ (Rakhlin et al. [2017]). However, there always exists a gap between $\mathbf{w}_{t,m}$ and $\mathbf{w}^{*}$ since only a few iterations can be executed to guarantee the computing efficiency. In this chapter, it is captured as model training loss $\epsilon_{t,m}^{M}$ in (1.11). To evaluate its impact on the model accuracy, the following lemma is provided.

Lemma 1.2 *When $F(\mathbf{w}, D)$ is L-Lipchitz and λ-strongly convex, and $\nabla F(\mathbf{w}, D)$ is with finite scale, i.e., $\mathbb{E}\{\|\nabla F(\mathbf{w}, D)\|^{2}\} \leqslant G^{2}$, an upper bound of model training loss $\epsilon_{t,m}^{M}$ can be written as*

$$\epsilon_{t,m}^{M} \leqslant \bar{\epsilon}_{t,m}^{M} = \frac{4G^{2}L^{2}}{\lambda^{2} N_{t,m}}. \tag{1.14}$$

By substituting (1.11), (1.12), (1.13), and (1.14) into (1.10), an upper bound of $\mathcal{L}_{t}$ can be derived, which is given by the following corollary.

Corollary 1.1 An upper bound of model accuracy loss $\mathcal{L}_{t}$ given by (1.9), which is denoted as $\bar{\mathcal{L}}_{t}$, can be expressed as

$$\bar{\mathcal{L}}_{t} = \sum_{m=1}^{M} \frac{2N_{m}}{N} \left((1 - z_{t,m})\mathcal{L}_{t-1} + 3z_{t,m} \left(L^{2} \, tr \left(h_{t,m}^{2} \mathbf{D}_{t,m} \mathbf{A}_{t,m} \mathbf{A}_{t,m}^{H} \mathbf{D}_{t,m}^{H} \right. \right. \right.$$

$$+ a_{t,m} \mathbf{D}_{t,m} \mathbf{D}_{t,m}^{H} + \mathbf{I}_{d} - h_{t,m} \mathbf{D}_{t,m} \mathbf{A}_{t,m} - h_{t,m}^{H} \mathbf{A}_{t,m}^{H} \mathbf{D}_{t,m}^{H} \Big)$$

$$+ \frac{128(d \log N_{t,m} + 4)}{N_{t,m}} + \frac{4G^{2}L^{2}}{\lambda^{2} N_{t,m}} \bigg) \bigg). \tag{1.15}$$

It is worth mentioning that it is unnecessary to obtain $F(\mathbf{w}_{t,F})$ and $G(\mathbf{w}^*)$ with high computation costs based on the following schemes. First, we can obtain $F(\mathbf{w}_{t-1,F})$ without additional computation overhead since the global loss can be computed as a linear combination of local loss $F(\mathbf{w}_{t,F}, D_{u_m})$ with respect to each participant, and the local loss can be obtained by U_m via local model training. Second, the true value of $G(\mathbf{w}^*)$ is known for some specific learning tasks, such as the supervised multiclassification tasks. As introduced in McMahan et al. [2017], since all image data can be labeled correctly when the global optimal parameter setting $\mathbf{w}^*$ can be obtained, the corresponding minimum expected risk can be set as $G(\mathbf{w}^*) = 0$ for the supervised multiclassification tasks. Moreover, when the true value of $G(\mathbf{w}^*)$ is unknown, zero can be treated as a lower bound of the expected risk function, and we also have $F(\mathbf{w}_{t-1,F}) \geqslant G(\mathbf{w}^*)$ since $\mathbf{w}^*$ is global optimal solution. By far, an upper bound of $\mathcal{L}_{t-1}$ can be derived as $\mathcal{L}_{t-1} = \mathbb{E}\left\{\| F(\mathbf{w}_{t-1,F}) - G(\mathbf{w}^*) \|^2\right\} \leqslant \| F(\mathbf{w}_{t-1,F}) \|^2 = \overline{\mathcal{L}}_{t-1}$, and $\overline{\mathcal{L}}_{t-1}$ still can capture the relationship between the model accuracy loss and the selection strategies of participants and data samples, which can be substituted into (1.15) instead of $\mathcal{L}_{t-1}$.

1.2.5 Problem Formulation

1.2.5.1 Objective Function

Since the resource of MEC systems is restricted, a sophisticated trade-off between the model accuracy and cost should be kept to guarantee that a high-quality model can be generated in a cost-efficient way. In this chapter, two categories of costs are considered, which are the computation and communication costs, respectively.

Each data element is processed by the same learning model, which means the computation cost is proportional to the size of employed training dataset. In accordance with (Zhao et al. [2020], Chen et al. [2021]), we model the computation cost as a linear function of $N_{t,m}$. The compression and communication cost is modeled as the mutual information between $\mathbf{x}_{t,m}$ and $\overline{\mathbf{w}}_{t,m}$ since the mutual information can characterize the capacity of wireless link for a compressed signal to be successfully recovered (Zhang et al. [2007]). This standard rate-distortion theoretic metric (El Gamal and Kim [2011]) in the analysis of compression design can provide useful insight to deploy FL with successful compression schemes. Then the model cost of FL can be expressed as

$$
C_t = \sum_{m=1}^{M} \left\{ \beta_1 N_{t,m} + \beta_2 \left[\log \det \left(\mathbf{A}_{t,m} \mathbf{A}_{t,m}^H + \mathbf{\Omega}_{t,m} \right) - \log \det \left(\mathbf{\Omega}_{t,m} \right) \right] \right\} z_{t,m},
$$

(1.16)

where β_1 and β_2 denote the weights of computation and communication costs, respectively.

In this chapter, our target is to jointly minimize the model accuracy loss and the costs of FL, and the objective function can be modeled as

$$Q_t = w_1 \overline{\mathcal{L}}_t + w_2 C_t, \tag{1.17}$$

where w_1 and w_2 denote the weights of $\overline{\mathcal{L}}_t$ and C_t, respectively.

1.2.5.2 Energy Consumption Constraint

Due to the limitation of battery volume, the energy consumption of each user should be restricted independently. During the tth iteration, the user energy consumption is mainly caused by local model training and update result feedback. In particular, the energy consumption of local model training grows linearly with respect to the size of training dataset $\mathcal{D}_{t,m}$, and the transmit energy consumption can be derived based on (1.5). Then the energy consumption constraint can be expressed as follows:

$$\mathrm{tr}\left(\mathbf{A}_{t,m}\mathbf{A}_{t,m}^H + \mathbf{\Omega}_{t,m}\right) + \alpha N_{t,m} \leqslant P_m, \tag{1.18}$$

where α denotes the coefficient of energy consumption caused by processing a single data element in the local model training phase, and P_m is the maximum energy consumption for the tth iteration of FL.

1.2.5.3 User Selection Constraint

As introduced in Section 1.1, only a part of users are selected to feedback their update results. To guarantee the convergence performance and computing efficiency, the maximum number of selected users in the tth iteration is set as K_t, $K_t \leqslant M$. Recalling (1.3), the following constraint of user selection is established:

$$\sum_{m=1}^{M} z_{t,m} \leqslant K_t. \tag{1.19}$$

1.2.5.4 Data Volume Constraint of Local Training Datasets

By employing sophisticated sample selection strategies, the impact of sample selection loss can be mitigated significantly. In particular, the local training data-set $\mathcal{D}_{t,m}$ is selected randomly from $\mathcal{D}_m$, and thus its data volume should follow the constraint

$$0 \leqslant N_{t,m} \leqslant N_m, N_{t,m} \in \mathbb{N}, \tag{1.20}$$

where N_m denotes the data volume of $\mathcal{D}_m$, and $\mathbb{N}$ denotes the set of nonnegative integers. In this chapter, we focus on the minimization of model accuracy loss and cost of FL in the MEC systems, which can be captured by the upper bound given by (1.15). Therefore, the optimization problem can be formulated as follows:

$$\min_{\mathbf{A}_m, \mathbf{D}_m, z_{t,m}, N_{t,m}} \quad Q_t \text{ in (1.17), } s.t. \text{ (1.18), (1.19), (1.20).} \tag{1.21}$$

1.3 A Joint Optimization Algorithm

The established optimization problem given by (1.21) is a nonlinear and non-convex problem. In this section, to obtain a tractable solution, it is first decoupled as three independent subproblems, and an iterative optimization algorithm is designed. Then we prove that the proposed algorithm can approach a stationary optimal solution.

1.3.1 Compression Optimization

As shown in (1.15) and (1.16), $\mathbf{A}_{t,m}$ and $\mathbf{D}_{t,m}$ only determine the accuracy loss and cost of communications, and thus an equivalent form of the objective function can be expressed as follows by removing the terms that are not related with $\mathbf{A}_{t,m}$ and $\mathbf{D}_{t,m}$:

$$\overline{Q}_1 = \sum_{m=1}^{M} \overline{Q}_{1,m} = \sum_{m=1}^{M} \left(w_1 z_{t,m} \frac{6N_m}{N} L^2 G_{1,m} + w_2 z_{t,m} \beta_2 G_{2,m} \right), \tag{1.22}$$

where

$$G_{1,m} = \operatorname{tr} \left(h_{t,m}^2 \mathbf{D}_{t,m} \mathbf{A}_{t,m} \mathbf{A}_{t,m}^H \mathbf{D}_{t,m}^H + a_{t,m} \mathbf{D}_{t,m} \mathbf{D}_{t,m}^H \right.$$
$$\left. - (h_{t,m} \mathbf{D}_{t,m} \mathbf{A}_{t,m} + h_{t,m}^H \mathbf{A}_{t,m}^H \mathbf{D}_{t,m}^H) \right), \tag{1.23}$$

$$G_{2,m} = \log \det (\mathbf{A}_{t,m} \mathbf{A}_{t,m}^H + \mathbf{\Omega}_{t,m}). \tag{1.24}$$

Equation (1.22) indicates that the compression of each user can be optimized independently by minimizing $\overline{Q}_{1,m}$, which is restricted by an individual energy consumption constraint. Therefore, the optimization of compression and decompression matrices can be transformed into M-dependent subproblems, which can be expressed as

$$\min_{\mathbf{A}_{t,m}, \mathbf{D}_{t,m}} \overline{Q}_{1,m} \text{ in (1.22)}$$

$$s.t. \ G_{3,m} = \operatorname{tr} \left(\mathbf{A}_{t,m} \mathbf{A}_{t,m}^H + \mathbf{\Omega}_{t,m} \right) - (P_m - \alpha N_{t,m}) \leqslant 0. \tag{1.25}$$

We first consider solving $\mathbf{A}_{t,m}$ and $\mathbf{D}_{t,m}$ independently, and then propose Algorithm 1.1 to optimize them jointly.

1.3.1.1 Optimization of $\mathbf{A}_{t,m}$

The optimization problem of $\mathbf{A}_{t,m}$ is identical with (1.25) when $\mathbf{D}_{t,m}$ is fixed. First, we verify the convexity of $G_{1,m}$, $G_{2,m}$, and $G_{3,m}$, which can be provided by the following lemma. Since $\overline{Q}_{1,m}$ $G_{1,m}$, $G_{2,m}$, and $G_{3,m}$ can be treated as functions with respect to $\mathbf{A}_{t,m}$, they are denoted as $\overline{Q}_{1,m}(\mathbf{A}_{t,m})$, $G_{1,m}(\mathbf{A}_{t,m})$, $G_{2,m}(\mathbf{A}_{t,m})$, and $G_{3,m}(\mathbf{A}_{t,m})$ in this part, respectively.

Algorithm 1.1 Optimization of $\mathbf{A}_{t,m}$ and $\mathbf{D}_{t,m}$

1: **Initialization**: $\bar{\mathbf{A}}_{t,m}^{(0)}$, the maximum number of iterations K_1 and K_A, and the convergence threshold ξ_1 and ξ_A.

2:

3: **Repeat**: For $k_a = 0$ to K_A

4: *Step 1. Optimize* $\mathbf{A}_{t,m}$.

  - **Repeat**: For $k_1 = 0$ to K_1
    - Update $\bar{\mathbf{A}}_{t,m}^{(k_1)}$ by solving (1.27).
  - **Termination**: When $k_1 > K_1$ or $\|\bar{\mathbf{A}}_{t,m}^{(k_1-1)} - \bar{\mathbf{A}}_{t,m}^{(k_1)}\| \leqslant \xi_1$.
    - Return $\mathbf{A}_{t,m}^{(k_a)} = \bar{\mathbf{A}}_{t,m}^{(k_1)}$.

  Step 2. Optimize $\mathbf{D}_{t,m}$.

  - Calculate $\mathbf{D}_{t,m}^{(k_a)}$ based on (1.30).

  Step 3. Update $\bar{Q}_1^{(k_a)}$ *based on* (1.22).

5: **Termination**: When $k_a > K_A$ or $\|\bar{Q}_1^{(k_a)} - \bar{Q}_1^{(k_a-1)}\| \leqslant \xi_A$.

  - Return $\mathbf{A}_{t,m}^{\text{opt}} = \mathbf{A}_{t,m}^{(k_a)}$ and $\mathbf{D}_{t,m}^{\text{opt}} = \mathbf{D}_{t,m}^{(k_a)}$.

Lemma 1.3 $G_{1,m}(\mathbf{A}_{t,m})$ *and* $G_{3,m}(\mathbf{A}_{t,m})$ *given by* (1.23) *are both convex, while* $G_{2,m}(\mathbf{A}_{t,m})$ *given by* (1.24) *is concave.*

Lemma 1.3 shows that (1.25) is a convex–concave procedure problem, where both the objective function and constraints can be treated as a difference of two convex functions (Stephen and Lieven [2004]). It can be solved by using majorization minimization algorithm. The key idea is to replace the concave terms, i.e., $G_{2,m}(\mathbf{A}_{t,m})$ in (1.25), by using successive convex approximation, which can transform the original optimization problem into a convex form. As shown in Algorithm 1.1, a stationary solution of (1.25) can be obtained in an iterative approach. Without loss of generality, we focus on the k_1th iteration. By using first-order Taylor expansion, $G_{2,m}(\mathbf{A}_{t,m})$ can be approximated in a convex form, which can be expressed as follows:

$$
\begin{aligned}
\overline{G}_{2,m}&(\mathbf{A}_{t,m}^{(k_1)}; \overline{\mathbf{A}}_{t,m}^{(k_1-1)}) \\
&= \overline{G}_{2,m}(\overline{\mathbf{A}}_{t,m}^{(k_1-1)}) + \nabla_{\mathbf{A}_{t,m}} \overline{G}_{2,m}(\overline{\mathbf{A}}_m^{(k_1-1)})^H (\mathbf{A}_{t,m}^{(k_1)} - \overline{\mathbf{A}}_{t,m}^{(k_1-1)}) \\
&= \text{tr}((\overline{\mathbf{A}}_{t,m}^{(k_1-1)})^H (\overline{\mathbf{A}}_{t,m}^{(k_1-1)} (\overline{\mathbf{A}}_{t,m}^{(k_1-1)})^H + \Omega_{t,m})^{-1} (\mathbf{A}_{t,m}^{(k_1)} - \overline{\mathbf{A}}_{t,m}^{(k_1-1)})) \\
&\quad + \log \det (\overline{\mathbf{A}}_{t,m}^{(k_1-1)} (\overline{\mathbf{A}}_{t,m}^{(k_1-1)})^H + \Omega_{t,m}),
\end{aligned} \tag{1.26}
$$

where $\mathbf{A}_{t,m}^{(k_1)}$ denotes the compression matrix of the k_1th iteration, and $\overline{\mathbf{A}}_{t,m}^{(k_1-1)}$ is update result of the $(k_1 - 1)$-th iteration. Based on (1.26), an approximated problem of (1.25) can be established as follows:

$$\min_{\mathbf{A}_{t,m}^{(k_1)}} \overline{Q}_{1,m}^{\text{app}}(\mathbf{A}_{t,m}^{(k_1)}) = w_1 z_{t,m} \frac{6N_m}{N} L^2 G_{1,m}(\mathbf{A}_{t,m}^{(k_1)}) + w_2 z_{t,m} \beta_2 \overline{G}_{2,m}(\mathbf{A}_{t,m}^{(k_1)}; \overline{\mathbf{A}}_{t,m}^{(k_1-1)})$$

$$\text{s.t. } G_{3,m}(\mathbf{A}_{t,m}^{(k_1)}) \leqslant 0. \tag{1.27}$$

The approximated problem given by (1.27) is a convex problem satisfying the KKT condition, which can be solved efficiently by using the optimization package (Stephen and Lieven [2004]).

In Algorithm 1.1, $\mathbf{A}_{t,m}$ can be optimized iteratively, which can be updated by solving (1.27). To verify the convergence of *Step 1* in Algorithm 1.1, the following theorem with respect to the descent of $\overline{Q}_{1,m}$ is provided.

Theorem 1.1 *Denoting $\overline{\mathbf{A}}_{t,m}^{(k_1)}$ as the optimal solution of (1.27) for the k_1th iteration, the following inequality with respect to $\overline{\mathbf{A}}_{t,m}^{(k_1)}$ and $\overline{\mathbf{A}}_{t,m}^{(k_1-1)}$ can be established:*

$$\overline{Q}_{1,m}(\overline{\mathbf{A}}_{t,m}^{(k_1)}) \leqslant \overline{Q}_{1,m}(\overline{\mathbf{A}}_{t,m}^{(k_1-1)}). \tag{1.28}$$

Theorem 1.1 indicates that $\overline{Q}_{1,m}(\overline{\mathbf{A}}_{t,m}^{(k_1)})$ keeps decreasing as k_1 increases. Moreover, a tractable lower bound of $\overline{Q}_{1,m}(\mathbf{A}_{t,m}^{(k_1)})$ is 0. Therefore, *Step 1* of Algorithm 1.1 can converge to a stationary point.

1.3.1.2 Optimization of $\mathbf{D}_{t,m}$
When the compression matrix $\mathbf{A}_{t,m}$ is fixed, the optimization $\mathbf{D}_{t,m}$ is equivalent to solve the following unconstraint problem:

$$\min_{\mathbf{D}_{t,m}} G_{1,m} \text{ in } (1.23). \tag{1.29}$$

To provide a tractable solution of (1.29), the following lemma with respect to the convexity of $G_{1,m}$ is provided.

Lemma 1.4 $G_{1,m}$ *given by (1.23) is a convex function with respect to $\mathbf{D}_{t,m}$.*

Based on Lemma 1.4, the optimal solution of (1.29) can be derived straightforwardly by solving the equation $\nabla_{\mathbf{D}_{t,m}} G_{1,m} = \mathbf{0}$ and can be expressed as

$$\mathbf{D}_{t,m} = \left(h_{t,m} \mathbf{A}_{t,m}\right)^H \left(h_{t,m}^2 \mathbf{A}_{t,m} \mathbf{A}_{t,m}^H + a_{t,m} \mathbf{I}_d\right)^{-1}. \tag{1.30}$$

1.3.2 Joint Optimization of $\mathbf{A}_{t,m}$ and $\mathbf{D}_{t,m}$

As shown in Algorithm 1.1, $\mathbf{A}_{t,m}$ and $\mathbf{D}_{t,m}$ can be updated iteratively. During each iteration, $\mathbf{A}_{t,m}$ can be optimized by solving (1.27), and $\mathbf{D}_{t,m}$ is updated based on (1.30). As introduced previously, $\overline{Q}_1$ given by (1.22) keeps decreasing as k_a increases, since $\overline{\mathbf{A}}_{t,m}^{(k_1)}$ and $\mathbf{D}_{t,m}^{(k_a)}$ are updated based on the optimal solution. Therefore, the convergence of Algorithm 1.1 can be guaranteed.

1.3.3 Optimization of Sample Selection

When we focus on the optimization of sample selection, Q_t given by (1.17) is equivalent to the following function with respect to $N_{t,m}$:

$$\overline{Q}_2 = \sum_{m=1}^{M} \overline{Q}_{2,m} = \sum_{m=1}^{M} \frac{1}{N_{t,m}} \left(r_1 \log N_{t,m} + r_2 + w_2 z_{t,m} \beta_1 N_{t,m}^2 \right), \tag{1.31}$$

where $r_1 = \frac{768 N_m}{N} w_1 z_{t,m} d$, $r_2 = \frac{6 N_m}{N} z_{t,m} w_1 (4 + \frac{4 G^2 L^2}{\lambda^2})$. Moreover, recalling (1.21), the constraints with respect to $N_{t,m}$, which are given by (1.18) and (1.20), can be rewritten as follows:

$$1 \leqslant N_{t,m} \leqslant \overline{N}_m, \text{ and } \overline{N}_m = \min \left\{ N_m, \left\lfloor \frac{1}{\alpha} (P_m - \text{tr}(\mathbf{A}_{t,m}\mathbf{A}_{t,m}^H + \mathbf{\Omega}_{t,m})) \right\rfloor \right\}, \tag{1.32}$$

where $\lfloor \cdot \rfloor$ denotes the floor function with respect to its argument. Similar to (1.25), the sample selection of each user can be optimized independently, and then the optimization problem of sample selection can be established as follows:

$$\min_{N_{t,m}} \overline{Q}_{2,m} \text{ in } (1.31), \text{ s.t. } (1.32). \tag{1.33}$$

To solve (1.33) efficiently, we first consider a relaxed problem, where $N_{t,m}$ can be treated as a continuous variable. It is a fractional programming problem, and its objective function is nonlinear and nonconvex with respect to $N_{t,m}$. Due to Theorem 1 in Zhao et al. [2020], $\overline{Q}_{2,m}$ given by (1.31) can be transformed into a linear form, and its minimum value can be achieved if and only if the following constraint is satisfied:

$$\min_{N_{t,m}} f\left(N_{t,m}\right) - \rho^* N_{t,m} = f\left(N_{t,m}^*\right) - \rho^* N_{t,m}^* = 0, \tag{1.34}$$

where

$$f\left(N_{t,m}\right) = r_1 \log N_{t,m} + r_2 + w_2 \beta_1 N_{t,m}^2, \tag{1.35}$$

ρ^* denotes the minimum value of $\overline{Q}_{2,m}$, and $N_{t,m}^*$ is the corresponding optimal solution. Based on (1.34), the following optimization problem should be studied to obtain the optimal solution of (1.33):

$$\min_{N_{t,m}} \tilde{Q}_{2,m}\left(N_{t,m}\right) = f\left(N_{t,m}\right) - \rho^* N_{t,m}, \ s.t. \ (1.32). \tag{1.36}$$

Since $\tilde{Q}_{2,m}$ is continuous and derivable with respect to $N_{t,m}$, the optimal solution of (1.36) locates at either the stationary point or the boundary point, which can be expressed as

$$N_{t,m}^* = \arg\ \min\ \left\{ \tilde{Q}_{2,m}\left(1\right),\ \tilde{Q}_{2,m}\left(\overline{N}_{t,m}\right),\ \tilde{Q}_{2,m}\left(\tilde{N}_{t,m}\right) \right\}, \tag{1.37}$$

where $\tilde{N}_{t,m}$ denotes the solution of equation $\nabla_{N_{t,m}} \tilde{Q}_{2,m}\left(N_{t,m}\right) = 0$.

Based on (1.34) and (1.37), Algorithm 1.2 can be designed to approach the optimal solution of (1.32) iteratively. In particular, during the k_2th iteration, $N_{t,m}$ is first updated by solving the following problem:

$$\min_{N_{t,m}^{(k_2)}} \tilde{Q}_{2,m}\left(N_{t,m}^{(k_2)}\right) = f\left(N_{t,m}^{(k_2)}\right) - \rho^{(k_2-1)} N_{t,m}^{(k_2)} \ s.t. \ (1.32), \tag{1.38}$$

where $N_{t,m}^{(k_2)}$ denotes the update results of $N_{t,m}$ for the k_2th iteration, and $\rho^{(k_2-1)}$ is the value of $\overline{Q}_{2,m}$ for the $(k_2 - 1)$-th iteration. The optimal solution of (1.38) can be obtained straightforwardly based on (1.37). Then $\rho^{(k_2)}$ can be updated as follows:

$$\rho^{(k_2)} = \frac{f(N_{t,m}^{(k_2)})}{N_{t,m}^{(k_2)}}. \tag{1.39}$$

To evaluate the equivalency of (1.32) and (1.38), $q^{(k_2)}$ should be calculated as follows:

$$q^{(k_2)} = f\left(N_{t,m}^{(k_2)}\right) - \rho^{(k_2-1)} N_{t,m}^{(k_2)}. \tag{1.40}$$

Based on Theorem 3 in Zhao et al. [2020], it can be proved that $q^{(k_2)}$ approaches 0 monotonically as k_2 increases, which means that Algorithm 1.2 can approach the optimal solution of the original fractional optimization problem.

Denoting $N_{t,m}^*$ as the final update result that satisfies the termination constraints, the optimal solution of (1.32) can be expressed as follows by rounding:

$$N_{t,m}^{\text{opt}} = \begin{cases} \lfloor N_{t,m}^* \rfloor, \ \text{if} \ \lceil N_{t,m}^* \rceil > \overline{N}_m, \\ \arg\ \min\ \{\overline{Q}_{2,m}(\lfloor N_{t,m}^* \rfloor), \ Q(\lceil N_{t,m}^* \rceil)\}, \ \text{if} \ \lfloor N_{t,m}^* \rfloor, \ 1 \leqslant \lceil N_{t,m}^* \rceil \leqslant \overline{N}_m. \\ \lceil N_{t,m}^* \rceil, \ \text{if} \ \lfloor N_{t,m}^* \rfloor < 1, \end{cases} \tag{1.41}$$

where $\lceil \cdot \rceil$ is the ceiling function with respect to its argument.

Algorithm 1.2 Optimization of $N_{t,m}$

Initialization: $N_{t,m}^{(0)}$, $\rho^{(0)}$, the maximum number of iterations K_N, and the convergence threshold ξ_N.

Repeat: For $k_2 = 1$ to K_N
- Update $N_{t,m}^{(k_2)}$ by solving (1.38),
- Update $\rho^{(k_2)}$ based on (1.39),
- Update $q^{(k_2)}$ based on (1.40).

Termination: When $k_2 > K_N$ or $\|q^{(k_2)} - q^{(k_2-1)}\| \leqslant \xi_N$.

Return $N_{t,m}^* = N_{t,m}^{(k_2)}$, and obtain $\mathbf{N}_{t,m}^{\text{opt}}$ based on (1.41).

1.3.4 Optimization of User Selection

Based on (1.21), by fixing $\mathbf{A}_{t,m}$, $\mathbf{D}_{t,m}$, and $N_{t,m}$, the subproblem of user selection can be written as follows:

$$\min_{z_{t,m}} Q_t = 2w_1 \mathcal{L}_{t-1} - \sum_{m=1}^{M} v_{t,m} z_{t,m},$$

$$s.t. \ (1.19), \tag{1.42}$$

where

$$v_{t,m} = w_1 \frac{2N_m}{N} \left(\mathcal{L}_{t-1} - 3\epsilon_{t,m}^C - \frac{384(d \log N_{t,m} + 4)}{N_{t,m}} - \frac{12G^2 L^2}{\lambda^2 N_{t,m}} \right)$$
$$- w_2 \left(\beta_1 N_{t,m} + \beta_2 \left(\log \det (\mathbf{A}_{t,m} \mathbf{A}_{t,m}^H + \mathbf{\Omega}_{t,m}) - \log \det (\mathbf{\Omega}_{t,m}) \right) \right). \tag{1.43}$$

$v_{t,m}$ given by (1.43) denotes the utility of U_m for the tth iteration of FL, which in practical is positive by choosing suitable w_1 and w_2. Therefore, (1.42) is a 0-1 backpack problem and can be solved efficiently by using dynamic programming. When $v_{t,m}$ is negative, in order to minimize Q_t, U_m could never be selected, which means $z_{t,m} = 0$. After removing users with $v_{t,m} \leqslant 0$ from set of candidate users, the remaining problem turns into a standard 0-1 backpack problem, which can by solved by dynamic programming.

1.3.5 A Joint Optimization Algorithm

Since all the subproblems can be solved efficiently, a joint optimization algorithm can be proposed by optimizing $\mathbf{A}_{t,m}$, $\mathbf{D}_{t,m}$, $z_{t,m}$, and $N_{t,m}$ iteratively. As shown in Algorithm 1.3, during each iteration, the update results of $\mathbf{A}_{t,m}$ and $\mathbf{D}_{t,m}$ can be

---

Algorithm 1.3 The Joint Optimization Algorithm

Initialization: $A_{t,m}$, $D_{t,m}$, $N_{t,m}$ and $z_{t,m}$, the maximum number of iterations K_J, and the convergence threshold ξ_Q.

Repeat: For $0 \leqslant k_3 \leqslant K_J$
- Update $A_{t,m}^{k_3}$ and $D_{t,m}^{k_3}$ by Algorithm 1;
- Update $N_{t,m}^{k_3}$ by Algorithm 2;
- Update $z_{t,m}^{k_3}$ by solving (1.42);
- Update $Q_t^{k_3}$;

Termination: When $k_3 > K_J$ or $\|Q_t^{k_3} - Q_t^{k_3-1}\| \leqslant \xi_Q$.

---

obtained by employing Algorithm 1.1, $N_{t,m}$ can be updated based on Algorithm 1.2, and $z_{t,m}$ can be obtained by solving (1.42).

As introduced previously, the optimal solution of each subproblem can be obtained, which can guarantee that Q_t given by (1.17) keeps decreasing as the iteration index k_3 increases. Moreover, 0 can be treated as a fixed lower bound of Q_t, and thus Algorithm 1.3 converges to the optimal solution of (1.21) with limited iterations. The final update results can be obtained when it achieves the maximum iteration time, or satisfies the accuracy requirement. Due to the non-convexity of the formulated problem, the proposed iterative solution can only guarantee the convergence to the stationary point for one time. To overcome this defect, in our work, we use random initialization and run multiple simulations. Moreover, our proposed algorithm can be executed in parallel with multiple initialization values to search multiple stationary points individually. The best result selected among these stationary points can approach to the optimal solution. Therefore, we have the following corollary.

1.4 Simulation and Experiment Results

In this section, both the numerical simulation and experiment results are provided to evaluate the performance of our proposed optimization algorithm. Unlike the conventional works that focus on the data center networks, we study the implementation of FL at the edge of wireless networks, where each base station can associate with dozens of users at most. Therefore, we set the number of users is as $M = 20$ and 50, the ratio of maximum number of selected users as $K_t/M = 0.2$. The maximum iteration indexes of Algorithm 1.3 and federated global averaging are set as $K_J = 30$ and $T = 50$, respectively, and the learning rate is $\eta = 0.01$. We consider multiclassification learning task and MNIST dataset (Xiao et al. [2017]), which is a commonly used image dataset. The entire training dataset consists of 60,000 training images and 10,000 testing images, which can be classified into 10 different digits. The following two cases with respect to local training datasets are considered, which are named as balanced and unbalanced dataset cases, respectively: (i) In the balanced dataset case, each client randomly and independently sample from the entire training dataset and $N_{t,m}$ is identically set as 600; (ii) In the unbalanced dataset case, each client independently and randomly sample two different digits from the training dataset and $N_{t,m}$ is set as a random integer variable that follows uniform distribution with a mean value of 600 for fair comparison, namely $N_{t,m} \sim \mathcal{U}[100, 1100]$ and $\mathbb{E}\{N_{t,m}\} = 600$, which means all clients will only have two kinds of labels at most and have different size of local dataset.

In this part, the performance of Algorithm 1.3 is testified practically by employing MNIST dataset. Though the model training loss cannot directly account for the models with nonconvex empirical risk functions in theory, our optimization framework can still provide some insightful results for neural networks. It is verified by our experiment results that our proposed algorithm helps to improve the model performance in both multilayer perceptron (MLP) and convolutional neural network (CNN) case. The MLP and CNN models are employed for local learning model generation, and the detailed model structure is described as follows: (i) A three-layer MLP model is employed, which consists of an input layer with 784 neurons, a hidden layer with 64 neurons, and an output layer based on softmax classifier and (ii) the same CNN model as LeNet-5 in Lecun et al. [1998] is employed which is constructed by seven layers.

In Figure 1.2(a) and 1.2(b), the experiment results of model accuracy are provided, where the MLP and CNN models are employed for local model training, respectively. To evaluate the performance of Algorithm 1.3, the centralized learning scheme with ideal transmission condition is employed as a comparable scheme. To the best of our knowledge, it is the state-of-the-art scheme that can achieve the best accuracy performance. For fair comparison, we set $M = 50$, $K_t = 10$, and $SNR = 10$ dB for all algorithms, and random user selection and fixed sample selection with $N_{t,m} = 100$ are employed for unoptimized FL. For centralized learning, fixed sample selection with $N_t = 1000$ is employed. As shown in Figure 1.2(a), the test accuracy can be improved from 89.37% to 94.04% for the balanced dataset case, while it can be improved from 85.52% to 88.86% for the unbalanced dataset case. The experiment result based on the CNN model shows a similar tendency, which is provided in Figure 1.2(b). It shows that the test accuracy can be improved from 91.28% to 95.75% for the balanced dataset case, while it can be improved from 85.39% to 90.89% for the unbalanced dataset case. Moreover, the performance of FL with balanced data can approach the centralized learning scheme after being optimized by using Algorithm 1.3, which demonstrates that FL can generate high-quality model in a cost-efficient way.

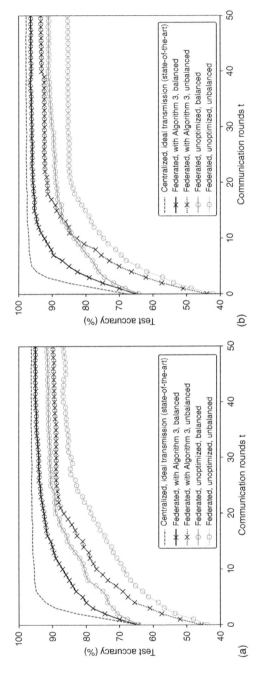

Figure 1.2 The test accuracy of optimized FL with Algorithm 1.3 trained by MLP and CNN. (a) Local training with MLP model. (b) Local training with CNN model.

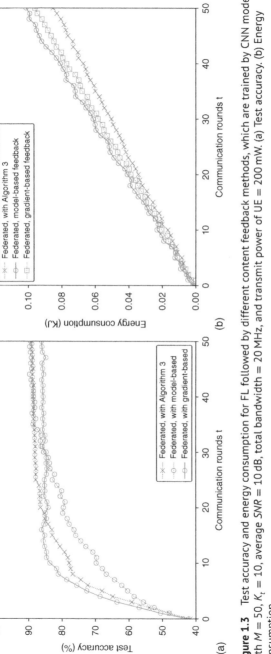

Figure 1.3 Test accuracy and energy consumption for FL followed by different content feedback methods, which are trained by CNN model with $M = 50$, $K_t = 10$, average $SNR = 10$ dB, total bandwidth $= 20$ MHz, and transmit power of UE $= 200$ mW. (a) Test accuracy. (b) Energy consumption.

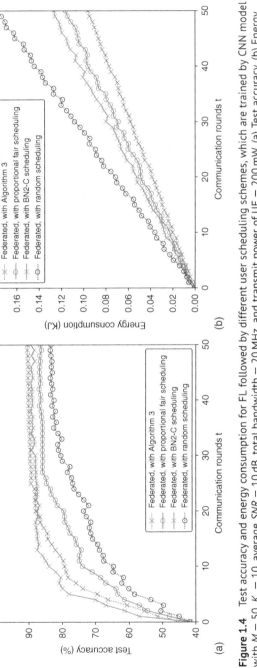

Figure 1.4 Test accuracy and energy consumption for FL followed by different user scheduling schemes, which are trained by CNN model with $M = 50$, $K_t = 10$, average $SNR = 10$ dB, total bandwidth = 20 MHz, and transmit power of UE = 200 mW. (a) Test accuracy. (b) Energy consumption.

To evaluate the impact of user scheduling, we choose conventional random scheduling as baseline and compare proportional fair in Yang et al. [2020], BN2-C in Amiri et al. [2020] with our proposed algorithm. In Yang et al. [2020], the server will select users with K_t largest SNR value, and in Amiri et al. [2020], the server will select users with K_t largest l_2-norm of local model update gradient from the user candidate set with best conditions. First, Figure 1.3 shows that our proposed algorithm causes the smallest energy consumption and obtains the highest test accuracy. It is reasonable because our user selection and sample optimization schemes are jointly designed to minimize model accuracy loss and cost; however, proportional fair, BN2-C mainly focus on improving leaning performance, leaving model cost optimization alone. Second, the BN2-C algorithm converges most quickly and our algorithm gets close to it.

To evaluate the feedback process, we compare the model-based feedback (Chen et al. [2021]) and gradient-based feedback (Ren et al. [2019]) with our proposed algorithm. In Chen et al. [2021], users will feedback the local model without compression, and the server will select users based on communication resource allocation. And in Ren et al. [2019], users will feedback the local model gradient without compression, and the server will select the UEs contributing to larger global loss function decay without considering communication conditions. First, Figure 1.4 shows that our proposed algorithm will cause the smallest energy consumption, and the test accuracy of our algorithm is close to model-based feedback without compression in Chen et al. [2021]. It is reasonable because (i)both our algorithm and Chen et al. [2021] consider communication optimization, however, (Ren et al. [2019]), does not consider the channel state information; (ii) model compression is implemented in our work, which will introduce a little accuracy loss but also reduce the energy consumption, especially for communication cost. Second, the gradient-based feedback algorithm in Ren et al. [2019] converges most quickly and our algorithm gets close to it.

Bibliography

Mohammad Mohammadi Amiri, Deniz Gündüz, Sanjeev R. Kulkarni, and H. Vincent Poor. Update aware device scheduling for federated learning at the wireless edge. In *2020 IEEE International Symposium on Information Theory (ISIT)*, pages 2598–2603, 2020.

Mingzhe Chen, Zhaohui Yang, Walid Saad, Changchuan Yin, H. Vincent Poor, and Shuguang Cui. A joint learning and communications framework for federated learning over wireless networks. *IEEE Transactions on Wireless Communications*, 20(1):269–283, 2021.

Luc Devroye, László Györfi, and Gábor Lugosi. *A probabilistic theory of pattern recognition*. New York: Springer-Verlag, 1996.

Abbas El Gamal and Young-Han Kim. *Network information theory*. Cambridge University Press, 2011.

Jakub Konecný, H. Brendan McMahan, Felix X. Yu, Peter Richtárik, Ananda Theertha Suresh, and Dave Bacon. Federated learning: Strategies for improving communication efficiency. In *Proceedings of the Advances in Neural Information Processing Systems (NeurIPS) Workshop on Private Multi-Party Machine Learning*, pages 5003–5009, Barcelona, Spain, 2016.

Y. Lecun, L. Bottou, Y. Bengio, and P. Haffner. Gradient-based learning applied to document recognition. *Proceedings of the IEEE*, 86(11):2278–2324, 1998.

Brendan McMahan, Eider Moore, Daniel Ramage, Seth Hampson, and Blaise Aguera y Arcas. Communication-efficient learning of deep networks from decentralized data. In *Proceedings of the International Conference on Artificial Intelligence and Statistics (AISTATS)*, pages 1273–1282, Florida, USA, 2017.

Alexander Rakhlin, Ohad Shamir, and Karthik Sridharan. Making gradient descent optimal for strongly convex stochastic optimization. In *Proceedings of the International Conference on International Conference on Machine Learning (ICML)*, pages 1571–1578, Edinburgh, Scotland, 2017.

Jinke Ren, Guanding Yu, and Guangyao Ding. Accelerating DNN training in wireless federated edge learning system. In *arXiv preprint*, 2019. URL https://arxiv.org/abs/1905.09712.

Shalev-Shwartz Shai, Ohad Shamir, Srebro Nathan, and Sridharan Karthik. Learnability, stability and uniform convergence. *Journal of Machine Learning Research*, 11:2635–2670, 2010.

Boyd Stephen and Vandenberghe Lieven. *Convex optimization*. Cambridge, UK: Cambridge University Press, 2004.

Han Xiao, Kashif Rasul, and Roland Vollgraf. Fashion-MNIST: A novel image dataset for benchmarking machine learning algorithms. In *arXiv preprint*, 2017. URL https://arxiv.org/abs/1708.07747.

H. Howard Yang, Zuozhu Liu, Q. S. Tony Quek, and H. Vincent Poor. Scheduling policies for federated learning in wireless networks. *IEEE Transactions on Communications*, 68(1):317–333, 2020.

Xin Zhang, Jun Chen, Stephen B. Wicker, and Toby Berger. Successive coding in multiuser information theory. *IEEE Transactions on Information Theory*, 53(6):2246–2254, 2007.

Zhongyuan Zhao, Shuqing Bu, Tiezhu Zhao, Zhenping Yin, Mugen Peng, Zhiguo Ding, and Tony Q. S. Quek. On the design of computation offloading in fog radio access networks. *IEEE Transactions on Vehicular Technology*, 68(7):7136–7149, 2020.

2

Federated Learning with non-IID data in Mobile Edge Computing Systems

Chenyuan Feng[1], Daquan Feng[1], Zhongyuan Zhao[2], Geyong Min[3], and Hancong Duan[4]

[1] Shenzhen Key Laboratory of Digital Creative Technology, The Guangdong Province Engineering Laboratory for Digital Creative Technology, The Guangdong-Hong Kong Joint Laboratory for Big Data Imaging and Communication, College of Electronics and Information Engineering, Shenzhen University, Shenzhen, Guangdong, China
[2] State Key Laboratory of Networking and Switching Technology, School of Information and Communication Engineering, Beijing University of Posts and Telecommunications, Beijing, China
[3] Department of Computer Science, College of Engineering Mathematics and Physical Sciences, College of Engineering Mathematics and Physical Sciences, Zhejiang University/University of Illinois at Urbana-Champaign Institute, Exeter, UK
[4] Information Systems Technology and Design Pillar, School of Computer Science and Engineering, University of Electronic Science and Technology of China, Sichuan, Chengdu, China

2.1   System Model

In this chapter, we consider the deployment of FL in wireless networks, which can be implemented via the interactions between a central server and multiple clients. In particular, the server S_B is equipped with a base station (BS), and $U_1, \ldots, U_M$ are M different users associated with this BS, which can act as clients participating FL. Each user U_m is equipped with a local processor S_m to enable local data processing and model training. Unlike the conventional centralized learning paradigms, FL is an efficient approach to train high-quality learning models without collecting raw data from the users (McMahan et al. [2017]). It allows the users to train local models based on its own data, and then these local models are aggregated at the server. Without loss of generality, we focus on the tth round interaction, and the following two key steps should be executed:

2.1.1   Local Model Training

Without loss of generality, we focus on a specific user U_m, $m = 1, \ldots, M$. A local training dataset D_m is with U_m, which can be employed for its local model training.

Federated Learning for Future Intelligent Wireless Networks, First Edition.
Edited by Yao Sun, Chaoqun You, Gang Feng, and Lei Zhang.

The update result can be expressed as follows:

$$\mathbf{w}^m_{(t-1)K+k} = \mathbf{w}^m_{(t-1)K+(k-1)} - \eta G(\mathbf{w}^m_{(t-1)K+(k-1)}, \mathbf{x}(B^m_{(t-1)K+k})), \tag{2.1}$$

where $\mathbf{w}^m_{(t-1)K+k}$ and $\mathbf{w}^m_{(t-1)K+(k-1)}$ denote the update results with respect to the kth and the $(k-1)$-th batches of training data at U_m during the tth round interaction, respectively, η denotes the learning rate, $B^m_{(t-1)K+k}$ is the training data batch employed for generating $\mathbf{w}^m_{(t-1)K+k}$, $B^m_{(t-1)K+k} \subset D_m$, and $G(\mathbf{w}, \mathbf{x}(B))$ denotes the gradient of the empirical risk function with respect to $\mathbf{w}$ based on the training data batch B, $\mathbf{x} \in B$. In this chapter, a mini batch-based training strategy is used, and K data batches are used for local model training in each round interaction. The data volume of each batch at U_m is fixed as N_m.

2.1.2 Federated Averaging

In this chapter, a synchronous federated averaging strategy is used. After training with K local data batches, the update result $\mathbf{w}^m_{tK}$ is transmitted to S_B to generate the global model of the tth interaction. By averaging the feedback results from all the users, the global model can be updated as follows:

$$\mathbf{w}^F_t = \sum_{m=1}^M a_m \mathbf{w}^m_{tK}, \tag{2.2}$$

where a_m denotes the average weight of $\mathbf{w}^m_{tK}$, $0 \leqslant a_m \leqslant 1$, and $\sum_{m=1}^M a_m = 1$. Then $\mathbf{w}^F_t$ is sent back to the users, and each user updates its local model accordingly.

The conventional deep learning paradigms are based on the IID data assumption, and thus the training datasets of all the users can be modeled as unbiased samples of IID random variables. However, due to the high dynamics, the collected data of users are neither independent nor identical, and thus we focus on a non-IID scenario. In particular, we assume that the local dataset D_m of U_m is generated based on a random vector $\mathbf{X}_m$, and its probability density function (PDF) can be denoted as $P(\mathbf{X}_m)$, $m = 1, \ldots, M$. Similarly, the global distribution is characterized by $\mathbf{X}_g$, and its PDF can be denoted as $P(\mathbf{X}_g)$, and $\mathbf{X}_1, \ldots, \mathbf{X}_M, \mathbf{X}_g$ are non-IID, i.e., $P(\mathbf{X}_i) \neq P(\mathbf{X}_j)$, $i, j \in \{1, \ldots, M, g\}$, $i \neq j$.

2.2 Performance Analysis and Averaging Design

Although the existing works claimed that FL is applicable for the non-IID data (McMahan et al. [2017]), its performance is severely skewed (Zhao et al. [2018]), and the impact of distribution divergence of non-IID data is still unknown.

To provide some insights, the accuracy performance of FL with non-IID data is analyzed by comparing the following two learning paradigms:

- *FL scheme with non-IID data S_F*: The learning model is generated based on FL. As introduced previously, the training datasets $D_1, \ldots, D_M$ are non-IID.
- *Centralized learning scheme with IID data S_C*: In S_C, the local collected datasets of $U_1, \ldots, U_M$ are required to be sent back to the server S_B, and a global training dataset D_g can be formulated, $D_g = \cup_{m=1}^{M} D_m$. Moreover, each batch of training dataset is assumed to be an IID sample of D_g, which follows the global distribution $P(\mathbf{X}_g)$.

To the best of our knowledge, the centralized learning with IID data is the state-of-the-art learning paradigm, which can achieve the best accuracy performance ([Sattler et al. [2020]]). To evaluate the accuracy loss caused by non-IID data in FL, the expected weight divergence with respect to S_F and S_C is used as an efficient metric, which can be defined as Zhao et al. [2018]

$$\delta(t) = \mathbb{E}\left\{\|\mathbf{w}_t^F - \mathbf{w}_t^C\|\right\}, \tag{2.3}$$

where $\mathbf{w}_t^F$ is defined by (2.2), $\mathbf{w}_t^C$ is defined similarly for S_C after the tth round of training, and $\|\cdot\|$ denotes the 1-norm of its augment. In this chapter, SGD method is also used to update $\mathbf{w}_t^C$ in S_C, which can be expressed as

$$\mathbf{w}_t^C = \mathbf{w}_{t-1}^C - \eta G(\mathbf{w}_{t-1}^C, \mathbf{x}(\mathcal{B}_t^C)), \quad t = 1, \ldots, TK, \tag{2.4}$$

where $\mathcal{B}_t^C$ is the training data batch selected randomly from D_g to update $\mathbf{w}_t^C$. To guarantee the fairness of comparison, a common loss function is employed for both S_F and S_C. Moreover, the size of $\mathcal{B}_t^C$ is set as $N = \sum_{m=1}^{M} N_m$ to ensure that the learning models of S_F and S_C are trained by employing the same volume of data.

In the existing learning models, the gradient of the empirical risk is employed to estimate the gradient of the expected loss. Therefore, the gradient of the empirical risk functions in (2.1) and (2.4) can be defined as follows:

$$G(\mathbf{w}, \mathbf{x}(B)) = \frac{1}{N} \sum_{j=1}^{N} \nabla l(\mathbf{w}, \mathbf{x}_j(B)), \tag{2.5}$$

where $l(\mathbf{w}, \mathbf{x}_j(B))$ denotes the loss function with respect to a specific data element $\mathbf{x}_j$ in B, $\nabla l(\mathbf{w}, \mathbf{x}_j(B))$ is its first-order derivative with respect to $\mathbf{w}$, and N is the size of training dataset B. To guarantee the model accuracy and provide some insightful analytical results, the following remark is first provided, which has been widely employed in the existing works (Wang et al. [2019]).

Remark 2.1 For any $\mathbf{w}_1$ and $\mathbf{w}_2$, the following characteristics of loss function $l(\mathbf{w}, \mathbf{x}_j(B))$ should be guaranteed:

1. $l(\mathbf{w}, \mathbf{x}_j(B))$ is ρ-Lipschitz, i.e., $\|l(\mathbf{w}_1, \mathbf{x}_j(B)) - l(\mathbf{w}_2, \mathbf{x}_j(B))\| \leqslant \rho\|\mathbf{w}_1 - \mathbf{w}_2\|$.

2. $l(\mathbf{w}, \mathbf{x}_j(B))$ is β-smooth, i.e., $\|\nabla l(\mathbf{w}_1, \mathbf{x}_j(B)) - \nabla l(\mathbf{w}_2, \mathbf{x}_j(B))\| \leqslant \beta\|\mathbf{w}_1 - \mathbf{w}_2\|$. em $l(\mathbf{w}, \mathbf{x}_j(B))$ is λ-strongly convex, i.e., $\langle \nabla l(\mathbf{w}_1, \mathbf{x}_j(B)) - \nabla l(\mathbf{w}_2, \mathbf{x}_j(B)),$ $\mathbf{w}_1 - \mathbf{w}_2 \rangle \geq \lambda\|\mathbf{w}_1 - \mathbf{w}_2\|^2$.
3. There exists an upper bound of $\nabla l(\mathbf{w}, \mathbf{x}_j(B))$, i.e., $\|\nabla l(\mathbf{w}, \mathbf{x}_j(B))\| \leqslant \mu$ for any $\mathbf{x}_j$.
4. Based on (4), the standard deviation of $\nabla l(\mathbf{w}, \mathbf{x}_j(B))$ is limited, i.e.,

$$s(\nabla l(\mathbf{w}, \mathbf{x}_j(B))) = \sqrt{\mathbb{E}\{\|\nabla l(\mathbf{w}, \mathbf{x}_j(B))\|^2\} - (\mathbb{E}\{\|\nabla l(\mathbf{w}, \mathbf{x}_j(B))\|\})^2} \leqslant \sigma.$$

2.2.1 The Analysis of Expected Weight Divergence

By substituting (2.2) into (2.3), $\delta(tK)$ can be expressed as follows:

$$\delta(tK) = \mathbb{E}\left\{ \left\|\sum_{m=1}^{M} a_m \mathbf{w}_{tK}^m - \mathbf{w}_{tK}^C\right\| \right\} \leqslant \mathbb{E}\left\{ \sum_{m=1}^{M} a_m \underbrace{\left\|\mathbf{w}_{tK}^m - \mathbf{w}_{tK}^C\right\|}_{\theta_{tK}^m} \right\}. \tag{2.6}$$

Due to the Lipschitz and smooth characteristics given Remark 2.1, (2.6) can provide a tractable upper bound of model accuracy loss caused by FL with non-IID data by comparing the model weight divergence with the centralized learning with IID data, no matter if they converge in the same pace. It has been widely employed as an efficient metric to evaluate the performance of learning models (Zhao et al. [2018]). To derive a tractable upper bound of $\delta(tK)$, a recurrence relation with respect to θ_{tK} and θ_{tK-1} is provided by the following lemma.

Lemma 2.1 *An upper bound of θ_{tK} given by (2.6) can be expressed as follows:*

$$\theta_{tK}^m \leqslant \sqrt{1 + \beta^2\eta^2 - 2\eta\lambda}\,\theta_{tK-1}^m + \eta\mathcal{L}_m, \tag{2.7}$$

where $\mathcal{L}_m = \left\| G(\mathbf{w}_{tK-1}^C, \mathbf{x}(\mathcal{B}_{tK}^m)) - G(\mathbf{w}_{tK-1}^C, \mathbf{x}(\mathcal{B}_{tK}^C)) \right\|.$

Based on Lemma 2.1, the upper bound of $\delta(tK)$ given by (2.6) can be derived as

$$\delta(tK) \leqslant \sum_{m=1}^{M} a_m \left(\sqrt{1 + \beta^2\eta^2 - 2\eta\lambda}\,\mathbb{E}\{\theta_{tK-1}^m\} + \eta\mathbb{E}\{\mathcal{L}_m\} \right), \tag{2.8}$$

As shown in (2.8), besides the weight divergence θ_{tK-1}^m, $\delta(tK)$ is also determined by the data distribution divergence $\mathcal{L}_m$. Recalling (2.7), $\mathcal{L}_m$ denotes the difference of loss values with respect to S_C by employing two different training data batches $\mathcal{B}_{tK}^m$ and $\mathcal{B}_{tK}^C$, respectively. As introduced previously, $\mathcal{B}_{tK}^m$ and $\mathcal{B}_{tK}^C$ are generated based on $P(\mathbf{X}_m)$ and $P(\mathbf{X}_g)$, respectively, which are non-IID. To provided some insightful results, $\mathcal{L}_m$ is derived as follows.

2.2.1.1 The Analysis of Expected Data Distribution Divergence $\mathbb{E}\{\mathcal{L}_m\}$

$G(\mathbf{w}_{tK-1}^C, \mathbf{x}(\mathcal{B}_{tK}^m))$, and $G(\mathbf{w}_{tK-1}^C, \mathbf{x}(\mathcal{B}_{tK}^C))$ in (2.7) are the gradients of the empirical risks of S_C with respect to $\mathcal{B}_{tK}^m$ and $\mathcal{B}_{tK}^C$, respectively, which are employed to estimate the gradients of the corresponding expected risks, i.e.,

$$\mathcal{G}(\mathbf{w}_n^C, \mathbf{X}_i) = \mathbb{E}_{\mathbf{x} \sim P(\mathbf{X}_i)} \{\nabla l(\mathbf{w}_n^C, \mathbf{x})\}, \ i \in \{1, \dots, M, C\}, \ n = 1, \dots, TK, \quad (2.9)$$

where $\mathcal{G}(\mathbf{w}_n^C, \mathbf{X}_i)$ denotes the gradient of the expected risk with respect to $\mathbf{X}_i$. To evaluate the estimation accuracy, $\mathbf{z}_{i,n}$ is defined to capture the gap between the gradients of the empirical risk $G(\mathbf{w}_n^C, \mathbf{x}(\mathcal{B}_n^i))$ and the gradient of the corresponding expected risk $\mathcal{G}(\mathbf{w}_n^C, \mathbf{X}_i)$, i.e.,

$$\mathbf{z}_{i,n} = G(\mathbf{w}_n^C, \mathbf{x}(\mathcal{B}_n^i)) - \mathcal{G}(\mathbf{w}_n^C, \mathbf{X}_i). \quad (2.10)$$

As shown in (2.5), $G(\mathbf{w}_n^C, \mathbf{x}(\mathcal{B}_n^i))$ is the mean of $\nabla l(\mathbf{w}_n^C, \mathbf{x}_1(\mathcal{B}_n^i)), \dots, \nabla l(\mathbf{w}_n^C, \mathbf{x}_{N_i}(\mathcal{B}_n^i))$ with respect to a data batch $\mathcal{B}_n^i$, which can be treated as a sample generated by identically independent sampling based on a random vector $\nabla l(\mathbf{w}_n^C, \mathbf{X}(\mathcal{B}_n^i))$. Moreover, $\mathcal{G}(\mathbf{w}_n^C, \mathbf{X}_i)$ denotes the expectation of $\nabla l(\mathbf{w}_n^C, \mathbf{X}(\mathcal{B}_n^i))$. Therefore, due to the central limit theorem, $\mathbf{z}_{i,n}$ follows q-dimension normal distribution when the batch size N_i is large, i.e.,

$$\mathbf{z}_{i,n} = \frac{1}{N_i} \sum_{n=1}^{N_i} \nabla l(\mathbf{w}_n^C, \mathbf{x}_n(\mathcal{B}_n^i)) - \mathbb{E}\{\nabla l(\mathbf{w}_n^C, \mathbf{x}(\mathcal{B}_n^i))\} \sim \mathcal{N}_q(\mathbf{0}, \Sigma_{i,n}/N_i), \quad (2.11)$$

where $\Sigma_{i,n}$ denotes the covariance matrix of $\nabla l(\mathbf{w}_n^C, \mathbf{x}(\mathcal{B}_n^i))$. And its PDF can be expressed as;

$$f(\mathbf{z}_{i,n}) = \frac{1}{(2\pi)^{q/2}\sqrt{\Sigma_{i,n}/N_i}} \exp\left(-\frac{N_i}{2}\mathbf{z}_{i,n}^T \Sigma_{i,n}^{-1} \mathbf{z}_{i,n}\right). \quad (2.12)$$

Based on (2.10), $\mathcal{L}_m$ given by (2.7) can be derived as

$$\mathcal{L}_m = \left\|(\mathbf{z}_{m,tK-1} + \mathcal{G}(\mathbf{w}_{tK-1}^C, \mathbf{X}_m)) - (\mathbf{z}_{C,tK-1} + \mathcal{G}(\mathbf{w}_{tK-1}^C, \mathbf{X}_C))\right\|. \quad (2.13)$$

Then an upper bound of $\mathbb{E}\{\mathcal{L}_m\}$ in (2.7) can be provided by the following theorem.

Theorem 2.1 *A tractable upper bound of* $\mathbb{E}\{\mathcal{L}_m\}$ *can be expressed as*

$$\mathbb{E}\{\mathcal{L}_m\} \leqslant \overline{\mathcal{L}}_m = 2\sigma\sqrt{\frac{2}{\pi N_m}} + \mu \int_{\mathbf{x}} \|P(\mathbf{X}_m) - P(\mathbf{X}_g)\| d\mathbf{x}, \quad (2.14)$$

where μ and σ follow the notations given by 3) and 4) of Remark 2.1, respectively.

2.2.1.2 An Upper Bound of $\delta(tK)$

Based on Lemma 2.1 and Theorem 2.1, $\mathbb{E}\{\theta_{tK-1}^m\}$ in (2.8) can be written as

$$\mathbb{E}\left\{\theta_{tK-1}^m\right\} \leqslant \left(\sqrt{1 + \eta^2\beta^2 - 2\eta\lambda}\right)^{K-1} \mathbb{E}\left\{\theta_{(t-1)K}^m\right\} + \sum_{i=1}^{K-1}\left(\sqrt{1 + \eta^2\beta^2 - 2\eta\lambda}\right)^i \eta\overline{\mathcal{L}}_m$$

$$\stackrel{(b)}{=} \left(\sqrt{1 + \eta^2\beta^2 - 2\eta\lambda}\right)^{K-1} \delta((t-1)K) + \sum_{i=1}^{K-1}\left(\sqrt{1 + \eta^2\beta^2 - 2\eta\lambda}\right)^i \eta\overline{\mathcal{L}}_m,$$

$$(2.15)$$

where (b) in (2.15) follows the fact that $\mathbf{w}_{(t-1)K}^m = \mathbf{w}_{(t-1)}^F$ due to the $(t-1)$-th round of federated averaging. Then a recurrence relationship of $\delta(tK)$ and $\delta((t-1)K)$ can be obtained by substituting (2.15) into (2.8), which can be expressed as follows:

$$\delta(tK) \leqslant \sum_{m=1}^{M} a_m \left(\left(\sqrt{1 + \eta^2\beta^2 - 2\eta\lambda}\right)^K \delta((t-1)K) + \sum_{i=0}^{K-1}\left(\sqrt{1 + \eta^2\beta^2 - 2\eta\lambda}\right)^i \eta\overline{\mathcal{L}}_m\right)$$

$$= \left(\sqrt{1 + \eta^2\beta^2 - 2\eta\lambda}\right)^K \delta((t-1)K) + \frac{\eta\left[1 - \left(\sqrt{1+\eta^2\beta^2-2\eta\lambda}\right)^K\right]}{1 - \sqrt{1+\eta^2\beta^2-2\eta\lambda}} \sum_{m=1}^{M} a_m \overline{\mathcal{L}}_m.$$

$$(2.16)$$

Similar to (2.15), a tractable upper bound of $\delta(tK)$ can be obtained based on (2.16). In particular, to guarantee the fairness of comparison, the initialized parameter vectors of S_F and S_C should be set identically, i.e., $\mathbf{w}_0^m = \mathbf{w}_0^C$, and thus $\delta(0) = 0$. Then the following corollary can be provided.

Corollary 2.1 *An upper bound of $\delta(tK)$ with respect to S_F and S_C, which is denoted as $\overline{\delta}(tK)$, can be expressed as follows:*

$$\delta(tK) \leqslant \overline{\delta}(tK) = \frac{\eta\left[1 - \left(\sqrt{1 + \eta^2\beta^2 - 2\eta\lambda}\right)^{tK}\right]}{1 - \sqrt{1 + \eta^2\beta^2 - 2\eta\lambda}} \sum_{m=1}^{M} a_m \overline{\mathcal{L}}_m, \qquad (2.17)$$

where $\overline{\mathcal{L}}_m$ is provided by (2.14), and the learning rate should satisfy the constraint $\eta \leqslant \frac{2\lambda}{\beta^2}$.

As shown in (2.17), to guarantee the convergence of SGD method, the learning rate should satisfy the constraint $\eta \leqslant \frac{2\lambda}{\beta^2}$.

2.2.2 Rethinking the Settings of Federated Averaging Weights

As shown in (2.17), $\overline{\delta}(tK)$ is mainly determined by $\overline{\mathcal{L}}_m$, and it can be lowered by reducing the data distribution divergence. Therefore, rather than setting the averaging weight simply based on the volume of training dataset, such as McMahan et al. [2017] and Wang et al. [2019], it should be designed based on $\overline{\mathcal{L}}_m$, which can jointly capture the impacts of distribution divergence and data volume. In particular, when the local model generated by U_m is with a large value of $\overline{\mathcal{L}}_m$, its federated averaging weight a_m should be small. By following the proportion weighting strategy employed in the existing federate learning schemes, a_m is defined as follows based on (2.17):

$$a_m = \frac{1/\overline{\mathcal{L}}_m^2}{\sum_{n=1}^{M}\left(1/\overline{\mathcal{L}}_n^2\right)}. \tag{2.18}$$

To obtain a_m, the risk of privacy leakage of U_m can be completely avoided by following reasons. First, to protect the privacy of U_m for sharing $P(\mathbf{X}_m)$, the order of all the elements in $P(\mathbf{X}_m)$ permutated, which can misguide the eavesdropper. The permutation of $P(\mathbf{X}_m)$ can be implemented by employing the interleaving process in wireless communications straightforwardly. Second, recalling eq. (2.18), a_m can be generated based on $\overline{\mathcal{L}}_1, \ldots, \overline{\mathcal{L}}_M$. Therefore, instead of sharing $P(\mathbf{X}_m)$, U_m can transmit $\overline{\mathcal{L}}_m$ to the BS for generating a_m, which cannot provide any specific information of $P(\mathbf{X}_m)$. To generate $\overline{\mathcal{L}}_m$ at U_m, it requires BS to broadcast the global distribution $P(\mathbf{X}_g)$ to all the users. In many cases, $P(\mathbf{X}_g)$ can be obtained based on historical experience without requiring $P(\mathbf{X}_m)$.

By substituting (2.18) into (2.17), the obtained upper bound of expected weight divergence can be derived as follows.

Corollary 2.2 *When the weights of federated averaging are set as (2.18), $\overline{\delta}(tK)$ given by Corollary 2.1 can be expressed as*

$$\overline{\delta}(tK) = \frac{\eta\left[1 - \left(\sqrt{1 + \eta^2\beta^2 - 2\eta\lambda}\right)^{tK}\right]}{1 - \sqrt{1 + \eta^2\beta^2 - 2\eta\lambda}} \frac{\sum_{m=1}^{M}(1/\overline{\mathcal{L}}_m)}{\sum_{m=1}^{M}(1/\overline{\mathcal{L}}_m^2)}. \tag{2.19}$$

Recalling (2.18), the data distribution $P(\mathbf{X}_m)$ of U_m is required to obtain the average weights. To protect the user privacy of sharing $P(\mathbf{X}_m)$, the order of all the elements of $\mathbf{X}_m$ can be permutated to misguide the eavesdropper. The permutation of $P(\mathbf{X}_m)$ is similar to the interleaving process. Moreover, instead of sharing $P(\mathbf{X}_m)$, U_m can transmit $\overline{\mathcal{L}}_m$ to the server for generating a_m, and the user privacy issue can be solved since $\overline{\mathcal{L}}_m$ does not carry any specific information of $P(\mathbf{X}_m)$. To generate $\overline{\mathcal{L}}_m$ at U_m, $P(\mathbf{X}_g)$ should be broadcasted to all the users, which can be obtained based on historical experience without knowing $P(\mathbf{X}_m)$ in many cases.

Next, we focus on a special case when the employed training data batches of all the users are IID, i.e., $P(\mathbf{X}_1) = \cdots = P(\mathbf{X}_M) = P(\mathbf{X}_g)$. Then a_m given by (2.18) can be written as

$$a_m^{IID} = \frac{1/(\sigma/N_m)}{\sum_{k=1}^{M}[1/(1/(\sigma/N_k))]} = \frac{N_m}{N}. \tag{2.20}$$

(2.20) indicates that a_m is identical with the conventional scheme in McMahan et al. [2017] when the IID data is employed. It shows that our proposed federated averaging scheme can achieve at least the same accuracy performance as the conventional approach, which can be verified by the experiment results.

2.3 Data Sharing Scheme

The analytical results given by Corollary 2.2 indicate that the data distribution divergence has a great impact on the model accuracy of FL. To harmonize the divergence of the global and local distributions, a data sharing scheme is studied in this section, and a joint optimization algorithm is designed to balance the accuracy performance and cost of FL with data sharing in wireless networks.

2.3.1 Data Sharing

As introduced in Zhao et al. [2018], we assume that a training dataset D_s, which follows the global distribution $P(\mathbf{X}_g)$, is with the centralized server S_B. The key idea is to share D_s among the users to mitigate the distribution divergence. The credibility of D_s can be guaranteed by the comprehensive security framework of MEC systems (Hu et al. [2015]), and thus the data collection and sharing procedures can be protected by using powerful authentication and identification mechanisms. Moreover, the contaminated and abnormal data samples can be cleaned via data preprocessing techniques ([Ma et al. [2021]), which can further improve the credibility of sharing data.

Without loss of generality, we focus on the data sharing procedure for U_m. In particular, a part of data elements are selected randomly from D_s, which constitute a sharing dataset D_s^m for U_m. Then, D_s^m is transmitted to U_m via the wireless channel. A local training dataset $\overline{D}_m$ is constituted by combining D_s^m and D_m, which are sampled by following the distributions $P(\mathbf{X}_m)$ and $P(\mathbf{X}_g)$, respectively. To generate $\mathbf{w}_t^m$, a data batch $\overline{B}_t^m = B_t^m \cup B_s^m$ is selected from $\overline{D}_m$, where B_t^m and B_s^m are sampled uniformly from D_s^m and D_m, and their batch sizes are fixed as N_m

and L_m, respectively.[1] Denoting $\overline{\mathbf{X}}_m$ as a random vector to capture the distribution of $\overline{B}_t^m$, its PDF can be derived as a linear combination of $P(\mathbf{X}_m)$ and $P(\mathbf{X}_g)$

$$P(\overline{\mathbf{X}}_m) = \frac{N_m}{N_m + L_m} P(\mathbf{X}_m) + \frac{L_m}{N_m + L_m} P(\mathbf{X}_g). \tag{2.21}$$

Based on (2.14) and (2.21), the upper bound of distribution divergence can be expressed as follows when the data sharing scheme is employed:

$$\tilde{\mathcal{L}}_m = 2\sigma \sqrt{\frac{2}{\pi(N_m + L_m)}} + \mu \int_{\mathbf{x}} \| P(\overline{\mathbf{X}}_m) - P(\mathbf{X}_g) \| d\mathbf{x}$$

$$= 2\sigma \sqrt{\frac{2}{\pi(N_m + L_m)}} + \frac{\mu N_m}{N_m + L_m} \int_{\mathbf{x}} \| P(\mathbf{X}_m) - P(\mathbf{X}_g) \| d\mathbf{x}. \tag{2.22}$$

As shown in (2.22), it indicates that the distribution divergence can be reduced by data sharing and so is the weight divergence due to (2.19).

As introduced in Zhao et al. [2018], the accuracy performance of FL with non-IID data can be significantly improved with only a small amount of sharing data, it is not necessary to broadcast all the samples of sharing dataset to protect the user privacy, and it is difficult for the eavesdroppers to infer any useful information from trivial sharing data. Moreover, the number of sharing data samples can be further reduced by sampling adaptively with the distribution divergence, which is much more cost efficient than broadcasting a common dataset.

To further improve security of data sharing, the original data can be protected by employing encryption and adding noise, such as in Hardy et al. [2017] and Wei et al. [2021]. Moreover, instead of using real user data, the sharing dataset can be constructed by employing forged data, which can be generated by using generative adversarial network (GAN)-based models based on global distribution (Ching et al. [2020]).

2.3.2 Problem Formation

To balance the trade-off between the accuracy performance and cost of FL with data sharing in wireless networks, it requires to jointly optimize the radio resource allocation, computation and data management, and user selection, which should be restricted by the communication and computation constraints.

---

1 D_s^m can be transmitted to U_m only once at the beginning, which can be fully reused in each iterations, and the sampled sharing data batch can be set as $B_s^m = D_s^m$. Please note that the size of B_s^m can be restricted as $N_m \gg L_m$, and the over fitting issue caused by reusing D_s^m can be overlooked. Therefore, the communication costs can be minimized by making full use of all the data samples of D_s^m.

2.3.2.1 Objective Function

During the data sharing procedure, the cost can be captured by the transmission latency and energy consumption. In particular, we assume that the sharing dataset is transmitted to each user via orthogonal channel, and thus the transmission latency can be written as

$$\tau_{m,1} = \frac{SL_m}{B \log(1 + |h_m|^2 P_m / \sigma^2)}, \tag{2.23}$$

where S denotes the size of each data element in bits, B denotes the bandwidth of the wireless channel between S_B and U_m, h_m captures the block channel fading of the wireless channel between B and U_m, P_m is the transmit power of U_m, and σ^2 denotes the power of noise. Then the energy consumption caused by data sharing can be expressed as

$$E_{m,1} = P_m \tau_{m,1} = \frac{SL_m P_m}{B \log(1 + |h_m|^2 P_m / \sigma^2)}. \tag{2.24}$$

During the FL phase, its cost is mainly determined by the computation cost caused by local model training. In each round of local model training, the model parameters of U_m are updated by employing a batch with $(N_M + L_m)$ samples, which are trained by K iterations to guarantee the convergence. Then, the corresponding computation latency and energy consumption for all t rounds of federated averaging can be expressed as

$$\tau_{m,2} = tKr_m(L_m + N_m)\tilde{\tau}_m = \frac{tK(L_m + N_m)r_m}{f_m},$$

$$E_{m,2} = tKr_m(L_m + N_m)\tilde{E}_m = tK(L_m + N_m)\kappa_m r_m f_m^2, \tag{2.25}$$

where $\tilde{\tau}_m = r_m / f_m$ and $\tilde{E}_m = \kappa_m r_m f_m^2$ denote the computation latency and energy consumption of training a single sample for one iteration at U_m, respectively. As introduced in Zhao et al. [2020], r_m is the number of CPU cycles needed for training per sample, f_m is the CPU frequency of local processor S_m, and κ_m is the effective switched capacitance of S_m.

Recalling (2.19), the model accuracy is jointly determined by distribution divergence and batch size of training data. When a local model is trained by employing a huge amount of data with large distribution divergence, it will keep the aggregated model from converging an stationary optimal point with respect to the global data distribution if it is selected to participate global averaging. Therefore, a sophisticated user selection strategy should be designed to reduce the distribution divergence.

In this chapter, a Boolean variable x_m is defined to indicate the selection status of U_m, i.e.,

$$x_m = \begin{cases} 1, & \text{if } U_m \text{ is selected} \\ 0, & \text{if } U_m \text{ is not selected.} \end{cases} \tag{2.26}$$

Based on (2.26), the weight divergence can be rewritten as follows (2.19):

$$\tilde{\delta}(tK) = \frac{\eta[1 - (\sqrt{1 + \eta^2\beta^2 - 2\eta\lambda})^{tK}]}{1 - \sqrt{1 + \eta^2\beta^2 - 2\eta\lambda}} \frac{\sum_{m=1}^{M} x_m}{\sum_{m=1}^{M} x_m / \tilde{\mathcal{L}}_m}, \tag{2.27}$$

where $\tilde{\mathcal{L}}_m$ follows the notation given by (2.22).

To characterize the tradeoff between the model accuracy and cost, the objective function can be defined as follows:

$$Q = c_1\tilde{\delta}(tK) + c_2 \sum_{m=1}^{M} x_m \left(E_{m,1} + E_{m,2}\right) + c_3 \max_{i=1,\dots,M} \left\{x_i(\tau_{i,1} + \tau_{i,2})\right\}, \tag{2.28}$$

where c_1, c_2, and c_3 denote the weight coefficients of weight divergence, energy consumption, and latency, respectively. Please note that data sharing and local model training of each user can be accomplished in parallel, and thus the total latency is determined by the maximum latency.

2.3.3 Optimization Constraints

As introduced in Zhao et al. [2020], the CPU cycle frequency of S_m can be adjusted by employing DVFS techniques, which should satisfy the following constraint:

$$0 \leqslant f_m \leqslant f_m^{\max}, \; m = 1, \dots, M, \tag{2.29}$$

where $f_{m,\max}$ is the maximum value of f_m. Moreover, the transmit power of data sharing should follow the constraint

$$\sum_{m=1}^{M} x_m P_m \leqslant P_{\max}, \tag{2.30}$$

where $P_{\max}$ is the maximum transmit power of the BS.

To guarantee the convergence speed and model accuracy, enough users should be selected to participate FL, which should satisfy the following constraint:

$$M_s \leqslant \sum_{m=1}^{M} x_m \leqslant M, \tag{2.31}$$

where M_s denotes the minimum number of selected users. Moreover, the constructed sharing datasets should follow the constraint

$$0 \leqslant L_m \leqslant L, \; m = 1, \dots, M. \tag{2.32}$$

Then, our studied optimization problem can be formulated as follows:

$$\min_{f_m, P_m, x_m, L_m} Q \text{ in (2.28)}$$

$$s.t. \text{ (2.29), (2.30), (2.31), (2.32).} \tag{2.33}$$

2.3.4 A Joint Optimization Algorithm

As shown in (2.33), our formulated optimization problem is nonlinear and non-convex. To provide an efficient method to solve this problem, it can be decoupled as four subproblems, and then a joint optimization algorithm can be designed to approach an optimal solution by solving the subproblems iteratively.

2.3.4.1 CPU Cycle Frequency Optimization Subproblem

When we focus on the optimization of f_m, the objective function given by (2.28) can be rewritten as follows:

$$Q_1 = c_2 \sum_{m=1}^{M} x_m \kappa_m a_m f_m^2 + c_3 \max_{i=1,\dots,M} \left\{ x_i \left(\frac{a_i}{f_i} + \tau_{i,1} \right) \right\}, \tag{2.34}$$

where $a_j = tKr_j(L_j + N_j)$. Then the CPU cycle frequency optimization subproblem can be expressed as

$$\min_{f_m} Q_1 \text{ in } (2.34)$$

$$s.t. \text{ (2.29).} \tag{2.35}$$

Denoting $\mathcal{U}_1$ as the set that consists of all the selected users, i.e., $\mathcal{U}_1 = \{U_{i_1}, \dots, U_{i_k} | x_{i_j} = 1, j = 1, \dots, k\}$, (2.35) can be transformed as k independent subproblems. In particular, when $\max_{i=1,\dots,M}\{x_i(a_i/f_i + \tau_{i,1})\} = a_{i_j}/f_{i_j} + \tau_{i_j,1}$, (2.35) can be expressed as

$$\min_{f_m} Q_1 = c_2 \sum_{m=1}^{M} x_m \kappa_m a_m f_m^2 + c_3 \left(\frac{a_{i_j}}{f_{i_j}} + \tau_{i_j,1} \right)$$

$$s.t. \text{ (2.29), } x_{i_m} \left(\frac{a_{i_m}}{f_{i_m}} + \tau_{i_m,1} \right) \leq \frac{a_{i_j}}{f_{i_j}} + \tau_{i_j,1}, \, i_m = 1, \dots, M, i_m \neq i_j. \tag{2.36}$$

Denoting $y_m = 1/f_m$, (2.36) can be rewritten as

$$\min_{y_m} Q_1 = c_2 \sum_{m=1}^{M} \frac{x_m \kappa_m a_m}{y_m^2} + c_3 \left(a_{i_j} y_{i_j} + \tau_{i_j,1} \right)$$

$$s.t. \, y_m \geq \frac{1}{f_m^{\max}}, \, x_{i_m} a_{i_m} y_{i_m} - a_{i_j} y_{i_j} \leq \tau_{i_j,1} - x_{i_m} \tau_{i_m,1}, \, i, i_m = 1, \dots, M, i_m \neq i_j.$$

$$\tag{2.37}$$

It can be verified that (2.37) is convex with respect to $y_1, \dots, y_M$. Therefore, the optimal solution of (2.36) can be obtained by solving (2.37) via Karush–Kuhn–Tucker (KKT) conditions, which can be denoted as $\mathbf{f}_{i_j}^* = [f_{1,i_j}^*, \dots, f_{M,i_j}^*]$. Then, the optimal solution of (2.35) can be expressed as

$$\mathbf{f}^* = [f_1^*, \dots, f_M^*] = \arg \min_{U_{i_j} \in \mathcal{U}_1} \left\{ Q(\mathbf{f}_{i_j}^*) \right\}, \tag{2.38}$$

where $Q_1(\mathbf{f}_{i_j}^*)$ is the value of Q_1 given by (2.34) when $[f_1, \dots, f_M] = \mathbf{f}_{i_j}^*$.

2.3.4.2 Transmit Power Allocation Subproblem

Recalling (2.28), the objective function can be transformed as follows when we focus on the optimization of transmit power allocation:

$$Q_2 = c_2 \sum_{m=1}^{M} x_m \frac{L_m S P_m}{B \log(1 + |h_m|^2 P_m / \sigma^2)}$$

$$+ c_3 \max_{m=1,\ldots,M} \left\{ x_i \left[\underbrace{\frac{S L_m}{B \log(1 + |h_m|^2 P_m / \sigma^2)} + \tau_{m,2}}_{z_m} \right] \right\}. \tag{2.39}$$

Then the transmit power allocation subproblem can be expressed as

$$\min_{P_m} \ Q_2 \text{ in (2.39)},$$

$$s.t. \ (2.30). \tag{2.40}$$

Similar to (2.35), (2.40) can be transformed as k independent subproblems, i.e., it can be expressed as follows when $\max_{m=1,\ldots,M}\{x_m z_m\} = z_j, \ U_j \in \mathcal{U}_1$:

$$\min_{P_m} \ Q_2 = c_2 \sum_{m=1}^{M} x_m \frac{L_m S P_m}{B \log(1 + |h_m|^2 P_m / \sigma^2)} + c_3 z_j,$$

$$s.t. \ (2.30), \ x_m z_m \leqslant z_j, \ m \neq j. \tag{2.41}$$

Please note that it is not necessary to allocate transmit power to the unselected users, and thus (2.41) can be rewritten as follows based on $P_m = \frac{1}{|h_m|^2/\sigma^2}[2^{\frac{L_m S}{B(z_m - \tau_{m,2})}} - 1]$:

$$\min_{z_m} \ Q_2 = c_2 \sum_{U_m \in \mathcal{U}_1} \frac{z_m - \tau_{m,2}}{|h_m|^2/\sigma^2} \left[2^{\frac{L_m S}{B(z_m - \tau_{m,2})}} - 1 \right] + c_3 z_j,$$

$$s.t. \ \sum_{U_m \in \mathcal{U}_1} \frac{1}{|h_m|^2/\sigma^2} \left[2^{\frac{L_m S}{B(z_m - \tau_{m,2})}} - 1 \right] \leqslant P_{\max}, \ z_m \leqslant z_j, \ U_m, U_j \in \mathcal{U}_1 \ m \neq j. \tag{2.42}$$

It can be verified that Q_2 in (2.42) is convex with respect to $z_1, \ldots, z_M$. To obtain the optimal solution, its Lagrangian function can be written as

$$\mathcal{F} = c_2 \sum_{U_m \in \mathcal{U}_1} \frac{z_m - \tau_{m,2}}{|h_m|^2/\sigma^2} (w_m - 1) + c_3 z_j + \eta_1 \left[\sum_{U_m \in \mathcal{U}_1} \frac{1}{|h_m|^2/\sigma^2} (w_m - 1) - P_{\max} \right]$$

$$+ \sum_{U_m, U_j \in \mathcal{U}_1, m \neq j} \lambda_n (z_m - z_j), \tag{2.43}$$

where $w_m = 2^{\frac{L_m S}{B(z_m - \tau_{m,2})}}$. Then its KKT conditions can be listed as follows:

$$\frac{\partial}{\partial z_m} F = 0, \quad \frac{\partial}{\partial z_j} F = 0, \quad z_m - z_j \leqslant 0,$$

$$\eta_1, \lambda_n \geq 0, \quad \lambda_n(z_m - z_j) = 0, \quad m \neq j, \tag{2.44}$$

$$\eta_1 \left[\sum_{U_m \in \mathcal{U}_1} \frac{1}{|h_m|^2 / \sigma^2} (w_m - 1) - P_{\max} \right] = 0, \quad U_m \in \mathcal{U}_1.$$

We focus on the first set of conditions given by (2.44), which can be expressed as

$$\frac{\partial}{\partial z_m} F = c_2 \frac{1}{|h_m|^2 / \sigma^2} \left[\underbrace{w_m - 1 - \frac{L_m S \ln 2}{B(z_m - \tau_{m,2})} w_m}_{g_m} - \frac{\eta_1 L_m S \ln 2}{|h_m|^2 / \sigma^2 B(z_m - \tau_{m,2})^2} w_m \right]$$

$$+ \lambda_n = 0, U_m \in \mathcal{U}_1, m \neq j. \tag{2.45}$$

Please note that g_m in (2.45) is strictly smaller than zero, which can be verified as follows based on $\frac{L_m S}{B(z_m - \tau_{m,2})} = \log_2 w_m$:

$$g_m = w_m - 1 - w_m \ln w_m < 0. \tag{2.46}$$

Based on (2.45) and (2.46), it can be proved that $\lambda_n < 0$. Therefore, to ensure that the KKT conditions $\lambda_n(z_m - z_j) = 0$ can be satisfied, it can be derived that $z_m = z_j$, $U_m, U_j \in \mathcal{U}_1$, $m \neq j$. Then (2.42) can be transformed as a univariate optimization problem, which can be expressed as

$$\min_{z_j} Q_2 = c_2 \sum_{U_m \in \mathcal{U}_1} \frac{(z_j - \tau_{m,2})}{|h_m|^2 / \sigma^2} \left[2^{\frac{L_m S}{B(z_j - \tau_{m,2})}} - 1 \right] + c_3 z_j,$$

$$s.t. \quad \sum_{U_m \in \mathcal{U}_1} \frac{1}{|h_m|^2 / \sigma^2} \left[2^{\frac{L_m S}{B(z_j - \tau_{m,2})}} - 1 \right] \leqslant P_{\max}. \tag{2.47}$$

It can be verified that (2.47) is convex with respect to z_j. Therefore, the optimal solution can be obtained straightforwardly based on KKT conditions, which can be denoted as z_j^*. Finally, the optimal solution of (2.40) can be expressed as follows by solving (2.47):

$$P_m^* = \begin{cases} \dfrac{(2^{S L_m / B(z_j^* - \tau_{m,2})} - 1)\sigma^2}{|h_m|^2}, & U_m \in \mathcal{U}_1 \\ 0, & U_m \notin \mathcal{U}_1 \end{cases}. \tag{2.48}$$

2.3.4.3 Sharing Dataset Optimization Subproblem

The performance of data sharing is determined by the size of B_s^m, i.e., L_m in (2.33), and thus the objective function of our studied subproblem can be rewritten as follows:

$$
Q_3 = c_1 A \sum_{m=1}^{M} x_m \underbrace{\left(\sum_{m=1}^{M} \frac{x_m}{2\sigma \sqrt{\frac{2}{\pi(N_m+L_m)}} + \frac{p_m N_m}{N_m+L_m}} \right)^{-1}}_{\tilde{\delta}(tK) \text{ in } (1.27)}
$$

$$
+ c_2 \sum_{m=1}^{M} q_m L_m + c_3 \max_{i=1,\dots,M} \{ s_i L_i + b_i \}, \tag{2.49}
$$

where

$$
A = \frac{\eta[1 - (\sqrt{1 + \eta^2 \beta^2 - 2\eta \lambda})^{tK}]}{1 - \sqrt{1 + \eta^2 \beta^2 - 2\eta \lambda}}, p_m = \mu \int_{\mathbf{x}} \|P(\mathbf{X}_m) - P(\mathbf{X}_g)\| d\mathbf{x},
$$

$$
b_i = \frac{tKSr_m N_m}{f_m}, q_m = \frac{SP_m}{B\log(1 + |h_m|^2 P_m/\sigma^2)} + \kappa_m tKSr_m f_m^2, \tag{2.50}
$$

$$
s_i = \frac{S}{B\log(1 + |h_m|^2 P_m/\sigma^2)} + \frac{tKSr_m}{f_m}.
$$

Then, the sharing dataset optimization subproblem can be expressed as

$$
\min_{L_1,\dots,L_M} \quad Q_3 \text{ in } (2.49)
$$

$$
s.t. \ (2.32). \tag{2.51}
$$

Similar to (2.35) and (2.40), (2.51) can be transformed as M independent subproblems equivalently, i.e., it can be rewritten as follows when $\max_{i=1,\dots,M} \{ r_i L_i + b_i \} = r_j L_j + b_j$:

$$
\min_{L_1,\dots,L_M} \quad Q_3 = c_1 \tilde{\delta}(tK) + \sum_{m=1}^{M} c_2 q_m L_m + c_3 r_j L_j
$$

$$
s.t. \ (2.32), \ r_j L_j \geq r_i L_i + b_i - b_j, \ i \neq j. \tag{2.52}
$$

Please note that Q_3 is neither linear nor convex with respect to $L_1, \dots, L_M$. To approach an optimal solution, L_m-s can be optimized iteratively, which can be updated by solving the following problem:

$$
\min_{L_m} \quad Q_3 = c_1 \tilde{\delta}(tK) + d_m L_m
$$

$$
s.t. \ L_m^{\min} \leqslant L_m \leqslant L_m^{\max}, \tag{2.53}
$$

where

$$
L_m^{\max} = \begin{cases} L, & m = j \\ \frac{q_j - b_i}{r_i}, & m \neq j, \end{cases}, \quad L_m^{\min} = \begin{cases} \max_{i \neq j} \left\{ \frac{q_i - b_j}{r_j} \right\}, & m = j \\ 0, & m \neq j \end{cases}, \tag{2.54}
$$

and

$$
d_m = \begin{cases} c_2 q_m + c_3 r_m, & m = j \\ c_2 q_m, & m \neq j \end{cases}, \tag{2.55}
$$

and $q_m = r_m L_m + b_m$. The optimal solution of (2.53) can be obtained by solving $\nabla_{L_m} Q_3 = 0$, which can be expressed as

$$
L_{j,m}^{\text{op}} = \begin{cases} \lceil L_m^{\min} \rceil, & \text{if } \lfloor \overline{L}_{j,m} \rfloor \leqslant L_m^{\min} \\ \arg \min_{\lfloor \overline{L}_{j,m} \rfloor, \lceil \overline{L}_{j,m} \rceil} \left\{ Q_3(\lfloor \overline{L}_{j,m} \rfloor), \, Q_3(\lceil \overline{L}_{j,m} \rceil) \right\}, & \text{if } L_m^{\min} < \lfloor \overline{L}_{j,m} \rfloor, \lceil \overline{L}_{j,m} \rceil < L_m^{\max}, \\ \lfloor L_m^{\max} \rfloor, & \text{if } \lceil \overline{L}_{j,m} \rceil \geqslant L_m^{\max} \end{cases}
$$

$$\tag{2.56}$$

where $\overline{L}_{j,m}$ is the solution of $\nabla_{L_m} Q_3 = 0$, $\lfloor \cdot \rfloor$ is the floor of its augment, and $\lceil \cdot \rceil$ is its ceiling.

Based on (2.52), (2.53), and (2.56), Algorithm 2.1 can be designed to obtain the solution of (2.51). In particular, it can be divided into two steps:

- First, the M equivalent problems given by (2.52) need to be solved independently. As shown in Algorithm 2.1, the optimal solution of (2.52) can be approached via an iterative way. During the ith iteration, L_m can be updated by solving (2.53). $L_{j,m}(i)$ denotes the update result of L_m for the jth problem formulated by (2.52), which can be obtained straightforwardly based on (2.56). Denoting $Q_{3,j}(i)$ as the value of Q_3 given by (2.49) for ith iteration when $\max_{i=1,\ldots,M} \{ r_i L_i + b_i \} = r_j L_j + b_j$, it keeps decreasing as i increases since (2.56) is the optimal solution of (2.53). Therefore, it can converge to the optimal solution of (2.52).
- The optimal solution of (2.51) is among $\mathbf{L}_1^{\text{op}}, \ldots, \mathbf{L}_M^{\text{op}}$, where $\mathbf{L}_j^{\text{op}}$ denotes the optimal solution of the jth problem. It can be searched based on the following criterion:

$$
\mathbf{L}^* = \arg \min_{\mathbf{L}_1^{\text{op}}, \ldots, \mathbf{L}_M^{\text{op}}} \left\{ Q_{3,j}^{\text{op}} \right\}, \tag{2.57}
$$

where $Q_{3,j}^{\text{op}}$ denotes the value of Q_3 when $\mathbf{L}_j^{\text{op}}$ is the solution of (2.51).

Algorithm 2.1 Optimization of Sharing Dataset

Initialization: $L_{1,m}(0) = \cdots L_{j,m}(0) \cdots = L_{M,m}(0) = L_m(0)$, $j, m = 1, \ldots, M$, the maximum number of iterations I_{max}, and the convergence threshold ξ.

Step1. Solve M equivalent optimization problems given by (2.52)

For the jth problem, i.e., $\max_{i=1,\ldots,M} \left\{ r_i L_i + b_i \right\} = r_j L_j + b_j$,
- **Repeat**: For the ith iteration, $1 \leqslant i \leqslant I_{max}$
  - Update $L_{j,m}(i)$ based on (2.56), $m = 1, \ldots, M$.
  - Update $Q_{3,j}(i)$ based on (2.49).
- **Termination**: When $i > I_{max}$ or $|Q_{3,j}(i) - Q_{3,j}(i-1)| \leqslant \xi$.
- **Return**: $\mathbf{L}_j^{op} = [L_{j,1}(i), \ldots, L_{j,M}(i)]$ and $Q_{3,j}^{op} = Q_{3,j}(i)$.

Step2. Search the optimal solution of (2.51) *based on* (2.57).

Return: $\mathbf{L}^{op}$ as the optimal solution of (2.51).

2.3.4.4 User Selection Optimization Subproblem

Recalling (2.33), the objective function can be rewritten as follows when we focus on the optimization of user selection:

$$Q_4 = c_1 A \frac{\sum_{m=1}^{M} x_m}{\sum_{m=1}^{M} x_m / \overline{\mathcal{L}}_m} + c_2 \sum_{m=1}^{M} E_m x_m + c_3 \max_{i=1,\ldots,M} \{ \tau_i x_i \}, \tag{2.58}$$

where A follows the notation given by (2.50), $E_m = E_{m,1} + E_{m,2}$, and $\tau_i = \tau_{i,1} + \tau_{i,2}$. Then, the user selection optimization subproblem can be expressed as

$$\min_{x_1,\ldots,x_M} \ Q_4 \text{ in } (2.58)$$

$$s.t. \ (2.31). \tag{2.59}$$

Similar to (2.51), (2.59) can be equivalently transformed as M independent optimization subproblems, i.e., it can be rewritten as follows when $\tau_j = \max_{i=1,\ldots,M} \{ \tau_i x_i \}$:

$$\min_{x_1,\ldots,x_M} \ \overline{Q}_4 \left(x_1, \ldots, x_M \right) = \frac{1}{\sum_{m=1}^{M} x_m / \overline{\mathcal{L}}_m} \left(c_1 A \sum_{m=1}^{M} x_m + c_2 \sum_{m=1}^{M} \sum_{n=1}^{M} \frac{E_m}{\overline{\mathcal{L}}_n} x_m x_n \right.$$

$$\left. + c_3 \sum_{m=1}^{M} \frac{\tau_j}{\overline{\mathcal{L}}_m} x_m \right)$$

$$s.t. \ (2.31), \ x_m \tau_m \leqslant \tau_j, \ x_j = 1. \tag{2.60}$$

As shown in (2.60), it can be modeled as a 0-1 fractional programming problem. Denoting q as the value of function $\overline{Q}_4(x_1, \ldots, x_M)$, we first focus on the following

---

Algorithm 2.2 Optimization of User Selection

Initialization: $q_1(0) = \cdots = q_M(0) = q_0$, the maximum number of iterations K_{max}, and the convergence threshold ε.

Step1. Solve M independent optimization subproblems given by (2.60)

For the jth problem, i.e., $\tau_j = \max_{i=1,\ldots,M}\{\tau_i x_i\}$
- **Repeat**: For the kth iteration, $1 \leqslant k \leqslant K_{max}$
  - Update $\mathbf{x}_j^*(k)$ by solving (2.64);
  - Update $q_j^*(k) = \bar{Q}_4(\mathbf{x}_j^*(k))$;
  - Update $g(q_j^*(k))$ based on (2.62).
- **Termination**: When $k > K_{max}$ or $|g(q_j^*(k))| < \varepsilon$.
- **Return**: $\mathbf{x}_j^* = \mathbf{x}_j^*(k)$, and $Q_{4,j} = q_j^*(k)$.

Step2. Find the optimal solution of (2.59).

Return: $\mathbf{x}^*$ as the final result based on (2.65).

---

optimization problem, where the objective function is a polynomial function with respect to $x_1, \ldots, x_M$:

$$
\begin{aligned}
\min_{x_1,\ldots,x_M} g(q) &= \left(c_1 A \sum_{m=1}^{M} x_m + c_2 \sum_{m=1}^{M}\sum_{n=1}^{M} \frac{E_m}{\mathcal{L}_n} x_m x_n + c_3 \sum_{m=1}^{M} \frac{\tau_j}{\mathcal{L}_m} x_m \right) \\
&\quad - q \left(\sum_{m=1}^{M} x_m / \bar{\mathcal{L}}_m \right) \\
&= c_2 \sum_{m=1}^{M}\sum_{n=1}^{M} \frac{E_m}{\mathcal{L}_n} x_m x_n + \sum_{m=1}^{M} \left(c_1 A + \frac{c_3 \tau_j - q}{\mathcal{L}_m} \right) x_m
\end{aligned}
\tag{2.61}
$$

s.t. (2.31), $x_m \tau_m \leqslant \tau_j, x_j = 1$.

To transform $g(q)$ as a linear function, it can be expressed as follows by denoting $y_{m,n} = x_m x_n$:

$$
g(q) = c_2 \sum_{m=1}^{M}\sum_{n=1}^{M} \frac{E_m}{\mathcal{L}_n} y_{m,n} + \sum_{m=1}^{M} \left(c_1 A + \frac{c_3 \tau_j - q}{\mathcal{L}_m} \right) x_m.
\tag{2.62}
$$

Moreover, the following constraints with respect to $y_{m,n}$ should be considered to guarantee that the transformed optimization problem is equivalent to (2.61):

$$
y_{m,n} \leqslant x_m, \ y_{m,n} \leqslant x_n, \ y_{m,n} \geqslant x_m + x_n - 1.
\tag{2.63}
$$

Then (2.61) can be written as follows:

$$
\min_{x_m, y_{m,n}} g(q) \text{ in (2.62), s.t. (2.31), (2.63), } x_m \tau_m \leqslant \tau_j, x_j = 1.
\tag{2.64}
$$

Algorithm 2.3 The Joint Optimization Algorithm

Initialization: $x_m(0)$, $P_m(0)$, $f_m(0)$, $L_m(0)$, the maximum number of iterations $L_{\max}$, and the convergence threshold ϵ.

Repeat: For the lth iteration, $1 \leqslant l \leqslant L_{\max}$

- Update $f_m(l)$ based on (2.38), $m = 1, \ldots, M$;
- Update $P_m(l)$ based on (2.48), $m = 1, \ldots, M$;
- Update $L_m(l)$ based on Algorithm 2.1, $m = 1, \ldots, M$;
- Update $x_m(l)$ based on Algorithm 2.2, $m = 1, \ldots, M$;
- Update $Q(l)$ based on (2.33).

Termination: When $l > L_{\max}$ or $|Q(l) - Q(l-1)| \leqslant \epsilon$.

Since (2.64) is a linear programming problem, its optimal solution can be obtained efficiently by using various calculation softwares, such as the optimization package provided by CVXPY.

Similar to Algorithm 2.1, the original optimization problem given by (2.59) can be solved based on the solutions of M independent subproblems given by (2.60). As shown in Algorithm 2.2, (2.60) can be solved by updating q and the solution of (2.61) iteratively. In particular, during the kth iteration, the optimal solution of (2.61) can be obtained by solving (2.64), which can be denoted as $\mathbf{x}_j^*(k) = [x_{j,1}^*(k) \cdots x_{j,M}^*(k)]$. Then, recalling (2.60), the value of q can be updated as $q^*(k) = \overline{Q}_4(\mathbf{x}_j^*(k))$. The final result $\mathbf{x}_j^*$ can be obtained until the threshold of accuracy can be satisfied or the maximum iterations are achieved.

Based on Theorem 1 given by Zhao et al. [2020], it can be proved that $\mathbf{x}_j^*$ is the optimal solution of (2.61) if and only if $g(q_j^*(k)) = 0$. By following the proof of Theorem 2 in Zhao et al. [2020], it can be verified that $g(q_j^*(k))$ monotonously approaches 0 as k increases. Therefore, $\mathbf{x}_j^*$ can converge to the optimal solution of (2.60). Then, the optimal solution of (2.59) can be selected among $\mathbf{x}_1^*, \ldots, \mathbf{x}_M^*$ to minimize the objective function given by (2.60), i.e.,

$$\mathbf{x}^* = \arg \min_{\mathbf{x}_1^*, \ldots, \mathbf{x}_M^*} \{Q_{4,j}\}, \tag{2.65}$$

where $Q_{4,j}$ denotes the value of Q_4 given by (2.58) with respect to $\mathbf{x}_j^*$ for the jth problem.

2.3.4.5 A Joint Optimization Algorithm

As shown in Algorithm 2.3, a joint optimization algorithm is designed to solve (2.33), where the solutions of four decoupled subproblems are updated iteratively. During the lth iteration, the optimal solutions of CPU cycle frequency and optimization and transmit power allocation subproblems can obtained based on

(2.38) and (2.48), respectively. Moreover, the sharing data optimization subproblem can be solved by using Algorithm 2.1, while the user selection subproblem can be optimized based on Algorithm 2.2. As introduced previously, the optimal solutions of all the subproblems can be obtained, which can guarantee that $Q(l)$ keeps decreasing as the iteration index l increases. Moreover, there exists a fixed minimum value of Q, which can be lower bounded by 0. Therefore, the convergence of Algorithm 2.3 can be guaranteed, which can approach a stationary solution of (2.33).

2.4 Simulation Results

In this section, the simulation results are provided to evaluate the performance of our proposed schemes. In this chapter, we consider employing FL for image classification in wireless networks. In the simulation part, we set $\eta = 0.001$, and SVM and classic CNN models, LeNet-5 in Lecun et al. [1998], are employed. Moreover, the datasets of users are sampled independently from MINST dataset. In particular, the datasets of users are generated by sampling each category of the entire dataset with different frequency. Therefore, the proportions of each category samples in the generated datasets are different with each other, which can characterize the non-IID distribution features of local datasets of users.

In Figure 2.1, the test accuracy of FL with our proposed averaging scheme given by (2.18) is provided, where the conventional averaging scheme in McMahan et al. [2017] is selected as a benchmark. In particular, the number of selected users is set as 5, and the sizes of training datasets of each user are as 1000, 600, 600, 400, and 400. We study two cases of local datasets, which are with different distribution divergences. Specifically, in our setting, the data distribution divergences among each local dataset and the global dataset are 0.2120, 0.0874, 0.1503, 0.2576, and 0.2928 for Case 1, and for Case 2, the data distribution divergences are 0.0424, 0.0446, 0.0554, 0.0506, and 0.0389. As shown in Figure 2.1a, SVM is employed as the local model of each user. Compared with the conventional averaging scheme, the test accuracy can be improved from 74.01% to 82.00% by using our proposed averaging scheme for Case 1 and from 79.63% to 84.25% for Case 2. In Figure 2.1b, CNN is employed for local model generation, and our proposed averaging scheme can improve test accuracy from 85.11% to 87.28% for Case 1 and from 93.51% to 94.25% for Case 2. The experiment results show that our proposed scheme can significantly improve the test accuracy of FL with non-IID data, especially when the distribution divergence is large. Moreover, its performance can approach the performance of centralized learning as the distribution divergence decreases.

As shown in Figure 2.2, the test accuracy of FL with our proposed averaging scheme is evaluated with both balanced and unbalanced datasets. For balanced

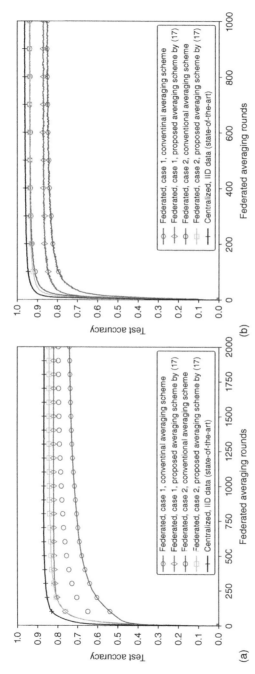

Figure 2.1 Test accuracy performance with different data distribution divergence. (a) Local training with SVM model. (b) Local training with CNN model.

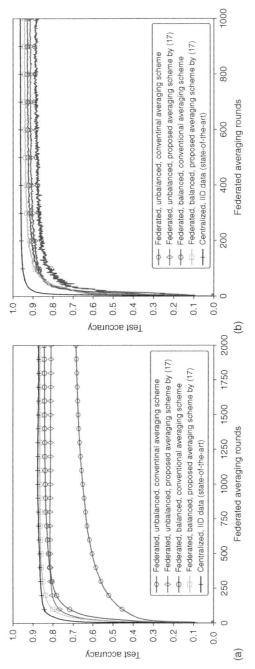

Figure 2.2 Test accuracy performance with balanced and unbalanced datasets. (a) Local training with SVM model. (b) Local training with CNN model.

case, the size of training dataset of each user is set as 600, and for unbalanced case, the sizes of each local dataset are set as 1000, 600, 600, 400, and 400. In Figure 2.2a, SVM is employed for local model generation. It shows that the test accuracy can be improved from 84.44% to 86.43% by using our proposed averaging scheme for the balanced dataset case and from 68.60% to 81.40% for the unbalanced dataset case. When the CNN model is deployed, Figure 2.2b shows that the test accuracy can be improved from 92.17% to 94.47% for the balanced dataset case and from 87.19% to 91.4% for the unbalanced dataset case. Therefore, the evaluation results indicate that our proposed averaging scheme can improve the accuracy performance of FL for both balanced and unbalanced data cases, and the performance gains can be enlarged when the unbalanced datasets are used. As shown in Figures 2.1 and 2.2, our proposed averaging scheme based on Corollary 2.1 can significantly improve the accuracy performance when the CNN models are employed. Therefore, the simulation results indicate that our analytical works can still be used to capture the relationship between distribution divergence and accuracy loss of nonconvex models.

As shown in Figures 2.3a,b, the test accuracy of all schemes converges as the number of total users grows, due to the increase of the amount of data participated in model training. Moreover, the test accuracy of FL with optimized data sharing scheme is second only to the conventional centralized learning scheme and out-performs the conventional FL and FL only with the FedAvg weighting schemes. As shown in Figure 2.3c–f, the total energy consumption and total latency of FL with optimized data sharing scheme are little larger than the FL only with FedAvg weighting scheme and is significantly less than the conventional centralized learning and conventional FL schemes. On the one hand, by comparing with the FL only with FedAvg weighting scheme, the FL with optimized data sharing scheme can further improve the learning performance with little cost of commu-nication and computation resources. In particular, compared with conventional FL schemes, the test accuracy trained by SVM can be improved by 161.2% and 57.6% when the number of total user is $M = 5$ and $M = 50$, respectively. On the other hand, compared with the conventional centralized learning scheme, the FL with optimized data sharing scheme can approach to the optimal learning performance with large reduction of cost of communication and computation resources. Though, our proposed data sharing scheme will produce additional overhead, and our optimized resource allocation and user selection schemes are helpful to reduce the model cost. In particular, compared with the conventional centralized learning scheme, although the test accuracy loss trained by SVM of FL with optimized data sharing scheme is 3.4% and 0.5% when $M = 5$ and $M = 50$, respectively, the cost of total energy consumption and total latency can be reduced significantly. Therefore, a sophisticated balance between communication and computation cost and performance of FL can be achieved, by employing our

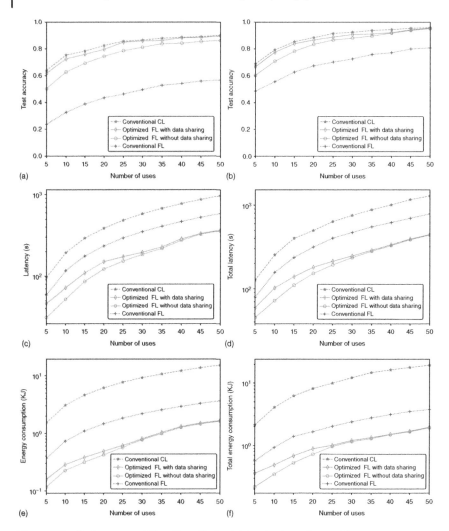

Figure 2.3 Performance of FL with optimized data sharing scheme, (a) test accuracy, (c) latency performance, (e) energy performance are trained via SVM and (b) test accuracy, (d) latency performance, (f) energy performance are trained via CNN.

proposed joint optimized data sharing scheme. Our proposed FedAvg weighting scheme can enhance the FL performance all the time, and our proposed data sharing scheme is more helpful when the number of users is small compared with user-dense case. It is reasonable since the large the sample is, the smaller the variance is, which means under ideal communication circumstance, the more clients participate in model aggregation, the smaller the weight difference of trained models between FL and centralized learning. Compared with SVM models, CNN can provide higher test accuracy but results in larger energy consumption and a little longer latency, and our proposed optimization algorithm enhances the performance of CNN model, which can offer some help to design an FL paradigm in MEC systems.

Bibliography

Cheng-Wei Ching, Tzu-Cheng Lin, Kung-Hao Chang, Chih-Chiung Yao, and Jian-Jhih Kuo. Model partition defense against gan attacks on collaborative learning via mobile edge computing. In *IEEE Global Communications Conference (GLOBECOM)*, pages 1–6, 2020.

Stephen Hardy, Wilko Henecka, Hamish Ivey-Law, Richard Nock, Giorgio Patrini, Guillaume Smith, and Brian Thorne. Private federated learning on vertically partitioned data via entity resolution and additively homomorphic encryption. *CoRR*, abs/1711.10677, 2017. URL http://arxiv.org/abs/1711.10677.

Yun Chao Hu, Milan Patel, Dario Sabella, Nurit Sprecher, and Valerie Young. Mobile edge computing - a key technology towards 5G, 2015. URL https://infotech.report/Resources/Whitepapers/f205849d-0109-4de3-8c47-be52f4e4fb27etsiwp11mecakeytechnologytowards5g.pdf.

Y. Lecun, L. Bottou, Y. Bengio, and P. Haffner. Gradient-based learning applied to document recognition. *Proceedings of the IEEE*, 86(11):2278–2324, 1998.

Lichuan Ma, Qingqi Pei, Lu Zhou, Haojin Zhu, Licheng Wang, and Yusheng Ji. Federated data cleaning: Collaborative and privacy-preserving data cleaning for edge intelligence. *IEEE Internet of Things Journal*, 8(8):6757–6770, 2021.

Brendan McMahan, Eider Moore, Daniel Ramage, Seth Hampson, and Blaise Aguera y Arcas. Communication-efficient learning of deep networks from decentralized data. In *Proceedings of the International Conference on Artificial Intelligence and Statistics (AISTATS)*, pages 1273–1282, Florida, USA, 2017.

Felix Sattler, Simon Wiedemann, Klaus-Robert Müller, and Wojciech Samek. Robust and communication-efficient federated learning from non-iid data. *IEEE Transactions on Neural Networks and Learning Systems*, 31(9):3400–3413, 2020.

Shiqiang Wang, Tiffany Tuor, Theodoros Salonidis, Kin K. Leung, Christian Makaya, Ting He, and Kevin Chan. Adaptive federated learning in resource constrained edge computing systems. *IEEE Journal on Selected Areas in Communications*, 37(6):1205–1221, 2019.

Kang Wei, Jun Li, Ming Ding, Chuan Ma, Hang Su, Bo Zhang, and H. Vincent Poor. User-level privacy-preserving federated learning: Analysis and performance optimization. *IEEE Transactions on Mobile Computing*, 21(9):33881–34011, 2021.

Yue Zhao, Meng Li, Liangzhen Lai, Naveen Suda, Damon Civin, and Vikas Chandra. Federated learning with non-iid data. In *arXiv preprint*, 2018. URL https://arxiv.org/abs/1806.00582.

Zhongyuan Zhao, Shuqing Bu, Tiezhu Zhao, Zhenping Yin, Mugen Peng, Zhiguo Ding, and Tony Q. S. Quek. On the design of computation offloading in fog radio access networks. *IEEE Transactions on Vehicular Technology*, 68(7):7136–7149, 2020.

3

How Many Resources Are Needed to Support Wireless Edge Networks

Yi-Jing Liu[1], Gang Feng[1], Yao Sun[2], and Shuang Qin[1]

[1]National Key Lab on Wireless Communications, University of Electronic Science and Technology of China, Chengdu, Sichuan, China
[2]James Watt School of Engineering, University of Glasgow, Glasgow, Scotland, UK

3.1 Introduction

Edge network intelligence is boosted by the unprecedented computing capability of smart devices. Nowadays, more than 15 billion user equipments (UEs) have been equipped with artificial intelligence (AI)-enabled computing modules, such as AI chips and graphic processing units (GPUs) Jovanović [n.d.]. On the one hand, these UEs can be potentially deployed as computing nodes to support emerging services, such as AI tasks, which paves the way for applying AI in wireless edge networks. On the other hand, in the paradigm of machine learning (ML), the powerful computing capability on these UEs can decouple traditional ML from acquiring, storing, and training data in data centers as conventional approaches.

Federated learning (FL) has recently been widely acknowledged as one of the most significant enablers to bring edge network intelligence into reality, as it facilitates collaborative training of ML models, while guaranteeing data security and individual user privacy (Lim et al. [2020], Yang et al. [2019]). In FL, ML models are trained locally; therefore, raw data remains in the UEs. Specifically, FL uses an iterative approach that requires a number of global iterations to achieve a certain global model accuracy. In each global iteration, UEs perform a number of local iterations to reach a local model accuracy (Lim et al. [2020], Yang et al. [2019]). As a result, the implementation of FL in wireless edge networks can also reduce the costs of transmitting raw data and relieve the burden on backbone networks.

While FL offers the valuable benefits, it also brings many challenges, especially when being deployed in wireless edge networks. For example, both local training and global/local model transmission can be unsuccessful due to constrained computing resources, communication resources, and unstable transmission. Moreover, different from conventional ML approaches where raw datasets are sent to a central server, only the lightweight model parameters (i.e., weights and gradients) are exchanged in FL. Nevertheless, the communication costs of FL can be still fairly large and cannot be ignored. The experimental results in Jeong et al. [2018] show that transmitting the model of a 5-layer convolutional neural network used for MNIST (classification) consumes about 4.567MB bandwidth per global iteration for 28×28 pixel images, while transmitting the model of ResNet-110 used for CIFAR-10 (classification) consumes around 4.6MB bandwidth per global iteration for 32×32 pixel images (He and Annavaram [2020]). Therefore, before deploying FL-enabled wireless edge networks, we need to answer two fundamental questions: (i) How accurate of an ML model can be achieved by using FL and (ii) How many costs are incurred to guarantee certain required FL performance? Obviously, answering these two questions is of paramount importance for facilitating edge network intelligence. Therefore, we need to deeply understand the relationship between FL performance and consumed multidimensional resources.

In this chapter, we theoretically analyze how many resources are needed to support an FL-empowered wireless edge network by assuming spatial-temporal domain Poisson distribution. We first derive the distribution of signal-to-interference-plus-noise ratio (SINR), signal-noise ratio (SNR), model transmission success probability, and resource consumption. Then, we evaluate the impact of the amount of resources on FL performance. Numerical results validate the accuracy of our theoretical modeling and analysis.

3.2 System Model

We consider an FL-enabled wireless edge network that consists of a central base station (BS) and multiple UEs, as shown in Fig. 3.1. Specifically, UEs can be regarded as local computing nodes for training local models, while the server (e.g., edge servers) serves as the model aggregator co-located with the BS (Lim et al. [2020], Liu et al. [2020b], Yang et al. [2019]). To quantitatively elaborate on the FL-enabled wireless network models, we need to model the distribution of the UEs and arrival rate of the interfering UEs. As one of the most commonly used point processes, PPP model is used to model UEs distribution and/or arrival rate of the interfering UEs in wireless edge networks, where a huge amount of data has validated the accuracy of the model (Flint et al. [2017], Hunter et al. [2008], Weber et al. [2010], Sun et al. [2019]). Nevertheless, other point processes, such as Poisson cluster process (PCP) (Chun et al. [2015]) for some specific scenarios, can also be used in our analytical model.

Figure 3.1 FL-enabled
wireless edge networks.

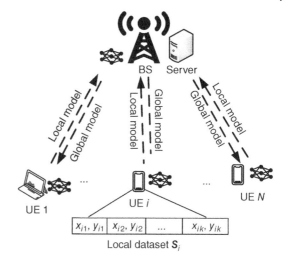

3.2.1 FL Model

3.2.1.1 Loss Function

Let random variable N denote the number of UEs that are geographically distributed as homogeneous PPP with intensity λ_i, where we use n to denote the value of N. Similarly, in the following, we use capital letter to denote random variables and the corresponding lower case to the value of the random variables. In FL, each global iteration is called a *communication round* (Liu et al. [2020b]), as shown in Fig. 3.2. A communication round consists of five phases including local model updating, local iterations (also called local training), local model transmission, global model updating, and global model transmission.

For a specific UE i, let S_i and S_i represent its local dataset and the number of data samples, where $S_i = \{x_{ik} \in \mathbb{R}^d, y_{ik} \in \mathbb{R}\}_{k=1}^{S_i}$. Moreover, let $f_k(w_i^r(t); x_{ik}, y_{ik})$ be a local loss function for data sample k, where $w_i^r(t)$ represents the model parameter of UE i at the tth local training during the rth communication round. The loss

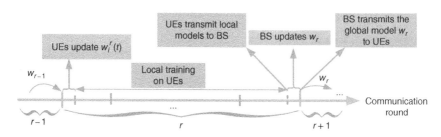

Figure 3.2 The process of a communication round.

function $f_k(w_i^r(t); x_{ik}, y_{ik})$ is different for various FL learning tasks (Hennig and Kutlukaya [2007]). For example, for a linear regression, the loss function is defined as $f_k(w_i^r(t); x_{ik}, y_{ik}) = \frac{1}{2}(x_{ik}^T w_i^r(t) - y_{ik})^2$. For neural network, the loss function can be mean squared error (i.e., $\frac{1}{n} \sum_{i=1}^n (y_{ik} - \hat{y}_{ik})^2$), where $\hat{y}_{ik}$ represents the predicted value of y_{ik}. Based on local loss function $f_k(w_i^r(t); x_{ik}, y_{ik})$, we define $F_i(w_i^r(t)): \mathbb{R}^m \rightarrow \mathbb{R}$ as a local loss function to capture the local training performance, which is as follows:

$$F_i(w_i^r(t)) \triangleq \frac{1}{S_i} \sum_{i \, k \in S_i} f_k(w_i^r(t); x_{ik}, y_{ik}). \tag{3.1}$$

Based on local loss function, we define $F(w_r)$ as the global loss function to measure the global training performance, which is expressed by

$$F(w_r) \triangleq \frac{\sum_{i=1}^n S_i F_i(w_i^r(t))}{S}, \tag{3.2}$$

where $S \triangleq \sum_{i=1}^n S_i$. The goal of the BS is to derive a vector w_r^* satisfying $w_r^* \triangleq \arg_{w_r} \min F(w_r)$.

3.2.1.2 Updating Model

As shown in Fig. 3.2, model updating includes the local model updating and global model updating. In the following, we present the details of the local and global model updating, respectively.

(1) *Local Model Updating*: The local model updating is performed based on a local learning algorithm, such as gradient descent (GD) and actor-critic (AC). Specifically, let t represent the index of local iterations and τ represent the total number of local iterations during a communication round. If $0 < t \leq \tau$, the local model $w_i^r(t)$ is updated by $w_i^r(t) = w_r - \eta \nabla F_i(w_i^r(t-1))$, while $w_i^r(t) = w_r$ if $t = 0$. Moreover, $\eta \geq 0$ is the step size, and w_r represents the global model at the rth communication round.

(2) *Global Model Updating*: After τ local iterations, i.e., $t = \tau$, UEs achieves a certain local model accuracy and send the local models $w_i^r(t)$ to the server. Then the server performs global aggregation according to

$$w_r = \frac{\sum_{i=1}^n S_i w_i^r(t)}{S}, t = \tau. \tag{3.3}$$

3.2.2 Computing Resource Consumption Model

For UE i, let Z_i denote its computing capacity in cycles/s. c_i (cycles/sample) represents the number of CPU cycles required for computing one data sample at UE i. T_i represents the local training time needed for one local iteration. Therefore, similar to that in Yang et al. [2020b], the computing resources consumption during one local iteration for UE i is given by $Z_i = \frac{c_i S_i}{T_i}$.

3.2.3 Communication Resource Consumption Model

3.2.3.1 Uplink

For UE i, the time for transmitting the local model $w_i^r(t)$ (uplink direction) is represented by $T_{up}^{i,r}$. Since the dimensions of local models are fixed for all UEs that participate in local training, the data size of all local models is constant and is represented by s (Yang et al. [2020b]). The transmission rate of UE i on the wireless channel to the BS during the rth communication round is denoted by $R_{up}^{i,r}$. Therefore, we have $\frac{s}{R_{up}^{i,r}} = T_{up}^{i,r}$, where

$$R_{up}^{i,r} = b_{up}^{i,r} \log_2(1 + SINR_{up}(D_1, N_I, \mathbf{D}_2)), \tag{3.4}$$

where $b_{up}^{i,r}$ denotes the amount of the consumed bandwidth for transmitting the local model $w_i^r(t)$. In addition, let $SINR_{up}(D_1, N_I, \mathbf{D}_2) = \frac{P_{up} G(D_1)}{\sum_{j=1}^{N_I} PG(D_2^{(j)}) + \delta^2}$ represent the signal-to-interference-plus-noise ratio (SINR). D_1 represents the distance between the UE and the BS, $\mathbf{D}_2 = [D_2^{(1)}, D_2^{(2)}, \ldots, D_2^{(j)}, \ldots, D_2^{(N_I)}]$ denotes the distance vector for all interfering UEs of UE i, N_I represents the number of interfering UEs with $N_I \leq N$, δ^2 denotes the noise power, P_{up} represents the transmit power of the UE, and $G(\cdot)$ represents the large-scale channel gain between the BS and UEs. Indeed, the channel gain model could be either large scale (e.g., path loss) or small scale (e.g., Rayleigh fading, Rician fading), which only affects SINR. Furthermore, local model transmission is successful only if $SINR_{up}(D_1, N_I, \mathbf{D}_2) > \beta_{up}$, where β_{up} denotes the SINR threshold.

3.2.3.2 DownLink

The transmission time for transmitting the global model w_r (downlink direction) is denoted by $T_{down}^{i,r}$. From Eq. (3.3), we see that the dimensions of the global model w_r are equal to that of local models. Therefore, the data size of the global model that the BS sends to each UE is also equal to s (Chen et al. [2021]). Let $R_{down}^{i,r}$ represents the transmission rate of the BS during the rth communication round. Therefore, we have $\frac{s}{R_{down}^{i,r}} = T_{down}^{i,r}$, where

$$R_{down}^{i,r} = b_{down}^{i,r} \log_2(1 + SNR_{down}(D_1)), \tag{3.5}$$

where $b_{down}^{i,r}$ denotes the consumed bandwidth for transmitting the global model. In addition, let SNR_{down} be signal-to-noise ratio (SNR), given by $SNR_{down}(D_1) = \frac{P_{down} G(D_1)}{\delta^2}$. Specifically, P_{down} represents the transmit power of the BS allocated to all UEs. Furthermore, the global model transmission is successful only if $SNR_{down}(D_1) > \beta_{down}$, where β_{down} denotes the SNR threshold.

3.3 Wireless Bandwidth and Computing Resources Consumed for Supporting FL-Enabled Wireless Edge Networks

A certain amount of wireless bandwidth is consumed in the uplink/downlink, when models are exchanged between the BS and the UEs. Specifically, in the uplink, the UEs send their local models to the BS via some channel partitioning schemes, such as orthogonal frequency division multiplexing (OFDM). In the downlink, the BS sends the global model to individual UEs. Indeed, the wireless transmission environment of the uplink/downlink will affect the transmission process of the local/global model, and thus affect the global aggregation and local training. In this section, we theoretically analyze SINR, SNR, as well as wireless bandwidth consumed in the uplink/downlink to support FL-empowered wireless edge networks.

3.3.1 SINR Analysis (Uplink Direction)

3.3.1.1 Probability Density Function (PDF) of SINR

To derive the probability density function (PDF) of SINR, we separately investigate the signal power and interference. As the UEs are geographically distributed as homogeneous PPP model with intensity λ_i, the number of UEs N is a variable of Poisson distribution with density parameter $\pi(r_0)^2\lambda_i$, where r_0 denotes the radius of the BS coverage. We assume that the transmit power P_{up} is always fixed for each UE. Therefore, for a specific UE i, the signal power is also a random variable, i.e., $S_{up} = P_{up}G(D_1)$, as it only relates to the distance D_1.

Corollary 3.1 *The PDF of the distance D_1 between a specific UE and the serving BS is* $f_{D_1}(d_1) = \frac{2d_1}{(r_0)^2}$.

Proof: As the PDF of location (X, Y) for UE i is $f(X, Y) = \frac{1}{\pi(r_0)^2}$, the CDF of distance $D_1 = d_1$ is $F_{D_1}(d_1) = \iint\limits_{X^2+Y^2\leq(d_1^2)} \frac{1}{\pi(r_0)^2}dXdY = \frac{(d_1)^2}{(r_0)^2}$. Therefore, the PDF of $D_1 = d_1$ is $f_{D_1}(d_1) = F'_{D_1}(d_1) = \frac{2d_1}{(r_0)^2}$. $\qquad\square$

Therefore, we can obtain the PDF of signal power (i.e., $f_S(S = P_{up}G(D_1)) = f_{D_1}(d_1)$) for deriving the closed-form expression of $\Pr(SINR_{up} > \beta_{up})$. Next, we derive the distribution of the received interference in the uplink. Note that only the transmitting UEs located in the interfering area with radius d_0 can contribute to the interference. Let $N_A(N_A \geq N_I)$ represent the number of UEs within the interfering area, which is also a variable of Poisson distribution with density parameter $\pi(d_0)^2 \lambda_i$. Moreover, the transmission time for the UEs is represented by t_{up}, where the transmitting UEs during time $[-t_{up}, t_{up}]$ can contribute to interference. Therefore, the number of interfering UEs is distributed as PPP with parameter $2t_{up}\lambda_a$, where λ_a represents the arrival rate of interfering UEs. Therefore, the interference probability of a transmitting UE during time $[-t_{up}, t_{up}]$ can be expressed by $\Pr(active) = 1 - \exp\{-2t_{up}\lambda_a\}$.

Therefore, the probability of the number of interfering UEs $N_I = n_I$ given $N_A = n_A$ is given by

$$\Pr(N_I = n_I | N_A = n_A) = C_{n_A}^{n_I} (1 - \exp\{1 - 2t_{up}\lambda_a\})^{n_I} \cdot (\exp\{1 - 2t_{up}\lambda_a\})^{n_A - n_I}, \tag{3.6}$$

where $C_{n_A}^{n_I}$ represents the combination number. Therefore, the PDF of N_I is given by

$$f_{N_I}(n_I) = \Pr(N_I = n_I) = \sum_{n_A = n_I}^{N} \Pr(N_I = n_I | N_A = n_A) \cdot \Pr(N_A = n_A), \tag{3.7}$$

where $\Pr(N_A = n_A) = \frac{(\pi(d_0)^2 \lambda_i)^{n_A}}{n_A!} \exp\{-\pi(d_0)^2 \lambda_i\}$. Based on 3.1, we can derive the PDF of interference I_i that is generated by UE i, i.e., $f_{I_i}(I_i = P_{up} G(d_2^{(i)})) = f_{D^{(i)}}(d_2^{(i)}) = \frac{2d_2^{(i)}}{d_0^2}$. As the total interference $I(N_I, \mathbf{D}_2)$ is affected by the number of interfering UEs N_I and the distance of the interfering UEs $\mathbf{D}_2$, we derive the PDF of $I(N_I, \mathbf{D}_2)$, as follows:

$$f_I(N_I = n_I, \mathbf{D}_2 = \mathbf{d}_2) = f_{N_I}(n_I) \Pr(\mathbf{D}_2 = \mathbf{d}_2 | N_I = n_I)$$

$$= f_{N_I}(n_I) \left(\frac{2}{(d_0)^2}\right)^{n_I} \prod_{n=1}^{n_I} d_2^{(n)}. \tag{3.8}$$

Therefore, the PDF of SINR can be given by

$$f_{SINR_{up}}(D_1 = d_1, N_I = n_I, \mathbf{D}_2 = \mathbf{d}_2) = f_{D_1}(d_1) f_I(N_I = n_I, \mathbf{D}_2 = \mathbf{d}_2), \tag{3.9}$$

3.3.1.2 Transmission Success Probability of Local Models

Local model transmission is successful only if $SINR_{up} > \beta_{up}$. Therefore, the transmission success probability of local models is expressed by

$$\Pr(SINR_{up} > \beta_{up}) = \iiint_{\mathcal{A}} f_{SINR_{up}} d\mathcal{A}, \tag{3.10}$$

in which $\mathcal{A}$ denotes the area of $(D_1, N_I, \mathbf{D}_2)$ that satisfies $SINR_{up}(D_1, N_I, \mathbf{D}_2) > \beta_{up}$. As we can obtain $f_{SINR_{up}}$ in Eq. (3.10), we only need to find the interfering area $\mathcal{A}$.

For the distance between the UE and the serving BS, intuitively $f_{SINR_{up}} < \beta_{up}$, when $D_1 > d_0$ (Sun et al. [2019]). Therefore, the satisfying range of D_1 is $(0, d_0]$.

Therefore, we can obtain the number of interfering UEs N_I and the location of these interfering UEs when $D_1 = d_1$ is given. Let $\overline{n}_I$ denote the mean of random variable N_I. Based on the UE distribution and interfering UE arrival models, we can obtain $\overline{n}_I$ as follows:

$$
\begin{aligned}
\overline{n}_I &\triangleq E(N_A)\Pr(active) \\
&= \pi(d_0)^2 \lambda_i \left(1 - \exp\{1 - 2t_{up}\lambda_I\}\right).
\end{aligned}
\tag{3.11}
$$

Therefore, SINR is only related to D_1 and $\mathbf{D}_2$, given as $SINR_{up}(D_1, \mathbf{D}_2) = \frac{P_{up}G(D_1)}{\sum_{i=1}^{\overline{n}_I} I_i + \delta^2}$, where $I_i = P_{up}G(D_2^{(i)})$ represents the interference generated by UE i. Therefore, we have

$$
\Pr(SINR_{up} > \beta_{up}) = \Pr\left(\sum_{i=1}^{\overline{n}_I} I_i < \frac{P_{up}G(D_1)}{\beta_{up}} - \delta^2\right).
\tag{3.12}
$$

In a typical FL framework, the number of UEs participating in local model training is fairly large, i.e., at least hundreds of UEs (Wang et al. [2021]). Therefore, based on the central limit theorem, $\sum_{i=1}^{\overline{n}_I} I_i$ follows a *normal distribution* $N(\mu_I, \sigma_I^2)$ (Sun et al. [2019], Hsu and Robbins [1947]). Furthermore, we have $\mu_I = \overline{n}_I E(I_i)$ and $\sigma_I = \sqrt{\overline{n}_I}D(I_i)$, which are the mean and variance of I_i, respectively (Hsu and Robbins [1947]), given by,

$$
\begin{aligned}
\mu_I &= \overline{n}_I E(I_i) = \overline{n}_I \int_{d_2^{(i)}=d_{min}}^{d_0} P_{up}g(d_2^{(i)})f_{D^{(i)}}(d_2^{(i)})d(d_2^{(i)}) \\
&= \overline{n}_I \int_{d_2^{(i)}=d_{min}}^{d_0} P_{up}g(d_2^{(i)}) \cdot \frac{2d_2^{(i)}}{(d_0)^2}d(d_2^{(i)}) \\
&= \frac{2P_{up}\overline{n}_I}{(d_0)^2} \int_{d_2^{(i)}=d_{min}}^{d_0} g(d_2^{(i)})(d_2^{(i)})d(d_2^{(i)}) \\
&\overset{(b)}{=} \frac{2P_{up}\overline{n}_I}{(d_0)^2} \int_{d_2^{(i)}=d_{min}}^{d_0} (d_2^{(i)})dG(d_2^{(i)}) \\
&= \frac{2P_{up}\overline{n}_I}{(d_0)^2} \left[(d_2^{(i)})G(d_2^{(i)})|_{d_2^{(i)}=d_{min}}^{d_0} - \int_{d_2^{(i)}=d_{min}}^{d_0} G(d_2^{(i)})d(d_2^{(i)})\right],
\end{aligned}
$$

where $\int_{d_2^{(i)}=d_{min}}^{d_0} G(d_2^{(i)})d(d_2^{(i)}) = G_2(d_0) - G_2(d_{min})$. Therefore, μ_I can be given by

$$
\mu_I = \frac{2P_{up}\overline{n}_I}{(d_0)^2} \left\{[d_0 G(d_0) - d_{min}G(d_{min})] - [G_2(d_0) - G_2(d_{min})]\right\},
$$

where d_{min} is the minimum distance between UEs and BS, and we define $g(\cdot) = G'$ and $G(\cdot) = G'_2(\cdot)$. Therefore, σ_I can be given by

$$\sigma_I = \sqrt{\overline{n}_I} D(I_i) = \sqrt{\overline{n}_I}[E(I_i^2) - E^2(I_i)]$$

$$= \sqrt{\overline{n}_I} \left[\int_{d_2^{(i)}=d_{min}}^{d_0} 2P_{up}^2 g^2(d_2^{(i)}) \frac{d_2^{(i)}}{(d_0)^2} d(d_2^{(i)}) - \left(\frac{\mu_I}{n_I}\right)^2 \right].$$

Let $Y = \frac{I-\mu_I}{\sigma_I}$ and $I = \sum_{i=1}^{\overline{n}_I} I_i$ and $I \sim N(\mu_I, \sigma_I)$. Therefore, we have $Y \sim N(0, 1)$, where

$$\Pr\left(\sum_{i=1}^{\overline{n}_I} I_i < \frac{P_{up}G(d_1)}{\beta_{up}} - \delta^2\right) = \Pr\left(Y < \frac{\frac{P_{up}G(d_1)}{\beta_{up}} - \delta^2 - \mu_I}{\sigma_I}\right) = \Phi(\xi(d_1)),$$

$$(3.13)$$

where $\xi(d_1) = \frac{\frac{P_{up}G(d_1)}{\beta_{up}} - \delta^2 - \mu_I}{\sigma_I}$ and Φ denote the cumulative distribution function of standard normal distribution. Therefore, we have

$$\Pr(SINR_{up} > \beta_{up}) = \iiint_{\mathcal{A}} f_{SINR_{up}} d\mathcal{A}$$

$$= \int_{d_1=d_{min}}^{d_0} f_{D_1}(d_1)\Phi(\xi(d_1))d(d_1). \qquad (3.14)$$

3.3.2 SNR Analysis (Downlink Direction)

As $SNR_{down}(D_1) = P_{down}G(D_1)/\delta^2$, we can derive the PDF of the signal power, as $f_{S_{down}}(S_{down} = P_{down}G(D_1)) = f_{D_1}(d_1)$, when transmit power P_{down} and noise level δ^2 are given. Therefore, given $G(d'_{min}) = \delta^2\beta_{down}/P_{down}$, where we assume $G(d_1)$ monotonically increases, we have

$$\Pr(SNR_{down} > \beta_{down}) = \Pr\left(\frac{P_{down}G(D_1)}{\delta^2} > \beta_{down}\right)$$

$$= \int_{d_1=d'_{min}}^{r_0} f_{D_1}(d_1)d(d_1). \qquad (3.15)$$

3.3.3 Wireless Bandwidth Needed for Transmitting Local/Global Models

Based on Eq. (3.4), in the rth communication round, the bandwidth needed for transmitting the local model $w_i^r(t)$ is given by $b_{up}^{i,r} = \frac{s}{T_{up}^{i,r}\log_2(1+SINR_{up}(D_1,N_I,D_2))}$. As s and $T_{up}^{i,r}$ are fixed, the PDF of $b_{up}^{i,r}$ for UE i in the uplink is equal to $f_{SINR_{up}}$. Therefore, the mean of bandwidth for all UEs transmitting local models during K communication rounds is given by

$$\overline{B}_{up} = K \cdot \sum_{i=1}^{n} b_{up}^{i,r} f_{SINR_{up}}. \tag{3.16}$$

Similarly, the mean of bandwidth for transmitting the global models during K communication rounds is expressed by

$$\overline{B}_{down} = K \cdot \sum_{i=1}^{n} b_{down}^{i,r} f_{SNR_{down}}. \tag{3.17}$$

3.3.4 Computing Resources Needed for Training Local Models

We assume the processing capacity of all UEs is constant in cycles/sample. In addition, as different ML models pose different degrees of complexity, all UEs are assumed to train the same FL task in our analytical model, where the local ML models have the same size and structure. Therefore, the total computing resources consumption is affected by the number of training UEs and the amount of datasets on the UEs. On the one hand, the number of training UEs is affected by the wireless transmission of the global model. On the other hand, many of the existing studies explicitly indicate that the amount of different datasets distributed on the UEs is imbalanced, as the data is collected directly and stored persistently (Hsu et al. [2019], McMahan et al. [2017]). Note here data imbalance means the different amount of local datasets, instead of different dataset contents, so our analytical model is based on i.i.d. data, where non-i.i.d data case is left for future work. Therefore, in this section, we theoretically analyze computing resources needed for training local models from the perspective of SNR and imbalanced datasets.

Let the amount of datasets on the UEs follow the *normal distribution* (Gao et al. [2020], Liu et al. [2020a]), i.e., $S_i \sim N(\mu_i, \sigma_i^2)$, where μ_i or/and σ_i^2 can be different for specific UEs. Indeed, we can also use other distributions such as Beta distribution and Gamma distribution in our analytical model. Moreover, as the computing resources consumption of UE i for one local iteration is $Z_i = \frac{c_i S_i}{T_i}$, the PDF of Z_i is equal to $f_{S_i}(s_i)$, i.e., $f_{Z_i}(z_i) = \frac{1}{\sqrt{2\pi}\sigma_i} \exp\left(-\frac{(s_i - \mu_i)^2}{2\sigma_i^2}\right)$.

For a specific UE i, UE i successfully receive the global model only if $SNR_{down} > \beta_{down}$. In other words, UE i will continue to perform local training in the next communication round and consume certain computing resources. Let $\hat{Z} = \{\hat{Z}_1, \hat{Z}_2, \ldots, \hat{Z}_i, \ldots, \hat{Z}_n\}$ indicate the certain computing resources consumed by all UEs, where the value of $\hat{Z}_i$ is set to z_i if UE i successfully receives the global model and 0 otherwise. Therefore, we can obtain the PDF of $\hat{Z}_i$ as follows:

$$f_{\hat{Z}_i}(\hat{Z}_i = z_i) = \Pr(\hat{Z}_i = z_i | SNR_{down} > \beta_{down})$$
$$= f_{Z_i}(z_i) \Pr(SNR_{down} > \beta_{down}). \tag{3.18}$$

Therefore, based on Eq. (3.18), we can obtain the mean of computing resources consumed by all UEs for one local iteration, which is given by $\overline{C}_{UE} = \sum_{i=1}^{n} z_i f_{\hat{Z}_i}(\hat{Z}_i = z_i)$. Therefore, the total computing resources consumed for local model training is given by

$$C_{\text{total}} = \tau K \overline{C}_{UE}, \tag{3.19}$$

where τ and K denote the total number of local trainings and communication rounds, respectively. Armed with the above preparation, we are now starting to analyze how the resources affect the FL performance by evaluating local and global model accuracy.

3.4 The Relationship between FL Performance and Consumed Resources

Indeed, the unsuccessful transmission of local models in the uplink affects the aggregation of the global model, while the unsuccessful transmission in the downlink affects the updating and training of local models. Therefore, in this section, we analyze how the computing and communication resources affect the FL performance by evaluating both the local and global model accuracy.

3.4.1 Local Model Accuracy

In an FL framework, no matter what local ML algorithm is used, each UE solves the local optimization problem for local training (Yang et al. [2020a,b], Dinh et al. [2020]), i.e.,

$$\min_{h_i \in \mathbb{R}^d} G_i(w_r, h_i) \triangleq F_i(w_r + h_i) - (\nabla F_i(w_r) - \zeta \nabla F(w_r))^{\mathrm{T}} h_i, \tag{3.20}$$

where ζ is constant and h_i denotes the difference between the global model and the local model for UE i. Without loss of generality, we use the GD algorithm to update local models, as it can achieve the required high accuracy and facilitate the convergence analysis (Yang et al. [2020b]), as follows:

$$h_i^{(r)(t+1)} = h_i^{(r)(t)} - \xi \nabla G_i(w_r, h_i^{(r)(t)}), \tag{3.21}$$

where ξ denotes the step size and $h_i^{(r)(t)}$ represents the value of h_i at the tth local iteration with given global model vector w_r. Moreover, $\nabla G_i(w_r, h_i^{(r)(t)})$ represents the gradient of $G_i(w_r, h_i)$. In addition, $w_i^r(t) = w_r + h_i^{(r)(t)}$ denotes the local model of UE i at the tth local iteration. For a small step ξ, we can derive a set of solutions $h_i^{(r)(0)}, \cdots, h_i^{(r)(t)}, \cdots, h_i^{(r)(\tau)}$, which meets

$$G_i(w_r, h_i^{(r)(0)}) \geq \cdots \geq G_i(w_r, h_i^{(r)(t)}) \geq \cdots \geq G_i(w_r, h_i^{(r)(\tau)}). \tag{3.22}$$

To provide the convergence condition for the GD method, we introduce local model accuracy loss ϵ_l (Yang et al. [2020b], Dinh et al. [2020]), as follows:

$$G_i(w_r, h_i^{(r)(t)}) - G_i(w_r, h_i^{(r)*}) \leq \epsilon_l(G_i(w_r, h_i^{(r)(0)}) - G_i(w_r, h_i^{(r)*})), \qquad (3.23)$$

where $h_i^{(r)*}$ denotes the optimal solution of problem (3.19). Note that each UE aims to solve the local optimization problem with a target local model accuracy $1 - \epsilon_l$. To achieve the local model accuracy $1 - \epsilon_l$ and the global model accuracy loss $1 - \epsilon_g$ in the following, we first make the following three assumptions on the local loss function $F_i(w)$, as that in (Yang et al. [2020a,b], Chen et al. [2021]).

• Assumption 1: Function $F_i(w)$ is L-Lipschitz, i.e., $\forall w, w' \in \mathbb{R}^d, \|\nabla F_i(w) - \nabla F_i(w')\| \leq L \|w - w'\|$.

• Assumption 2: Function $F_i(w)$ is γ-strongly convex, i.e., $\forall w, w' \in \mathbb{R}^d, F_i(w) \geq F_i(w') + \langle \nabla F_i(w'), (w - w') \rangle + \frac{\gamma}{2} \|w - w'\|^2$.

• Assumption 3: $F_i(w)$ is twice continuously differentiable. And $\gamma I \leq \nabla^2 F_i(w) \leq LI$.

Based on the three assumptions, we can derive the lower bound on the number of local iterations during each communication round, which is shown as Proposition 2.

Corollary 3.2 *Local model accuracy loss ϵ_l is achieved if $\xi < \frac{2}{L}$ and run the GD method*

$$\tau \geq \left\lceil \frac{2}{(2 - L\xi)\xi\gamma} \ln \frac{1}{\epsilon_l} \right\rceil$$

iterations during each communication round at each UE that participates in local training.

Proof: See Section 3. □

The lower bound in Corollary 3.2 reflects the growing trend of the number of local iterations with respect to the local model accuracy, which can approximate the consumption of computing resources for training local models.

3.4.2 Global Model Accuracy

In FL algorithms, a global model accuracy is also needed. For a specific FL task, we define ϵ_g as its global model accuracy loss (the global model accuracy is $1 - \epsilon_g$), which is given by

$$F(w_r(\hat{S}, SINR_{up})) - F(w_r^*) \leq \epsilon_g(F(w_1) - F(w_r^*)), \qquad (3.24)$$

where w_r^* denotes the actual optimal solution. Moreover, we provide the following Corollary 3.3 about the number of communication rounds for achieving the global model accuracy $1 - \epsilon_g$.

Algorithm 3.1 FL algorithm

Input: The required local model accuracy loss ϵ_l, the required global model accuracy loss ϵ_g.

output: The global model w_r, the number of local iterations τ, and the number of communication rounds K.

1: Initialization: local models $w_i^1(0) = 0$, the global model $g_1 = 0$.

2: **for** $r = 1, 2, \dots$ **do**

3:    Each UE calculates $\nabla F_i(w_r)$.

4:    Each UE sends $\nabla F_i(w_r)$ to the BS.

5:    The BS calculates $\nabla F(w_r) = \frac{\sum_{i=1}^{n} \hat{S}_i \nabla F_i(w_r)}{\sum_{i=1}^{n} \hat{S}_i}$.

6:    The BS broadcasts $\nabla F(w_r)$ to each UE.

7:    **Parallel** Each UE $i = 1, 2, \dots, n$

8:    Initialization: the local iteration number $t = 0$, and set $h_i^{(r)(0)} = 0$.

9:    **Repeat**

10:    Every V steps set $h_i^{(r)*} = h_i^{(r)(t)}$.

11:    Update $h_i^{(r)(t+1)} = h_i^{(r)(t)} - \xi \nabla G_i(w_r, h_i^{(r)(t)})$.

12:    Set $w_i^r(t) = w_r + h_i^{(r)(t)}$.

13:    **if** $\frac{G_i(w_r, h_i^{(r)(t)}) - G_i(w_r, h_i^{(r)*})}{(G_i(w_r, h_i^{(r)(0)}) - G_i(w_r, h_i^{(r)*}))} > \epsilon_l$ **then**

14:       Set $t = t + 1$

15:    **else**

16:       Each UE i sends $w_i^r(t)$ to the BS.

17:    **end if**

18:    The BS calculates $w_r = \frac{\sum_{i=1}^{n} \hat{S}_i w_i^r(t)}{\sum_{i=1}^{n} \hat{S}_i}$

19:    The BS sends w_r to all UEs.

20:    **if** $\frac{F(w_r(\hat{S}, SINR_{up})) - F(g^*)}{F(g_0) - F(g^*)} < \epsilon_g$ **then**

21:       Break;

22:    **end if**

23: **end for**

24: Set $\tau = t$, $K = r$.

25: **output** w_r, τ, K.

Corollary 3.3 *Global model accuracy $1 - \epsilon_g$ is achieved if the number of communication rounds K meets*

$$K \geq \left\lceil \frac{2L^2 \ln \frac{1}{\epsilon_g}}{(1 - \epsilon_l)\gamma^2 \zeta} \right\rceil,$$

when running FL algorithm shown as Algorithm 3.1 with $0 < \zeta < \frac{\gamma}{L}$

Proof: See Section 3.9.   □

Note that it is challenging to derive a closed-form expression of the global model during each communication round, due to the dynamic nature of the wireless channel and the uncertain nature of multiple random variables. Therefore, we assume the amount of datasets on each UE is fixed to facilitate the proof of Corollary 3.3. In addition, from Corollary 3.2 and Corollary 3.3, we can see that there is a trade-off between the number of communication rounds and the number of local iterations characterized by ϵ_l: small ϵ_l leads to large τ, yet small K, from which we can jointly approximate the communication and computing resources consumed by training FL tasks. The details can be found in the Section 3.5.

3.5 Discussions of Three Cases

Generally, the resources used for training FL tasks in wireless edge networks should be limited as (i) Communication and computing resources are constrained and precious and (ii) resource consumption quickly increases with the widespread use of smart UEs. In this section, we discuss three specific cases for different sufficiency of communication and computing resources in wireless edge networks. Furthermore, we derive the explicit expression of the model accuracy under FL framework, as a function of the amount of the consumed resources based on the sufficiency of respective communication and computing resources.

3.5.1 Case 1: Sufficient Communication Resources and Computing Resources

When both communication resources and computing resources are sufficient, we can approximate the amount of communication/computing resources based on Corollary 3.2 and Corollary 3.3. Specifically, the bandwidth needed for transmitting local models (uplink direction) should meet

$$\overline{B}_{\text{up}} \geq \left\lceil \frac{2L^2 \ln \frac{1}{\epsilon_g}}{(1-\epsilon_l)\gamma^2\zeta} \right\rceil \cdot \sum_{i=1}^{n} b_{\text{up}}^{i,r} f_{\text{SINR}_{\text{up}}}$$

$$= \left\lceil \frac{2L^2 \ln \frac{1}{\epsilon_g}}{(1-\epsilon_l)\gamma^2\zeta} \right\rceil \cdot \sum_{i=1}^{n} \frac{s}{T_{\text{up}}^{i,r}\log_2 \left(1 + \frac{P_{\text{up}}G(d_1)}{\sum_{j=1}^{N_I} PG(d_2^{(j)})+\delta^2}\right)}$$

$$\cdot \left(\frac{2d_1}{(d_0)^2}\right) \cdot \sum_{n_A=n_I}^{N} C_{n_A}^{n_I}(1 - \exp\{1 - 2t_{\text{up}}\lambda_a\})^{n_I}$$

$$\cdot (\exp\{1 - 2t_{up}\lambda_a\})^{n_A - n_I} \cdot \frac{(\pi(d_0)^2 \lambda_i)^{n_A}}{n_A!}$$

$$\cdot \exp\{-\pi(d_0)^2 \lambda_i\} \cdot \left(\frac{2}{(d_0)^2}\right)^{n_I} \prod_{n=1}^{n_I} d_2^{(n)}, \tag{3.24}$$

Similarly, we can obtain the bandwidth needed for transmitting the global model (downlink direction), given by

$$
\begin{aligned}
\overline{B}_{down} &\geq \left\lceil \frac{2L^2 \ln\frac{1}{\epsilon_g}}{(1 - \epsilon_l)\gamma^2 \zeta} \right\rceil \cdot \sum_{i=1}^{n} b_{down}^{i,r} f_{SNR_{down}} \\
&= \left\lceil \frac{2L^2 \ln\frac{1}{\epsilon_g}}{(1 - \epsilon_l)\gamma^2 \zeta} \right\rceil \cdot \sum_{i=1}^{n} \frac{s}{T_{down}^{i,r} \log_2\left(1 + \frac{P_{down} G(d_1)}{\delta^2}\right)} \cdot \frac{2d_1}{r_0^2}.
\end{aligned}
\tag{3.25}
$$

Based on Eq. (3.19), given local accuracy ϵ_l with $\xi < \frac{2}{L}$, the total amount of computing resources should satisfy,

$$
\begin{aligned}
C_{total} &\geq \left\lceil \frac{2}{(2 - L\xi)\xi\gamma} \ln\frac{1}{\epsilon_l} \right\rceil \cdot \left\lceil \frac{2L^2 \ln\frac{1}{\epsilon_g}}{(1 - \epsilon_l)\gamma^2 \zeta} \right\rceil \\
&\cdot \sum_{i=1}^{n} \frac{c_i S_i}{T_i} \cdot \frac{1}{\sqrt{2\pi}\sigma_i} \exp\left(-\frac{(S_i - \mu_i)^2}{2\sigma_i^2}\right) \cdot \frac{(d_0)^2 - (d_{min})^2}{(r_0)^2}.
\end{aligned}
\tag{3.26}
$$

3.5.2 Case 2: Sufficient Computing Resources and Insufficient Communication Resources

When computing resources are sufficient, while communication resources are insufficient, we aim to reduce bandwidth consumption by reducing the number of communication rounds. In this case, the number of local iterations still follows Corollary 3.2, as computing resources are sufficient. However, Corollary 3.3 may not be satisfied due to the lack of communication resources, which decreases the number of communication rounds K. As a result, the required global model accuracy cannot be achieved. Specifically, the maximal number of communication rounds K_{max} is constrained by communication resources, i.e., $K_{max} = \lfloor \min\{\frac{B_{down}^{max}}{\overline{B}_{down}}, \frac{B_{up}^{max}}{\overline{B}_{up}}\} \rfloor$, where B_{up}^{max} and B_{down}^{max} represent the maximal available bandwidth that can be used for FL on the uplink and the downlink, respectively, $\overline{B}_{up}^r = \sum_{i=1}^{n} b_{up}^{i,r} f_{SINR_{up}}$ and $\overline{B}_{down}^r = \sum_{i=1}^{n} b_{down}^{i,r} f_{SINR_{down}}$ represent the mean bandwidth consumption on the uplink and downlink during one communication round, respectively. To achieve the required global accuracy

$1 - \epsilon_g$, when the number of communication round is constrained, based on Section 3.9, we first give the following relationship:

$$F\left(g_r\left(\hat{S}, SINR_{\text{up}}\right)\right) - F\left(g^*\right) \leq \exp\left(-K\left(\frac{(1-\epsilon_l)\gamma^2\zeta}{2L^2}\right)\right) \cdot$$
$$\left(F(g_0) - F(g^*)\right),$$
(3.27)

where we can reasonably expect that the real achieved global accuracy loss $\tilde{\epsilon}_g$ is given by

$$\tilde{\epsilon}_g = \exp\left(-K\left(\frac{(1-\tilde{\epsilon}_l)\gamma^2\zeta}{2L^2}\right)\right).$$
(3.28)

Therefore, $K = \lceil\frac{2L^2\ln\frac{1}{\tilde{\epsilon}_g}}{(1-\tilde{\epsilon}_l)\gamma^2\zeta}\rceil$, where $\tilde{\epsilon}_l$ represents the realistic local model accuracy loss. Moreover, we can observe that when $\tilde{\epsilon}_g$ is fixed, K decreases as $\tilde{\epsilon}_l$ decreases. If we want to reduce K and thus to reduce bandwidth consumption, while keeping $\tilde{\epsilon}_g$ unchanged, we should decrease $\tilde{\epsilon}_l$ by increasing the number of local iterations. As a result, the computing resource consumption will increase. In fact, there exists a trade-off to some extent between the communication resources and computing resources for achieving a certain ML model accuracy. In addition, from the aspect of communication resources, the number of communication rounds K should meet $K \leq K_{\max}$. Therefore, we have $\tilde{\epsilon}_l \leq \lfloor 1 - \frac{2L^2\ln\frac{1}{\tilde{\epsilon}_g}}{K_{\max}\gamma^2\zeta}\rfloor$.

Therefore, based on Corollary 3.2, the number of local iterations should satisfy

$$\tau \geq \left\lceil \frac{2}{(2-L\xi)\xi\gamma}\ln\frac{K_{\max}\gamma^2\zeta}{K_{\max}\gamma^2\zeta - 2L^2\ln\frac{1}{\tilde{\epsilon}_g}} \right\rceil.$$
(3.29)

As a result, the total computing resources needed for the FL task is given by

$$C_{\text{total}} \geq \sum_{i=1}^{n}\frac{c_i S_i}{T_i} \cdot \frac{1}{\sqrt{2\pi}\sigma_i}\exp\left(-\frac{(s_i - \mu_i)^2}{2\sigma_i^2}\right) \cdot \frac{(d_0)^2 - (d_{\min})^2}{(r_0)^2}$$
$$\cdot \left\lceil \frac{2}{(2-L\xi)\xi\gamma}\ln\frac{K_{\max}\gamma^2\zeta}{K_{\max}\gamma^2\zeta - 2L^2\ln\frac{1}{\tilde{\epsilon}_g}} \right\rceil$$
$$\cdot \left\lceil \frac{2L^2\ln\frac{1}{\tilde{\epsilon}_g}}{(1-\tilde{\epsilon}_l)\gamma^2\zeta} \right\rceil.$$
(3.30)

3.5.3 Case 3: Sufficient Communication Resources and Insufficient Computing Resources

When communication resources are sufficient, while computing resources are insufficient, we aim to reduce the consumption of computing resources by reducing the number of local iterations. As communication resources are sufficient, the number of communication rounds still follows Corollary 3.3, while Corollary 3.2 may not be satisfied as the contrained computing resources decreases the number of local iterations. As a result, the required local model accuracy cannot be achieved. In addition, from the aspect of computing resources, the number of local iterations should satisfy $\tau \cdot \overline{C}_{\mathrm{UE}} \leq \sum_{i=1}^{n} C_i$, where C_i denotes the maximal computing resources used for local training on UE i. To achieve the required local accuracy although the number of local iterations are constrained, we give the following relationship based on Section 3.8,

$$
G_i\left(w_r, h_i^{(r)(t)}\right) - G_i\left(w_r, h_i^{(r)^*}\right) \leq \exp\left(-\tau \frac{(2 - L\xi)\xi\gamma}{2}\right)
$$
$$
\cdot \left(G_i\left(w_r, h_i^{(r)(0)}\right) - G_i\left(w_r, h_i^{(r)^*}\right)\right), \tag{3.31}
$$

from which we can reasonably expect that the real local model accuracy loss $\tilde{\epsilon}_l$ is given by $\tilde{\epsilon}_l = \exp\left(-\tau \frac{(2 - L\xi)\xi\gamma}{2}\right)$, based on which we can derive the number of local iterations, i.e., $\tau = \lceil \frac{2\ln\frac{1}{\tilde{\epsilon}_l}}{(2 - L\xi)\xi\gamma} \rceil$.

Therefore, when $\xi < \frac{2}{L}$, $\tilde{\epsilon}_l \geq \exp\left(\frac{(L\xi - 2)\xi\gamma \sum_{i=1}^{n} C_i}{2\overline{C}_{\mathrm{UE}}}\right)$, from which we can see that $\tilde{\epsilon}_l$ increases when the total amount of available computing resource decreases. Moreover, based on $\overline{C}_{\mathrm{UE}}$ in Section 3.3, we can derive the lower bound of the number of communication rounds, which is given by

$$
K \geq \left\lceil \frac{2L^2 \ln\frac{1}{\epsilon_g}}{\left(1 - \exp\left(\frac{(L\xi - 2)\xi\gamma \sum_{i=1}^{n} C_i}{2\sum_{i=1}^{n} \frac{c_i S_i}{T_i} \cdot \frac{1}{\sqrt{2\pi}\sigma_i} \exp\left(-\frac{(s_i - \mu_i)^2}{2\sigma_i^2}\right)}\right)\right) \gamma^2 \zeta} \right\rceil. \tag{3.32}
$$

Therefore, the bandwidth needed for transmitting local models and the global model in the uplink and the downlink are, respectively, given by

$$
\overline{B}_{\mathrm{up}} \geq \left\lceil \frac{2L^2 \ln\frac{1}{\epsilon_g}}{\left(1 - \exp\left(\frac{(L\xi - 2)\xi\gamma \sum_{i=1}^{n} C_i}{2\sum_{i=1}^{n} \frac{c_i S_i}{T_i} \cdot \frac{1}{\sqrt{2\pi}\sigma_i} \exp\left(-\frac{(s_i - \mu_i)^2}{2\sigma_i^2}\right)}\right)\right) \gamma^2 \zeta} \right\rceil
$$
$$
\cdot \sum_{i=1}^{n} \frac{S}{T_{\mathrm{up}}^{i,r} \log_2\left(1 + \frac{P_{\mathrm{up}} G(d_1)}{\sum_{j=1}^{N_I} P G(d_2^{(j)}) + \delta^2}\right)}
$$

$$\cdot \left(\frac{2d_1}{(d_0)^2}\right) \cdot \sum_{n_\mathcal{A}=n_l}^{N} C_{n_\mathcal{A}}^{n_l} (1 - \exp\{1 - 2t_{\text{up}}\lambda_a\})^{n_l}$$

$$\cdot (\exp\{1 - 2t_{\text{up}}\lambda_a\})^{n_\mathcal{A} - n_l} \cdot \frac{(\pi(d_0)^2 \lambda_i)^{n_\mathcal{A}}}{n_\mathcal{A}!}$$

$$\exp\{-\pi(d_0)^2\lambda_i\} \cdot \left(\frac{2}{(d_0)^2}\right)^{n_l} \prod_{n=1}^{n_l} d_2^{(n)}, \tag{3.33}$$

$$\overline{B}_{\text{down}} \geq \left[\frac{2L^2 \ln\frac{1}{\epsilon_g}}{\left(1 - \exp\left(\frac{(L\xi - 2)\xi\gamma \sum_{i=1}^{n} C_i}{2\sum_{i=1}^{n} \frac{c_i S_i}{T_i} \cdot \frac{1}{\sqrt{2\pi}\sigma_i} \exp\left(-\frac{(s_i - \mu_i)^2}{2\sigma_i^2}\right)}\right)\right)\gamma^2\zeta}\right] \tag{3.34}$$

$$\cdot \sum_{i=1}^{n} \frac{s}{T_{\text{down}}^{i,r} \log_2(1 + \frac{P_{\text{down}} G(d_1)}{\delta^2})} \cdot \frac{2d_1}{(r_0)^2}.$$

Therefore, based on the analysis aforementioned, we provide Corollary 3.4 about the resource consumption for the three cases discussed above.

Corollary 3.4

(1) *Case 1-Sufficient Communication and Computing Resources: to achieve the required model accuracy ϵ_g and ϵ_l, the consumption of bandwidth (uplink direction) is given by* $\overline{B}_{\text{up}} \geq \lceil \frac{2L^2 \ln\frac{1}{\epsilon_g}}{(1-\epsilon_l)\gamma^2\zeta} \rceil \cdot \left(\frac{2}{(d_0)^2}\right)^{n_l} \prod_{n=1}^{n_l} d_2^{(n)} \cdot$ $\sum_{i=1}^{n} \frac{s}{T_{\text{up}}^{i,r} \log_2(1 + \frac{P_{\text{up}} G(d_1)}{\sum_{j=1}^{N_l} PG(d_2^{(j)}) + \delta^2})} \cdot (\frac{2d_1}{(d_0)^2}) \cdot \sum_{n_\mathcal{A}=n_l}^{N} C_{n_\mathcal{A}}^{n_l} (1 - \exp\{1 - 2t_{\text{up}}\lambda_a\})^{n_l} \cdot$ $(\exp\{1 - 2t_{\text{up}}\lambda_a\})^{n_\mathcal{A} - n_l} \cdot \frac{(\pi(d_0)^2\lambda_i)^{n_\mathcal{A}}}{n_\mathcal{A}!} \exp\{-\pi(d_0)^2\lambda_i\},$ *the consumption of bandwidth (downlink direction) is given by* $\overline{B}_{\text{down}} \geq \lceil \frac{2L^2 \ln\frac{1}{\epsilon_g}}{(1-\epsilon_l)\gamma^2\zeta} \rceil \cdot \frac{2d_1}{r_0^2} \cdot$ $\sum_{i=1}^{n} \frac{s}{T_{\text{down}}^{i,r} \log_2(1 + \frac{P_{\text{down}} G(d_1)}{\delta^2})},$ *and the consumption of computing resources needed for local training is given by* $C_{\text{total}} \geq \lceil \frac{2}{(2-L\xi)\xi\gamma} \ln\frac{1}{\epsilon_l} \rceil \cdot \lceil \frac{2L^2 \ln\frac{1}{\epsilon_g}}{(1-\epsilon_l)\gamma^2\zeta} \rceil \cdot \sum_{i=1}^{n} \frac{c_i S_i}{T_i} \cdot$ $\frac{1}{\sqrt{2\pi}\sigma_i} \exp\left(-\frac{(s_i - \mu_i)^2}{2\sigma_i^2}\right) \cdot \frac{(d_0)^2 - (d_{\min})^2}{(r_0)^2}.$

(2) *Case 2-Sufficient Computing Resources and Insufficient Communication Resources: to achieve the required global model accuracy ϵ_g, the consumption of computing resources should satisfy* $C_{\text{total}} \geq \sum_{i=1}^{n} \frac{c_i S_i}{T_i} \cdot \frac{1}{\sqrt{2\pi}\sigma_i} \exp\left(-\frac{(s_i - \mu_i)^2}{2\sigma_i^2}\right) \cdot$ $\frac{(d_0)^2 - (d_{\min})^2}{(r_0)^2} \cdot \lceil \frac{2}{(2-L\xi)\xi\gamma} \ln\frac{K_{\max}\gamma^2\zeta}{K_{\max}\gamma^2\zeta - 2L^2\ln\frac{1}{\epsilon_g}} \rceil \cdot \lceil \frac{2L^2 \ln\frac{1}{\epsilon_g}}{(1-\tilde{\epsilon}_l)\gamma^2\zeta} \rceil.$

(3) *Case 3-Sufficient Communication Resources and Insufficient Computing Resources: to achieve the required global model accuracy ϵ_g, the consumption of bandwidth (uplink direction) should meet*

$$\overline{B}_{up} \geq \left\lceil \frac{2L^2 \ln \frac{1}{\epsilon_g}}{\left(1 - \exp\left(\frac{(L\xi - 2)\xi\gamma \sum_{i=1}^{n} C_i}{2\sum_{i=1}^{n} \frac{c_i S_i}{T_i} \cdot \frac{1}{\sqrt{2\pi}\sigma_i} \exp\left(-\frac{(s_i - \mu_i)^2}{2\sigma_i^2}\right)}\right)\right)\gamma^2 \zeta} \right\rceil \cdot \sum_{i=1}^{n} \frac{s}{T_{up}^{i,r} \log_2(1 + \frac{P_{up} G(d_1)}{\sum_{j=1}^{N_I} P G(d_2^{(j)}) + \delta^2})} \cdot \left(\frac{2d_1}{(d_0)^2}\right) \cdot$$

$$\sum_{n_A = n_I}^{N} C_{n_A}^{n_I} (1 - \exp\{1 - 2t_{up}\lambda_a\})^{n_I} \cdot (\exp\{1 - 2t_{up}\lambda_a\})^{n_A - n_I} \cdot \frac{(\pi(d_0)^2 \lambda_i)^{n_A}}{n_A!}$$

$$\exp\{-\pi(d_0)^2 \lambda_i\} \cdot \left(\frac{2}{(d_0)^2}\right)^{n_I} \prod_{n=1}^{n_I} d_2^{(n)}, \text{ while the consumption of bandwidth}$$

(downlink direction) should meet

$$\overline{B}_{down} \geq \left\lceil \frac{2L^2 \ln \frac{1}{\epsilon_g}}{\left(1 - \exp\left(\frac{(L\xi - 2)\xi\gamma \sum_{i=1}^{n} C_i}{2\sum_{i=1}^{n} \frac{c_i S_i}{T_i} \cdot \frac{1}{\sqrt{2\pi}\sigma_i} \exp\left(-\frac{(s_i - \mu_i)^2}{2\sigma_i^2}\right)}\right)\right)\gamma^2 \zeta} \right\rceil \cdot \sum_{i=1}^{n} \frac{s}{T_{down}^{i,r} \log_2(1 + \frac{P_{down} G(d_1)}{\delta^2})} \cdot \frac{2d_1}{(r_0)^2} \cdot$$

3.6 Numerical Results and Discussion

In this section, we verify our analytical modeling using numerical simulations by (i) verifying the analytical results of transmission success probability (uplink and downlink) and the consumption of bandwidth and computing resources; (ii) measuring the performance of FL settings; and (iii) examining the trade-off between the computing resources and communication resources under FL framework.

3.6.1 Simulation Setting

We consider an FL-enabled wireless edge network cosisting of multiple UEs that are randomly generated and one central BS with a cloud server that serves as the FL model aggregator. The coverage of the BS is a circular area with a radius of 1 km. Similarly, the interfering area is also a circular area with a radius of 100 m, where the density of interfering UEs λ_a is set to 1 UE/m^2. The transmit power of UEs and the transimit power of the serving BS are set to 20 and 43 dBm respectively (Sun ct al. [2019]). Moreover, the noise power is set to -173 dBm (Sun et al. [2019]). t_{up} is randomly set within $[1, 3]$ ms. The path loss is modeled as $g(D_1) = 34 + 40 \log(D_1)$ (Liu et al. [2020b]). The number of CPU cycles is randomly set within $[1, 4] \times 10^4$ cycles/sample (Yang et al. [2020b]). μ_i and σ_i are randomly set within $[1000, 10, 000]$ and $[0.2, 0.5]$.

We consider using FL to solve the multiclass classification problem over MNIST datasets (LeCun [1998]) for model training, where all datasets of UEs are randomly splitted with $75:25$ ratio, for training and testing, respectively

(McMahan et al. [2017]). Moreover, we use a fully connected two-layer network built over PyTorch, where the size of input layer, hidden layer, and output layer is set to 784 (28 × 28), 40, and 10, respectively. The activation function is ReLU, as it can greatly accelerate the convergence of gradient descent and increase the number of the activated neurons (Li and Yuan [2017], Krizhevsky et al. [2012]). In addition, inspired by the hyperparameter analysis in Darken and Moody [1990] and Liu et al. [2019] and the corresponding experimental results, we set learning rate $\xi = 0.03$. Moreover, according to our neural network settings, the transmitted model size s is around 1.156 MB, when using 32-bit floating-point computing. $L = 1/10, \gamma = 1/10, \xi = 1/10, \zeta = 1/10$ (Yang et al. [2020b]).

3.6.2 Simulation Results

3.6.2.1 Verifying Analytical Results

First, we examine the local/global model transmission success probability with varying UE density. We randomly generated 30 specific point distributions for each UE density (0.1 intervals) based on PPP model. Note both the simulation results of $\Pr(\mathrm{SINR}_{\mathrm{up}} > \beta_{\mathrm{up}})$ and $\Pr(\mathrm{SNR}_{\mathrm{down}} > \beta_{\mathrm{down}})$ are averaged over these 30 different channel instances for each UE density. Fig. 3.3 and Fig. 3.4 show the probability $\Pr(\mathrm{SINR}_{\mathrm{up}} > \beta_{\mathrm{up}})$ (uplink direction) and $\Pr(\mathrm{SNR}_{\mathrm{down}} > \beta_{\mathrm{down}})$ (downlink direction), respectively, for both analytical and simulation results with varying UE density under different threshold parameters. The analytical results of $\Pr(\mathrm{SINR}_{\mathrm{up}} > \beta_{\mathrm{up}})$ and $\Pr(\mathrm{SNR}_{\mathrm{down}} > \beta_{\mathrm{down}})$ are computed based on Eqs. (3.13) and (3.14), respectively. From Fig. 3.3 and Fig. 3.4, we can see that the curves of analytical results match closely to simulations for both the uplink and the downlink. Moreover, we can see that the smaller threshold (β_{up} in Fig. 3.3, β_{down} in Fig. 3.4), the larger transmission success probability.

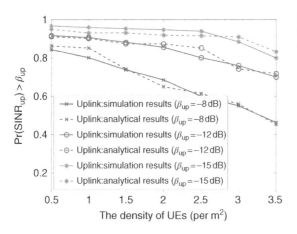

Figure 3.3 Comparison of the probability of successful transmission in the uplink.

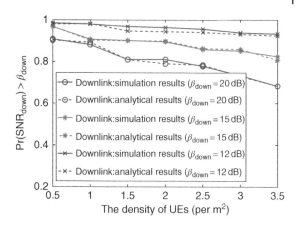

Figure 3.4 Comparison of the probability of successful transmission in the downlink.

Next, we examine the bandwidth consumption for uplink direction and downlink direction, respectively, for both analytical and simulation results with respect to the global accuracy loss ϵ_g. The UE density is randomly set within $[1, 2]$, where we randomly select 10 specific point distributions under the corresponding UE density. In this simulation experiment, we train the same FL task (i.e., classification on MNIST datasets) for each point distribution, where the simulation results are averaged over 10 point distributions. Fig. 3.5 and Fig. 3.6 show the bandwidth consumption for uplink direction and downlink direction changes with the global accuracy loss, respectively. From Fig. 3.5 and Fig. 3.6, we can see that the curves of analytical results match closely to simulations for both the uplink and the downlink. Moreover, both the bandwidth consumption in the uplink and the downlink decrease with global accuracy loss. Moreover, we can also find that the lower local accuracy leads to more bandwidth consumption to guarantee a specific global

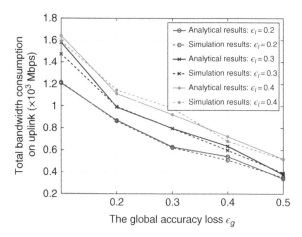

Figure 3.5 Comparison of bandwidth consumption in the uplink.

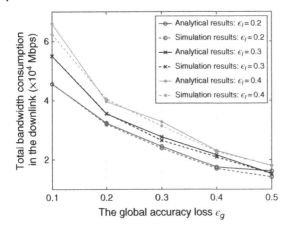

Figure 3.6 Comparison of bandwidth consumption in the downlink.

Legend:
—⊖— Analytical results: $\epsilon_l = 0.2$
- ⊖ - Simulation results: $\epsilon_l = 0.2$
—✳— Analytical results: $\epsilon_l = 0.3$
- ✳ - Simulation results: $\epsilon_l = 0.3$
—●— Analytical results: $\epsilon_l = 0.4$
- ● - Simulation results: $\epsilon_l = 0.4$

Y-axis: Total bandwidth consumption in the downlink ($\times 10^4$ Mbps)
X-axis: The global accuracy loss ϵ_g

accuracy. Specifically, as shown in Fig. 3.5, the amount of bandwidth (uplink direction) consumed to guarantee $\epsilon_l = 0.3$ is 0.144×10^3 Mbps more than that to guarantee $\epsilon_l = 0.2$ on average, while the amount of bandwidth consumed to guarantee $\epsilon_l = 0.4$ is 0.089×10^3 Mbps more than that to guarantee $\epsilon_l = 0.3$ on average. As shown in Fig. 3.6, the amount of bandwidth (downlink direction) consumed to guarantee $\epsilon_l = 0.3$ is 0.54×10^4 Mbps more than that to guarantee $\epsilon_l = 0.2$ on average, while the amount of bandwidth to guarantee $\epsilon_l = 0.4$ is 0.28×10^4 Mbps more than that to guarantee $\epsilon_l = 0.3$ on average. This is because the lower local accuracy needs more communication rounds to aggregate the local models to achieve a certain global accuracy, and thus consumes more bandwidth.

In the following, we examine the for both analytical and simulation results with respect to the density of UEs. Specifically, the analytical results of computing resource consumption are computed based on Eq. (3.19), while the simulation results are averaged over 10 randomly generated point distributions for each UE density. Fig. 3.7 shows the computing resource consumption changes with the density of UEs. From Fig. 3.7, we can see that the consumption of computing resource increases in the beginning and then decreases with the density of UEs. Specifically, the consumption of computing resource increases with UE density, when it is approximately below 2 (i.e., $\lambda_i \leq 2$). This is because the number of UEs that participate in local training increases with UE density. When approximately $\lambda_i \geq 2$, the consumption of computing resource decreases with UE density, as poor SNR causes that some UEs fail in successfully receiving the global model. As a result, the number of UEs that participate in local training decreases in the next communication round, and thus the consumption of computing resource decreases. Moreover, we also find that achieving higher local accuracy consumes more computing resources. Specifically, the amount of computing resources consumed to guarantee $\epsilon_l = 0.2$ is 0.27×10^{12} cycles/s more than that to guarantee

Figure 3.7 Comparison of the computing resource consumption.

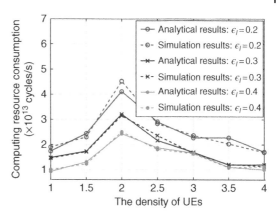

$\epsilon_l = 0.3$ on average, while the amount of computing resources consumed to guarantee $\epsilon_l = 0.3$ is 15×10^{12} cycles/s more than that to guarantee $\epsilon_l = 0.4$ on average.

3.6.2.2 Measuring the Performance of FL Settings

First, we examine the convergence property by using simulation experiments. In this simulation experiment, we randomly set the UE density within $[1, 2]$ and randomly generate data points (the same settings for the following simulations). Moreover, we set $\epsilon_g = 0.2$, $\beta_{up} = -15$ dB, and $\beta_{down} = 15$ dB. As shown in Fig. 3.8, we randomly choose 3 UEs to observe the changes of the local optimization function, from which we can see that the local optimization function converges in

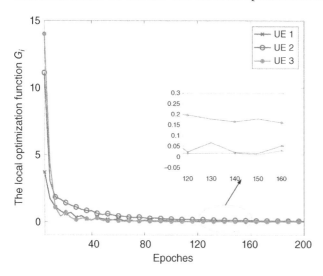

Figure 3.8 Local training during each communication round.

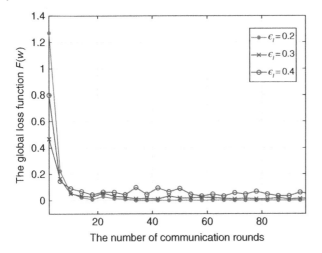

Figure 3.9 Convergence.

about 40 epochs. In addition, as shown in Fig. 3.9, we can observe that the global loss function convergences in around 12 communication rounds.

Next, we examine the global accuracy loss with respect to the number of communication rounds when local accuracy loss is fixed (i.e., $\epsilon_l = 0.2$). In this simulation experiment, we still set $\beta_{\mathrm{up}} = -15\,\mathrm{dB}$ and $\beta_{\mathrm{down}} = 15\,\mathrm{dB}$. Fig. 3.10 shows the global model accuracy loss changes with respect to the number of communication rounds. From Fig. 3.10, we can see that the global model accuracy loss decreases with the number of communication rounds. Moreover, the difference between $\epsilon_l = 0.2$ and the actual global accuracy loss is always within 0.1 when

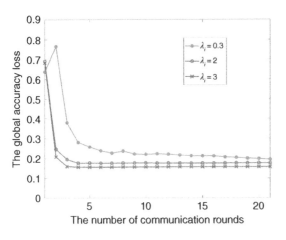

Figure 3.10 Comparison of the global accuracy loss.

Figure 3.11 The relationship between training accuracy and testing accuracy.

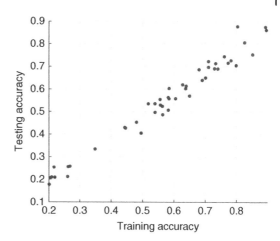

the learning convergences. Note that SINR and SNR practically affect the global aggregation and local training, respectively.

Next, we examine whether the well-trained model is effective for the test datasets. In this simulation experiment, the test datasets are drawn from the same distribution as the training data. We randomly select 3 UEs and calculate the testing accuracy every 2 communication rounds. As shown in Fig. 3.11, we can see that the testing accuracy increases with respect to training accuracy, where ϵ_l and $\hat{\epsilon}_l$ denote the local training accuracy and the testing training accuracy, respectively. Moreover, we can also see that the difference between the training accuracy and the testing accuracy is always within $[0, 0.1]$.

3.6.2.3 Examining the Trade-Off between the Computing and Communication Resources under FL Framework

First, we examine the relationship between the global model accuracy and bandwidth resources (uplink direction). We assume that the bandwidth resources in the downlink and computing resources are sufficient, and the required local model accuracy $(0.1, 0.2, 0.3)$ is fixed. From Fig. 3.12, we can see that the global model accuracy sharply increases in the beginning and then increases slowly with respect to the amount of bandwidth resources (uplink direction). The reason is that the number of local UEs that can participate in global aggregation quickly increases with the amount of bandwidth in the beginning. When the amount of bandwidth resources increases to be sufficient, it has little effect on the transmission success probability of local models (uplink direction), and thus the global model accuracy keeps fairly steady. Moreover, in the beginning, higher local model accuracy (lower ϵ_l) leads to a higher global model accuracy. Specifically, the global model accuracy when $\epsilon_l = 0.1$ is 8.5% higher than that when $\epsilon_l = 0.2$, while the global model accuracy when $\epsilon_l = 0.2$ is 17.2% higher than that when $\epsilon_l = 0.3$.

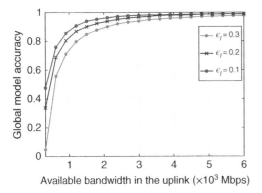

Figure 3.12 The relationship between available bandwidth in the uplink and global model accuracy.

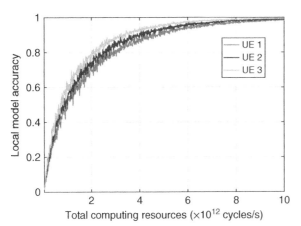

Figure 3.13 The relationship between available computing resources and local model accuracy.

After that we examine the relationship between the local model accuracy and computing resources, as shown in Fig. 3.13. We randomly select 3 different UEs and assume that the bandwidth resources in both the uplink and downlink are sufficient. From Fig. 3.13, we can see that the local model accuracy quickly increases in the beginning and then keeps fairly steady with respect to the amount of computing resources. This is because that more computing resources lead to more local iterations in the beginning, while sufficient computing resources have little effect on local iterations. Note the fluctuations of the curves in Fig. 3.13 are due to that the local model accuracy in Fig. 3.13 is recorded per local iteration, while the global accuracy in Fig. 3.12 is recorded per communication round composed of 30 local iterations.

Finally, we examine the trend of the global model accuracy with respect to the amount of computing resources as well as the amount of bandwidth resources (uplink direction). Fig. 3.14 shows the relationship among the global model

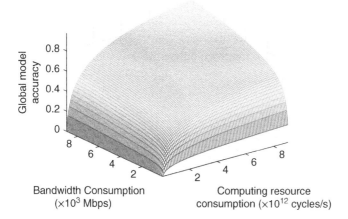

Figure 3.14 The trade-off between the computing resources and bandwidth.

accuracy, computing resources, and bandwidth resources. From Fig. 3.14, we can see that the trade-off between the amount of computing resources and the amount of bandwidth resources is verified. As shown in Fig. 3.14, both bandwidth and computing resources affect the global model accuracy, where we can flexibly adjust the amount of computing resources and/or the amount of bandwidth resources to guarantee the specific global model accuracy. Specifically, when the amount of bandwidth used for transmitting the local models is fixed, we can increase the global model accuracy by increasing the amount of computing resources. When the amount of computing resources used for local training is fixed, we can increase the global model accuracy by increasing the amount of bandwidth resources.

3.7 Conclusion

Wireless edge networks enabled by FL have been widely acknowledged as a very promising means to address a wide range of challenging network issues. In this chapter, we have theoretically analyzed how accurate of an ML model can be achieved by using FL and how many resources are consumed to guarantee a certain local/global accuracy. Specifically, we have derived the explicit expression of the model accuracy under FL framework, as a function of the amount of computing resources/communication resources for FL-enabled wireless edge networks. Numerical results validate the effectiveness of our theoretical modeling. The modeling and results can provide some fundamental understanding for the trade-off between the learning performance and consumed resources, which is useful for promoting FL empowered wireless network edge intelligence.

3.8 Proof of Corollary 3.2

First, based on the definition of $G_i(\cdot)$, we have $\nabla^2 G_i(\cdot) = \nabla^2 F_i(\cdot)$. Therefore, $\nabla^2 G_i(\cdot)$ also meets the L-smooth, γ-convex, and twice-differentiable assumptions, i.e.,

$$\|\nabla G_i(w) - \nabla G_i(w')\| \leq L\|w - w'\|,$$

$$\|\nabla G_i(w) - \nabla G_i(w')\| \geq \gamma\|w - w'\|,$$

$$\gamma I \leq \nabla^2 G_i(w) \leq LI$$

$$G_i(w) \geq G_i(w') + \nabla\left(G_i(w')\right)^T(w - w') + \frac{\gamma}{2}\|w - w'\|^2.$$

Then, when given w_r, we rewrite $G_i(\cdot)$ using the second-order Taylor expansion as

$$\begin{aligned}
G_i\left(w_r, h_i^{(r)(t+1)}\right) &= G_i\left(w_r, h_i^{(r)(t)}\right) + \left(h_i^{(r)(t+1)} - h_i^{(r)(t)}\right)^T \\
&\quad \cdot \nabla G_i\left(w_r, h_i^{(r)(t)}\right) + \frac{1}{2}\left(h_i^{(r)(t+1)} - h_i^{(r)(t)}\right)^T \\
&\quad \cdot \nabla^2 G_i\left(w_r, h_i^{(r)(t)}\right) \cdot \left(h_i^{(r)(t+1)} - h_i^{(r)(t)}\right).
\end{aligned}$$

As in GD method, we have

$$h_i^{(r)(t+1)} = h_i^{(r)(t)} - \xi\nabla G_i(w_r, h_i^{(r)(t)}).$$

Therefore, we have

$$\begin{aligned}
G_i(w_r, h_i^{(r)(t+1)}) &= G_i\left(w_r, h_i^{(r)(t)}\right) - \xi\left\|\nabla G_i\left(w_r, h_i^{(r)(t)}\right)\right\|^2 \\
&\quad + \frac{\xi^2}{2}\left\|\nabla G_i\left(w_r, h_i^{(r)(t)}\right)\right\|^2 \nabla^2 G_i\left(w_r, h_i^{(r)(t)}\right) \\
&\overset{(C.3)}{\leq} G_i\left(w_r, h_i^{(r)(t)}\right) - \xi\left\|\nabla G_i\left(w_r, h_i^{(r)(t)}\right)\right\|^2 \\
&\quad + \frac{L\xi^2}{2}\left\|\nabla G_i\left(w_r, h_i^{(r)(t)}\right)\right\|^2 \\
&= G_i\left(w_r, h_i^{(r)(t)}\right) - \frac{(2 - L\xi)\xi}{2} \cdot \left\|\nabla G_i\left(w_r, h_i^{(r)(t)}\right)\right\|^2,
\end{aligned}$$

Next we start to find the lower bound of $\|\nabla G_i\left(w_r, h_i^{(r)(t)}\right)\|^2$. For the optimal $h_i^{(r)^*}$ of $G_i(w_r, h_i^{(r)^*})$, we always have $\nabla G_i(w_r, h_i^{(r)^*}) = 0$. Therefore, we have

$$\begin{aligned}
\left\|\nabla G_i\left(w_r, h_i^{(r)(t)}\right)\right\|^2 &= \left\|\nabla G_i\left(w_r, h_i^{(r)(t)}\right) - \nabla G_i\left(w_r, h_i^{(r)^*}\right)\right\|^2 \\
&\overset{(C.2)}{\geq} \gamma\left\|\nabla G_i\left(w_r, h_i^{(r)(t)}\right) - \nabla G_i\left(w_r, h_i^{(r)^*}\right)\right\|
\end{aligned}$$

$$\cdot \left\| (w_r, h_i^{(r)(t)}) - (w_r, h_i^{(r)^*}) \right\|$$

$$\geq \gamma \left(\nabla G_i \left(w_r, h_i^{(r)(t)} \right) - \nabla G_i \left(w_r, h_i^{(r)^*} \right) \right)^{\mathrm{T}}$$

$$\cdot \left((w_r, h_i^{(r)(t)}) - (w_r, h_i^{(r)^*}) \right)$$

$$= \gamma \nabla G_i \left(w_r, h_i^{(r)(t)} \right)^{\mathrm{T}} \left((w_r, h_i^{(r)(t)}) - (w_r, h_i^{(r)^*}) \right)$$

$$\overset{(C.4)}{\geq} \gamma \left(G_i \left(w_r, h_i^{(r)(t)} \right) - G_i \left(w_r, h_i^{(r)^*} \right) \right).$$

Therefore, we have

$$G_i \left(w_r, h_i^{(r)(t+1)} \right) - G_i \left(w_r, h_i^{(r)^*} \right) \leq G_i \left(w_r, h_i^{(r)(t)} \right) - G_i \left(w_r, h_i^{(r)^*} \right)$$

$$- \frac{(2 - L\xi)\xi\gamma}{2} \left(G_i \left(w_r, h_i^{(r)(t)} \right) - G_i \left(w_r, h_i^{(r)^*} \right) \right)$$

$$\overset{(c)}{\leq} \left(1 - \frac{(2 - L\xi)\xi\gamma}{2} \right)^{(\tau+1)} \cdot \left(G_i \left(w_r, h_i^{(r)(0)} \right) - G_i \left(w_r, h_i^{(r)^*} \right) \right)$$

$$\leq \exp \left(-(\tau + 1) \frac{(2 - L\xi)\xi\gamma}{2} \right) \cdot \left(G_i \left(w_r, h_i^{(r)(0)} \right) - G_i \left(w_r, h_i^{(r)^*} \right) \right),$$

where (c) can be obtained from $1 - x \leq \exp(-x)$. Therefore, to ensure that $G_i(w_r, h_i^{(r)(t)}) - G_i(w_r, h_i^{(r)^*}) \leq \epsilon_l (G_i(w_r, h_i^{(r)(0)}) - G_i(w_r, h_i^{(r)^*}))$, we have

$$\epsilon_l \geq \exp \left(-(\tau) \frac{(2 - L\xi)\xi\gamma}{2} \right).$$

Therefore, when $\xi < \frac{2}{L}$ (as $(2 - L\xi)\xi\gamma > 0$), we have

$$\tau \geq \frac{2}{(2 - L\xi)\xi\gamma} \ln \left(\frac{1}{\epsilon_l} \right),$$

3.9 Proof of Corollary 3.3

Under the assumptions of $F_i(w)$, the following conditions hold on

$$\frac{1}{L} \| \nabla F_i(w_r + h_i) - \nabla F_i(w_r) \|^2 \leq \left(\nabla F_i(w_r + h_i) - \nabla F_i(w_r) \right)^{\mathrm{T}} h_i$$

$$\leq \frac{1}{\gamma} \| \nabla F_i(w_r + h_i) - \nabla F_i(w_r) \|^2,$$

$$\| \nabla F(w) \|^2 \geq \gamma \left(F(w) - F(w^*) \right).$$

Based on Lagrange median theorem, we always have a w such that

$$\nabla F_i(w_r + h_i) - \nabla F_i(w_r) = \nabla^2 F_i(w) h_i.$$

For the optimal solution of local optimization problem, we always have

$$F_i\left(w_r + h_i^{(r)*}\right) - \left(\nabla F_i(w_r) - \zeta\nabla F(w_r)\right)^{\mathrm{T}} h_i^{(r)*} = 0.$$

With the above equalities and inequalities, we now start to prove Corollary 3.3.

$$F\left(w_{r+1}\left(\hat{S}, SINR_{\mathrm{up}}\right)\right)$$

$$= F\left(w_r\left(\hat{S}, SINR_{\mathrm{up}}\right) + \frac{\sum_{i=1}^n \hat{S}_i h_i^{(r)(\tau)}}{\sum_{i=1}^n \hat{S}_i}\right)$$

$$\overset{(A1)}{\leq} F\left(w_r\left(\hat{S}, SINR_{\mathrm{up}}\right)\right) + \frac{1}{\sum_{i=1}^n \hat{S}_i}\sum_{i=1}^n \hat{S}_i \nabla F\left(w_r\left(\hat{S}, SINR_{\mathrm{up}}\right)\right)^{\mathrm{T}}$$

$$\cdot h_i^{(r)(\tau)} + + \frac{L}{2\left(\sum_{i=1}^n \hat{S}_i\right)^2}\left\|\sum_{i=1}^n \hat{S}_i h_i^{(r)(\tau)}\right\|^2.$$

As

$$G_i(w_r, h_i) = F_i(w_r + h_i) - \left(\nabla F_i(w_r) - \zeta\nabla F(w_r)\right)^{\mathrm{T}} h_i,$$

$$\left\|\frac{1}{\left(\sum_{i=1}^n \hat{S}_i\right)^2}\sum_{i=1}^n \hat{S}_i h_i^{(r)(\tau)}\right\|^2 \leq \left(\frac{1}{\sum_{i=1}^n \hat{S}_i}\sum_{i=1}^n \left\|\hat{S}_i h_i^{(r)(\tau)}\right\|\right)\left\|h_i^{(r)(\tau)}\right\|$$

$$\overset{(\hat{S}_i \geq 0)}{\leq} \frac{1}{\sum_{i=1}^n \hat{S}_i}\sum_{i=1}^n \hat{S}_i \left\|h_i^{(r)(\tau)}\right\|^2,$$

Therefore, we have

$$F\left(w_{r+1}\left(\hat{S}, SINR_{\mathrm{up}}\right)\right) \leq F\left(g_r\left(\hat{S}, SINR_{\mathrm{up}}\right)\right)$$

$$+ \frac{1}{\zeta\sum_{i=1}^n \hat{S}_i}\sum_{i=1}^n \hat{S}_i \{G_i\left(g_r\left(\hat{S}, SINR_{\mathrm{up}}\right), h_i^{(r)(\tau)}\right)$$

$$- F_i\left(w_r\left(\hat{S}, SINR_{\mathrm{up}}\right) + h_i^{(r)(\tau)}\right) + \nabla F_i\left(w_r\left(\hat{S}, SINR_{\mathrm{up}}\right)\right)h_i^{(r)(\tau)}\}$$

$$+ \frac{L}{2\left(\sum_{i=1}^n \hat{S}_i\right)^2}\left\|\sum_{i=1}^n \hat{S}_i h_i^{(r)(\tau)}\right\|^2 \overset{(A2)}{\leq} F\left(w_r\left(\hat{S}, SINR_{\mathrm{up}}\right)\right)$$

$$+ \frac{1}{\zeta\sum_{i=1}^n \hat{S}_i}\sum_{i=1}^n \hat{S}_i \{G_i\left(w_r\left(\hat{S}, SINR_{\mathrm{up}}\right), h_i^{(r)(\tau)}\right) - F_i\left(w_r\left(\hat{S}, SINR_{\mathrm{up}}\right)\right)$$

$$- \frac{\gamma}{2}\|h_i^{(r)(\tau)}\|^2\} + \frac{L}{2\left(\sum_{i=1}^n \hat{S}_i\right)^2}\left\|\sum_{i=1}^n \hat{S}_i h_i^{(r)(\tau)}\right\|^2 \overset{(D.7)}{\leq} F\left(w_r\left(\hat{S}, SINR_{\mathrm{up}}\right)\right)$$

$$+ \frac{1}{\zeta\sum_{i=1}^n \hat{S}_i}\sum_{i=1}^n \hat{S}_i \{G_i\left(w_r\left(\hat{S}, SINR_{\mathrm{up}}\right), h_i^{(r)(\tau)}\right)$$

$$-F_i\left(w_r\left(\hat{S}, SINR_{up}\right)\right) - \frac{\gamma - L\zeta}{2}\|h_i^{(r)(\tau)}\|^2\} \overset{(D.6)}{=} F\left(w_r\left(\hat{S}, SINR_{up}\right)\right)$$

$$+ \frac{1}{\zeta \sum_{i=1}^{n} \hat{S}_i} \sum_{i=1}^{n} \hat{S}_i \{G_i\left(w_r\left(\hat{S}, SINR_{up}\right), h_i^{(r)(\tau)}\right)$$

$$- G_i\left(w_r\left(\hat{S}, SINR_{up}\right), h_i^{(r)*}\right) - G_i\left(w_r\left(\hat{S}, SINR_{up}\right), 0\right)$$

$$+ G_i\left(w_r\left(\hat{S}, SINR_{up}\right), h_i^{(r)*}\right) - \frac{\gamma - L\zeta}{2}\|h_i^{(r)(\tau)}\|^2\}$$

$$\leq F\left(w_r\left(\hat{S}, SINR_{up}\right)\right) - \frac{1}{\zeta \sum_{i=1}^{n} \hat{S}_i} \sum_{i=1}^{n} \hat{S}_i \{\frac{\gamma - L\zeta}{2}\|h_i^{(r)(\tau)}\|^2$$

$$+ \left(1 - \epsilon_l\right) [G_i\left(w_r\left(\hat{S}, SINR_{up}\right), 0\right) - G_i\left(w_r\left(\hat{S}, SINR_{up}\right), h_i^{(r)*}\right)]\}.$$

Therefore, we can calculate $G_i\left(w_r, 0\right)$ and $G_i\left(w_r, h_i^{(r)*}\right)$ as follows:

$$G_i\left(w_r\left(\hat{S}, SINR_{up}\right), 0\right) = F_i\left(w_r(\hat{S}, SINR_{up})\right),$$

$$G_i\left(w_r\left(\hat{S}, SINR_{up}\right), h_i^{(r)*}\right) = F_i\left(w_r\left(\hat{S}, SINR_{up}\right) + h_i^{(r)*}\right)$$

$$- [\nabla F_i\left(w_r\left(\hat{S}, SINR_{up}\right)\right)$$

$$- \zeta \nabla F\left(w_r\left(\hat{S}, SINR_{up}\right)\right)]^{\mathrm{T}} h_i^{(r)*}.$$

Therefore, we have

$$F\left(w_{r+1}\left(\hat{S}, SINR_{up}\right)\right) \leq F\left(w_r\left(\hat{S}, SINR_{up}\right)\right)$$

$$- \frac{1}{\zeta \sum_{i=1}^{n} \hat{S}_i} \sum_{i=1}^{n} \hat{S}_i \{\frac{\gamma - L\zeta}{2}\|h_i^{(r)(\tau)}\|^2$$

$$+ \left(1 - \epsilon_l\right) \{F_i\left(w_r(\hat{S}, SINR_{up})\right) - F_i\left(w_r\left(\hat{S}, SINR_{up}\right) + h_i^{(r)*}\right)$$

$$+ [\nabla F_i\left(w_r\left(\hat{S}, SINR_{up}\right)\right) - \zeta \nabla F\left(w_r\left(\hat{S}, SINR_{up}\right)\right)]^{\mathrm{T}} h_i^{(r)*}\}\}$$

$$\overset{(D.4)}{=} F\left(w_r\left(\hat{S}, SINR_{up}\right)\right) - \frac{1}{\zeta \sum_{i=1}^{n} \hat{S}_i} \sum_{i=1}^{n} \hat{S}_i \{\frac{\gamma - L\zeta}{2}\|h_i^{(r)(\tau)}\|^2$$

$$+ \left(1 - \epsilon_l\right) \{F_i\left(w_r(\hat{S}, SINR_{up})\right) - F_i\left(w_r\left(\hat{S}, SINR_{up}\right) + h_i^{(r)*}\right)$$

$$+ \nabla F_i\left(w_r\left(\hat{S}, SINR_{up}\right) + h_i^{(r)*}\right)^{\mathrm{T}} h_i^{(r)*}\} \overset{A2}{\leq} F\left(w_r\left(\hat{S}, SINR_{up}\right)\right)$$

$$- \frac{1}{\zeta \sum_{i=1}^{n} \hat{S}_i} \sum_{i=1}^{n} \hat{S}_i \{\frac{\gamma - L\zeta}{2}\|h_i^{(r)(\tau)}\|^2 + \frac{(1 - \epsilon_l)\gamma}{2}\|h_i^{(r)*}\|^2\}$$

$$\leq F\left(w_r\left(\hat{S}, SINR_{up}\right)\right) - \frac{(1 - \epsilon_l)\gamma}{2\zeta \sum_{i=1}^{n} \hat{S}_i} \sum_{i=1}^{n} \hat{S}_i \|h_i^{(r)^*}\|^2$$

$$\overset{A1}{\leq} F\left(g_r\left(\hat{S}, SINR_{up}\right)\right) - \frac{(1 - \epsilon_l)\gamma}{2\zeta L^2 \sum_{i=1}^{n} \hat{S}_i}$$

$$\sum_{i=1}^{n} \hat{S}_i \|\nabla F_i\left(w_r\left(\hat{S}, SINR_{up}\right) + h_i^{(r)^*}\right)$$

$$- \nabla F_i\left(w_r\left(\hat{S}, SINR_{up}\right)\right)\|^2 \overset{D.4}{=} F\left(g_r\left(\hat{S}, SINR_{up}\right)\right)$$

$$- \frac{(1 - \epsilon_l)\gamma\zeta}{2L^2 \sum_{i=1}^{n} \hat{S}_i} \sum_{i=1}^{n} \hat{S}_i \|\nabla F\left(g_r\left(\hat{S}, SINR_{up}\right)\right)\|^2 \overset{D.2}{\leq} F\left(g_r\left(\hat{S}, SINR_{up}\right)\right)$$

$$- \frac{(1 - \epsilon_l)\gamma\zeta}{2L^2 \sum_{i=1}^{n} \hat{S}_i} \sum_{i=1}^{n} \hat{S}_i \gamma \left(F\left(g_r\left(\hat{S}, SINR_{up}\right) - F(g^*)\right)\right)$$

$$= F\left(g_r\left(\hat{S}, SINR_{up}\right)\right) - \frac{(1 - \epsilon_l)\gamma^2\zeta}{2L^2} \left(F\left(g_r\left(\hat{S}, SINR_{up}\right) - F(g^*)\right)\right).$$

Therefore, we have

$$F\left(w_{r+1}\left(\hat{S}, SINR_{up}\right)\right) - F(w^*) \leq F\left(w_r\left(\hat{S}, SINR_{up}\right)\right) - F(w^*)$$

$$- \frac{(1 - \epsilon_l)\gamma^2\zeta}{2L^2} \left(F\left(w_r\left(\hat{S}, SINR_{up}\right)\right) - F(w^*)\right)$$

$$= \left(1 - \frac{(1 - \epsilon_l)\gamma^2\zeta}{2L^2}\right) \left(F\left(w_r\left(\hat{S}, SINR_{up}\right)\right) - F(w^*)\right)$$

$$\leq \left(1 - \frac{(1 - \epsilon_l)\gamma^2\zeta}{2L^2}\right)^{(r+1)} \left(F(w_0) - F(w^*)\right)$$

$$\overset{(c)}{\leq} \exp\left(-(r + 1)\left(\frac{(1 - \epsilon_l)\gamma^2\zeta}{2L^2}\right)\right) \left(F(w_0) - F(w^*)\right).$$

To guarantee the global accuracy, i.e., $F(w_r) - F(w^*) \leq \epsilon_g(F(w_0) - F(w^*))$, we have

$$\epsilon_g \geq \exp\left(-r\frac{(1 - \epsilon_l)\gamma^2\zeta}{2L^2}\right).$$

Therefore, when $0 < \zeta < \frac{\gamma}{L}$, we have $r \geq \frac{2L^2 \ln\frac{1}{\epsilon_g}}{(1-\epsilon_l)\gamma^2\zeta}$.

References

M. Chen, Z. Yang, W. Saad, C. Yin, H. V. Poor, and S. Cui. A joint learning and communications framework for federated learning over wireless networks. *IEEE Transactions on Wireless Communications*, 20(1):269–283, 2021. doi: 10.1109/TWC.2020.3024629.

Young Jin Chun, Mazen O. Hasna, and Ali Ghrayeb. Modeling heterogeneous cellular networks interference using Poisson cluster processes. *IEEE Journal on Selected Areas in Communications*, 33(10):2182–2195, 2015. doi: 10.1109/JSAC.2015. 2435271.

Christian Darken and John E. Moody. Note on learning rate schedules for stochastic optimization. In *Proceedings of the 4th International Conference on Neural Information Processing Systems*, volume 91, pages 832–838, 1990.

Canh T. Dinh, Nguyen H. Tran, Minh N. H. Nguyen, Choong Seon Hong, Wei Bao, Albert Y. Zomaya, and Vincent Gramoli. Federated learning over wireless networks: Convergence analysis and resource allocation. *IEEE/ACM Transactions on Networking*, 29(1):398–409, 2020.

Ian Flint, Han-Bae Kong, Nicolas Privault, Ping Wang, and Dusit Niyato. Analysis of heterogeneous wireless networks using Poisson hard-core hole process. *IEEE Transactions on Wireless Communications*, 16(11):7152–7167, 2017.

Yansong Gao, Minki Kim, Sharif Abuadbba, Yeonjae Kim, Chandra Thapa, Kyuyeon Kim, Seyit A. Camtep, Hyoungshick Kim, and Surya Nepal. End-to-end evaluation of federated learning and split learning for Internet of Things. In *Proceedings of 2020 IEEE Computer Society International Symposium on Reliable Distributed Systems (SRDS)*, pages 91–100, 2020.

Chaoyang He and Murali Annavaram. Group knowledge transfer: Federated learning of large CNNs at the edge. In *Proceedings of Advances in Neural Information Processing Systems 33 (NeurIPS 2020)*, 2020.

Christian Hennig and Mahmut Kutlukaya. Some thoughts about the design of loss functions. *REVSTAT–Statistical Journal*, 5(1):19–39, 2007.

P. L. Hsu and H. Robbins. Complete convergence and the law of large numbers. *Proceedings of the National Academy of Sciences of the United States of America*, 33(2):25–31, 1947. ISSN 0027-8424. doi: 10.1073/pnas.33.2.25

Tzu-Ming Harry Hsu, Hang Qi, and Matthew Brown. Measuring the effects of non-identical data distribution for federated visual classification. *CoRR*, abs/1909.06335, 2019. URL http://arxiv.org/abs/1909.06335.

Andrew M. Hunter, Jeffrey G. Andrews, and Steven Weber. Transmission capacity of Ad Hoc networks with spatial diversity. *IEEE Transactions on Wireless Communications*, 7(12):5058–5071, 2008.

Eunjeong Jeong, Seungeun Oh, Hyesung Kim, Jihong Park, Mehdi Bennis, and Seong-Lyun Kim. Communication-efficient on-device machine learning: Federated distillation and augmentation under non-iid private data. *arXiv preprint arXiv:1811.11479*, 2018.

Bojan Jovanović. Internet of Things Statistics for 2021 –Taking Things Apart, May 5, 2023. URL https://dataprot.net/statistics/iot-statistics/.

Alex Krizhevsky, Ilya Sutskever, and Geoffrey E. Hinton. Imagenet classification with deep convolutional neural networks. *Advances in Neural Information Processing Systems 25 (NIPS 2012)*, pages 1097–1105, 2012.

Yann LeCun. The MNIST Database of Handwritten Digits. URL http://yann.lecun.com/exdb/mnist/, 1998.

Yuanzhi Li and Yang Yuan. Convergence analysis of two-layer neural networks with ReLU activation. In *Proceedings of the 31st International Conference on Neural Information Processing Systems*, pages 597–607, 2017.

Wei Yang Bryan Lim, Nguyen Cong Luong, Dinh Thai Hoang, Yutao Jiao, Ying-Chang Liang, Qiang Yang, Dusit Niyato, and Chunyan Miao. Federated learning in mobile edge networks: A comprehensive survey. *IEEE Communications Surveys & Tutorials*, 22(3):2031–2063, 2020.

Liyuan Liu, Haoming Jiang, Pengcheng He, Weizhu Chen, Xiaodong Liu, Jianfeng Gao, and Jiawei Han. On the variance of the adaptive learning rate and beyond. In *Proceedings of International Conference on Learning Representations*, 2019.

Yi Liu, Sahil Garg, Jiangtian Nie, Yang Zhang, Zehui Xiong, Jiawen Kang, and M. Shamim Hossain. Deep anomaly detection for time-series data in industrial IoT: A communication-efficient on-device federated learning approach. *IEEE Internet of Things Journal*, 8(8):6348–6358, 2020a.

Yi-Jing Liu, Gang Feng, Yao Sun, Shuang Qin, and Ying-Chang Liang. Device association for RAN slicing based on hybrid federated deep reinforcement learning. *IEEE Transactions on Vehicular Technology*, 69(12):15731–15745, 2020b.

Brendan McMahan, Eider Moore, Daniel Ramage, Seth Hampson, and Blaise Aguera y Arcas. Communication-efficient learning of deep networks from decentralized data. In *Proceedings of the 20th International Conference on Artificial Intelligence and Statistics*, volume 54, pages 1273–1282, 20–22 Apr 2017. URL http://proceedings.mlr.press/v54/mcmahan17a.html.

Y. Sun, L. Zhang, G. Feng, B. Yang, B. Cao, and M. A. Imran. Blockchain-enabled wireless Internet of Things: Performance analysis and optimal communication node deployment. *IEEE Internet of Things Journal*, 6(3):5791–5802, 2019. doi: 10.1109/JIOT.2019.2905743.

Jianyu Wang, Zachary Charles, Zheng Xu, Gauri Joshi, H. Brendan McMahan, Maruan Al-Shedivat, Galen Andrew, Salman Avestimehr, Katharine Daly, Deepesh Data, et al. A field Guide to Federated Optimization. *arXiv preprint arXiv:2107.06917*, 2021.

Steven Weber, Jeffrey G. Andrews, and Nihar Jindal. An overview of the transmission capacity of wireless networks. *IEEE Transactions on Communications*, 58(12):3593–3604, 2010.

Qiang Yang, Yang Liu, Tianjian Chen, and Yongxin Tong. Federated machine learning: Concept and applications. *ACM Transactions on Intelligent Systems and Technology (TIST)*, 10(2):1–19, 2019.

H. H. Yang, Z. Liu, T. Q. S. Quek, and H. V. Poor. Scheduling policies for federated learning in wireless networks. *IEEE Transactions on Communications*, 68(1):317–333, 2020a. doi: 10.1109/TCOMM.2019.2944169.

Zhaohui Yang, Mingzhe Chen, Walid Saad, Choong Seon Hong, and Mohammad Shikh-Bahaei. Energy efficient federated learning over wireless communication networks. *IEEE Transactions on Wireless Communications*, 20(3):1935–1949, 2020b.

4

Device Association Based on Federated Deep Reinforcement Learning for Radio Access Network Slicing

Yi-Jing Liu[1], Gang Feng[1], Yao Sun[2], and Shuang Qin[1]

[1]National Key Lab on Wireless Communications, University of Electronic Science and Technology of China, Sichuan, Chengdu, China
[2]James Watt School of Engineering, University of Glasgow, Scotland, Glasgow, UK

4.1 Introduction

Network slicing has become one of the most vital architectural technologies for 5G-and-beyond systems. Operators and companies can design, deploy, customize, and optimize multiple isolated network slices (NSs) on a common physical network infrastructure to support various applications with diverse quality of service (QoS) requirements in different application scenarios, i.e., enhanced mobile broadband (eMBB), massive machine-type communications (mMTC), and ultrareliable and low-latency communications (URLLC) (Ericsson AB [2017], Zhang [2019], Sun et al. [2019a]). The NS-based networks provide customized services efficiently and flexibly to meet the specific requirements of applications and corresponding service-level agreement. However, driven by the rapidly growing wireless applications with diversified service requirements, how to identify and classify service flows for accommodation by appropriate application-specific NS (i.e., device association including access control and handoff management) is still a challenging issue, especially in the radio access network (RAN) domain.

In RAN slicing, device association and relevant resource allocation are fundamentally different from that in traditional mobile networks due to the introduction of NSs. On the one hand, NSs are logically virtualized and isolated over shared physical networks. Therefore, both physical and virtual resource constraints should be considered to form a function chain/graph for specific services. On the other hand, according to the service requirements, smart devices need to select an appropriate NS, which may cover multiple access points (APs), i.e., base stations (BSs). Therefore, in NS-based networks, device association inherently includes NS selection, BS selection, handoff management, and associated resource allocation issues, which should be addressed jointly to improve network resource utilization while guaranteeing service quality. Moreover, due to the random nature of wireless condition and user mobility, the complexity of finding the optimal solution can be too high and the environmental changes may not be accurately described in the complex and dynamic scenarios. Fortunately, recently emerging reinforcement learning (RL) can efficiently solve such sequential decision problems by continuously interacting with the environments and thus obtain an optimal solution by using a trial and error learning process. Although RL works well in decision-making scenarios, the effectiveness of RL diminishes as the size of the state-action space becomes large (Zhang et al. [2019]). Then, deep reinforcement learning (DRL) emerges as a good alternative to solve the decision-making problems in the wireless system with large size of state-action space (Zhang et al. [2019]).

Although DRL is effective for solving the decision-making problems, a large amount of data is always necessary for training model to achieve satisfactory performance (Zhang et al. [2019], Lim et al. [2020]). With the dramatic growth of the heterogeneous data from geographically distributed smart devices, conventional centralized DRL algorithms may not be practically feasible as they require the data to be transferred and processed in a central entity, which definitely causes large latency in uploading a huge amount of raw data and consumes large amount of precious network bandwidth (Lim et al. [2020], Tran et al. [2019]). As a result, decentralized DRL algorithms that exploit local data are much more attractive. Furthermore, in light of the increasingly rigorous data security and device privacy concerns, an emerging decentralized approach, federated learning (FL) (Yang et al. [2019]), is introduced. In FL, smart devices train their own data locally and exploit the collaboration between them by aggregating their local machine learning models. Specifically, FL is classified into horizontally FL (hFL), vertically FL (vFL), and federated transfer learning (FTL) based on how data is distributed among various devices in the feature and sample space (Yang et al. [2019], Wang et al. [2019a]). Most of the current work focuses on hFL to share the sample space or vFL to share the feature space, such as Chen et al. [2019], Tran et al. [2019]. Indeed, in order to reduce the amount of required training samples and/or make more precise decisions, the combination of hFL and vFL, called hybrid FL, is intuitively advantageous (Yang et al. [2019]).

In this chapter, we propose an intelligent device association scheme for RAN slicing, called hybrid federated deep reinforcement learning (HDRL) scheme, with aim to improve network throughput while reducing handoff cost. Considering the large state-action space and the diversity of services, HDRL is designed to consist of two-layer model aggregations: (i) Horizontal aggregation: to select both appropriate BS and NS while guaranteeing the service quality of devices, we combine DRL and hFL to train local machine learning models on individual devices, and aggregate the models of the same type services (e.g., eMBB services) on BSs;

(ii) vertical aggregation: to design a global optimal NS-BS access selection for the devices with services of different types (e.g., eMBB and URLLC services), we aggregate the access features on the third encrypted party in vFL (Yang et al. [2019]), where we use Shapley value (Wang et al. [2019a]) which represents the average marginal contribution of a specific feature across all possible feature combinations, to evaluate aggregated access features. Numerical results show that in the typical scenarios, our proposed HDRL scheme for device association significantly outperforms some conventional solutions in terms of network throughput and communication efficiency.

4.2 System Model

4.2.1 Network Model

We consider a scenario where the NS-based mobile network is built upon a Software Define Network/Network Function Virtualization-enabled 5G network infrastructure, which consists of CN and RAN. As shown in Figure 4.1, the access

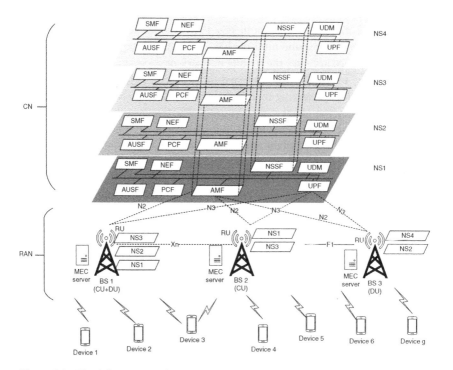

Figure 4.1 The NS-based mobile network model.

and mobility management function (AMF) is responsible for the connectivity and mobility management for associated devices with slices. Network slice instance selection for a device is triggered by the first contacted AMF. When the location of devices changes, the selected AMF may be changed to receive services and to enable mobility tracking. Specifically, if the AMF can serve the single network slice selection assistance information (S-NSSAI), the AMF remains the serving AMF for the devices. Otherwise, the network slice selection function (NSSF) will select the network slice instances and determine the target AMF set to serve the devices. Moreover, partial network functions can be shared between multiple slices, while others are slice specific. For example, in CN domain, AMF and NSSF can be shared among multiple slices, while user plane function (UPF) is slice specific. In cloud-RAN domain, distributed unit (DU) and radio unit (RU) can be shared if the "baseband processing functions" are implemented with physical devices (i.e., field programmable gate array [FPGA]) and "radio functions" are realized on specialized hardware. In general, centralized unit (CU) can be slice-specific because it realizes the "packet processing functions" as virtualized network functions.

4.2.2 RAN Slicing

We consider a multi-NSs and multi-BSs RAN slicing scenario, as shown as Figure 4.1. An NS can expand multiple BSs, and a BS can be covered by multiple NSs. When a device accesses the mobile network or experiences a handoff, both BS and NS need to be selected/reselected for provision seamless service for that device. Specifically, for serving mobile devices, the change of device association in this case is only the change of serving BS. For the case that a device moves out of the coverage of a specific NS, two methods can be used to guarantee the connections. One is to expand the coverage of the current serving NS by deploying it on more BSs. Another one is to change the device association to an exiting NS, which can provide the similar service thus to fulfill the QoS requirements. In this RAN slicing scenario, if the operator knows that a service is provided by some NSs, it must cover a specific region. In this region, if the required NS is not deployed in a specific BS or the QoS of devices cannot be guaranteed by a specific BS, a device cannot access the NS via this BS, but can via other BSs in this region to access.

Let $\mathcal{B}$, $\mathcal{N}$, and $\mathcal{D}$ denote the set of BSs, NSs, and devices, respectively. For a specific BS k, let $\mathcal{N}_k = \{j, \dots, g\}$ represent the set of NSs which are supported by it. For a specific NS j, let a four-tuple $\left(R_j, T_j, \Omega_j, \vec{W}_j\right)$ represent the state, where R_j and T_j represent the minimal transmission rate and the maximal latency that are provided by NS j to serve devices. Moreover, let Ω_j denote the bandwidth allocated to NS j in CN. $\vec{W}_j$ is a vector, which denotes the bandwidth allocation of NS j from all BSs. The kth element in $\vec{W}_j$ is denoted by $b_{j,k}$, which denotes the bandwidth resource allocated to NS j by BS k, where $b_{j,k} = 0$ means BS k is not covered by NS j.

4.2.3 Service Requirements

As the services required by devices vary with time, we assume the time is slotted, where the services remain fixed for the duration of one time slot and change from one slot to the next. Slotted time can be regarded as a sampled version of continuous time which consists of T time slots (fixed time intervals) (Wang et al. [2015], Mansouri et al. [2019]). During time slot $t \in \{1, \ldots, T\}$, we assume that a device requires only one service and remains connected to the same NS and BS. Let u represent the number of devices in the network. For a specific device $d_i \in D$, its service quality can be described by two metrics: the minimum transmission rate $\hat{r}_i^t$ and the maximum tolerated latency $\hat{d}_i^t$. Therefore, NS j can accommodate d_i only if $R_j \geq \hat{r}_i^t$ and $T_j \leq \hat{d}_i^t$.

Let $r_{i,t}^{j,k}$ represent the transmission rate of d_i that is served by NS j via BS k during time slot t. Moreover, let $w_{i,t}^{j,k}$ represent the wireless bandwidth that BS k allocates to d_i that is served by NS j during time slot t. As we focus the device association in the RAN slicing, the maximum delay in core network (including transport network) that NS j can provide to its serving devices is represented by T_j, which is a constant. Thus, the end-to-end delay can be expressed as $\hat{T}_{i,t}^{j,k} + T_j$, where $\hat{T}_{i,t}^{j,k} = q_i / r_{i,t}^{j,k}$ denotes the delay in the RAN and q_i represents the service flow data volume of d_i. Moreover, we use Shannon theory to calculate the transmission rate, i.e., $r_{i,t}^{j,k} = w_{i,t}^{j,k} \log_2\left(1 + SINR_{i,t}^k\right)$, where $SINR_{i,t}^k$ represents the signal-to-interference-plus-noise-ratio (SINR) between d_i and BS k during time slot t. Moreover, $SINR_{i,t}^k = \frac{p_{i,t}^k \cdot G_{i,t}^k}{\sum_{k \in B, k' \neq k} p_{i,t}^{k'} G_{i,t}^{k'} + \zeta^2}, t \in T$, where $p_{i,t}^k$ denotes the transmission power that BS k allocates to d_i, $G_{i,t}^k$ represents the channel gain between d_i and BS k, and ζ^2 represents the noise power level.

4.2.4 Handoff Cost

When the location of a device changes or the service quality of a device cannot be met, a handoff occurs to improve the experience of the user. Once a handoff happens, the device needs to reselect appropriate BS and NS. It is obvious that traditional reference signal received power (RSRP)-based handoff mechanisms (ETSI [2018]) are no longer applicable to RAN slicing due to a three-layer associate relationship device-BS-NS. Specifically, both the service type of NSs and the RSRP of BSs should be taken into account to guarantee the service quality when a handoff occurs. Therefore, different from the handoff in traditional mobile networks, there are three types of handoff we need to consider: switching NS only, switching BS only, and switching both NS and BS (Sun et al. [2019b]). Also, the amount of signaling data needed for a handoff is different for the three types. For example, switching NS only needs to exchange signaling in the same BS, while switching both NS and BS needs to exchange signaling between different BSs and NSs. Therefore, inspired by the idea of Sun et al. [2019b], we define the amount of signaling data for three types of handoff as (i) q_{NS}, the amount of signaling data needed for switching NS only; (ii) q_{BS}, the amount of signaling data needed for switching BS only; and (iii) q_{N-B}, the amount of signaling data needed for switching both NS and BS; with the relationship $q_{NS} < q_{BS} < q_{N-B}$. Intuitively, the amount of signaling data needed incurs corresponding signaling overhead in terms of bandwidth consumption for signaling exchange.

Furthermore, due to the bandwidth consumed by service flows and the bandwidth consumed by handoff may not be in the same order of magnitude, we define the handoff cost as follows (Lee et al. [2009]),

$$
\alpha^{HO} = \begin{cases} \dfrac{q_{NS}}{w_{NS}}, & \text{if switching NSs only,} \\[2ex] \dfrac{q_{BS}}{w_{BS}}, & \text{if switching BSs only,} \\[2ex] \dfrac{q_{N-B}}{w_{N-B}}, & \text{if switching both NSs and BSs,} \\[2ex] 0, & \text{otherwise.} \end{cases} \tag{4.1}
$$

where w_{NS} denotes the bandwidth consumed by the first type of handoff switching NS only, w_{BS} denotes the bandwidth consumed by switching BS only, and w_{N-B} represents the bandwidth consumed by switching both BS and NS.

4.3 Problem Formulation

4.3.1 Problem Statement

Given a set of devices which require services of different types, we investigate the device association problem for RAN slicing under network bandwidth, computing resource, and transmission rate constraints. Let a binary variable $x_{i,t}^{j,k}$ indicate whether the device d_i is served by NS j via BS k during time slot t or not: $x_{i,t}^{j,k} = 1$ yes and 0 otherwise. Therefore, multiplying the two variables $x_{i,t}^{j,k} x_{i,t-1}^{j',k'}$ in adjacent time slots indicates the handoff decision of d_i from time slot $t - 1$ to t. Note that if $x_{i,t}^{j,k} x_{i,t-1}^{j',k'} = 0$, we can only explain that device d_i is not served by NS j' via BS k'

during time slot $t-1$ or/and device d_i is not served by NS j via BS k during time slot t. Thus, we cannot judge whether handoff happens. However, it is much easier to judge if handoff happens when $x_{i,t}^{j,k} x_{i,t-1}^{j',k'} = 1$. When $x_{i,t}^{j,k} x_{i,t-1}^{j',k'} = 1$, we can judge whether handoff happens and derive the handoff types and corresponding α^{HO} from following four aspects: (i) $j \neq j'$, $k \neq k'$, switching both BS and NS; (ii) $j = j'$, $k \neq k'$, switching BS only; (iii) $j \neq j'$, $k = k'$, switching NS only; (iv) $j = j'$, $k = k'$, device d_i is served by NS j'/j via BS k'/k during both time slot $t-1$ and t. Thus, no handoff happens.

Therefore, with the aim to improve network throughput while reducing handoff cost, we define the communication efficiency of the network during time slot t as follows:

$$e_t = \sum_{i \in D} \left(\alpha_{i,t}^{flow} x_{i,t}^{j,k} - \alpha^{HO} x_{i,t}^{j,k} x_{i,t-1}^{j',k'} \right), \forall t \in [0, T], \tag{4.2}$$

where $\alpha_{i,t}^{flow} = \frac{q_i}{w_{i,t}^{j,k}}$ (Lee et al. [2009]), q_i denotes the service flow data volume of d_i, and $w_{i,t}^{j,k}$ denotes the wireless bandwidth that BS k allocates to d_i served by NS j.

In our model, $x_{i,t}^{j,k}$ is a decision variable, which represents the NS-BS access selection of d_i. Since the device association is indeed a sequential decision problem, we use the long-term network communication efficiency as the optimization objective in (4.3), with aim to improve network throughput while reducing handoff cost. Therefore, we formulate the device association problem as follows:

$$\max \lim_{T \to +\infty} E \left[\frac{1}{T} \sum_{t=1}^{T} e_t \right] \tag{4.3}$$

$$\text{s.t.} \sum_{k \in B i \in D} x_{i,t}^{j,k} r_{i,t}^{j,k} \leq \Omega_j, \forall j \in \mathcal{N}, t \in [0, T] \tag{4.3.1}$$

$$\sum_{i \in D} x_{i,t}^{j,k} w_{i,t}^{j,k} \leq b_{j,k}, \forall j \in \mathcal{N}, \forall k \in B, t \in [0, T] \tag{4.3.2}$$

$$\sum_{j \in \mathcal{N} k \in B} x_{i,t}^{j,k} r_{i,t}^{j,k} \geq \hat{r}_i^t, \forall i \in D, t \in [0, T] \tag{4.3.3}$$

$$\sum_{j \in \mathcal{N} k \in B} x_{i,t}^{j,k} R_j \geq \hat{r}_i^t, \forall i \in D, t \in [0, T] \tag{4.3.4}$$

$$\sum_{j \in \mathcal{N} k \in B} x_{i,t}^{j,k} \left(\hat{T}_{i,t}^{j,k} + T_j \right) \leq \hat{d}_i^t, \forall i \in D, t \in [0, T] \tag{4.3.5}$$

$$\sum_{j \in \mathcal{N} k \in B} x_{i,t}^{j,k} = 1, \forall i \in D, t \in [0, T] \tag{4.3.6}$$

$$x_{i,t}^{j,k} \in \{0,1\}, \forall i \in D, \forall j \in \mathcal{N}, \forall k \in B, t \in [0, T]. \tag{4.3.7}$$

In problem (4.3), (4.3.1) represents the constraint of wired link resource, where the total transmission rate offered by NS cannot exceed the link resource budget during any time slot t. (4.3.2) states the wireless bandwidth constraint. (4.3.3)–(4.3.5) state that the service quality of devices should be satisfied by its serving BS and NS even the access selection and network environment change. Specifically, (4.3.3) and (4.3.4) guarantee the transmission rate, and (4.3.5) guarantees the delay. Moreover, constraint (4.3.6) represents the access limitation, which means that a device can access only one NS via one BS during time slot t. The binary constraint on the decision variable is shown in (4.3.7).

Theorem 4.1 *Problem (4.3) with constraints (4.3.1)–(4.3.7) is non-deterministic polynomial (NP)-hard.*

Proof: A special case with fixed $w_{i,t}^{j,k}$ and $r_{i,t}^{j,k}$ in problem (4.3) can be mapped into a Multiple Choice Multidimensional Knapsack problem (MMKP) (Galeana-Zapien and Ferrus [2010]), which is NP-hard (Sbihi [2007]). When $w_{i,t}^{j,k}$ and $r_{i,t}^{j,k}$ change with time, problem (4.3) with constraints (4.3.1)–(4.3.7) is a dynamic MMKP (DMMKP). If DMMKP has solution in polynomial time, its corresponding MMKP should also have solution in polynomial time. Thus, DMMKP can reduce to MMKP. Therefore, problem (4.3) with constraints (4.3.1)–(4.3.7) is NP-hard. □

4.3.2 Markov Decision Process Modeling for Device Association

As Problem (4.3) is NP-hard, there is no polynomial-time algorithm for solving it. Meanwhile, in view of the dynamic nature of access conditions, we formulate the device association problem as a Markov decision process (MDP) model, which consists of four-tuple $\mathcal{M} = (S, \mathcal{A}, P, R)$. Specifically, S denotes the state space, $\mathcal{A}$ denotes the action space, P represents the transition probability between states, and R denotes the reward function. For a specific device, at the beginning of each time slot, the device should make a scheme to access appropriate BS and NS (action). This can change the state of access conditions, causing the network state to transit to another state. Through this action, the device can obtain a certain reward. In the following, we define the state, action, transition probability, and reward.

4.3.2.1 State

The current access conditions are used to describe the system state. Let S denote the set of network states for all devices. The number of NSs and BSs are represented by $|\mathcal{N}|$ and $|\mathcal{B}|$, respectively. For a specific device $d_i \in \mathcal{D}$, the state is denoted by $s_t^i = \{I_i, b_{1,1}^t, \ldots, b_{j,k}^t, \ldots, b_{|\mathcal{B}|,|\mathcal{N}|}^t\}$, where $s_t^i \in S, I_i = (j, k)$ denote the current NS and BS selection of d_i, and $b_{j,k}^t$ denotes the wireless bandwidth allocated to NS j from BS k with $b_{j,k}^t \leq b_{j,k}$. Note we filter out $b_{j,k}^t$ if NS j or BS k is not engaged for d_i.

4.3.2.2 Action

We remove the infeasible actions which cannot meet either the network resource limitation (4.3.1)–(4.3.2), the service quality of devices (4.3.3)–(4.3.5), or the access limitation (4.3.6)–(4.3.7). Let $\mathcal{A}$ represent the set of actions for all devices. For a specific device $d_i \in \mathcal{D}$, the action is denoted by $a_t^i = (j, k, w_{i,t}^{j,k})$, which means d_i consumes $w_{i,t}^{j,k}$ wireless bandwidth if it accesses to NS j via BS k at the beginning of time slot t.

4.3.2.3 Transition Probability

We assume that the transition probability of d_i is represented by $P = \{p_{s_t^i, s_{t+1}^i}^{a_t^i} | a_t^i \in \mathcal{A}, s_t^i, s_{t+1}^i \in S\}$, which denotes the probability that network state of d_i transits from s_t^i to s_{t+1}^i through action a_t^i.

4.3.2.4 Reward

To maximize the communication efficiency while considering the incurred communication cost, we define the reward as $r_t = e_t - u \cdot x_t \cdot \alpha_{i,k}^c$. Specifically, u denotes the number of devices, x_t denotes the number of communication rounds during $[1, t]$ time slots, and $\alpha_{i,k}^c$ represents the communication cost of each communication round in FL between d_i and BS k. More details about communication rounds and FL can be found in Section 4.4.

In the MDP for device association, a smart device can obtain an optimal long-term reward by continuously interacting with the network environments. But the effectiveness fades as the size of the state-action space becomes large (i.e., the state space of MDP for device association is a discrete space with $|\mathcal{B}| \cdot |\mathcal{N}| + 1$ dimensions, the action space is a discrete space with $|\mathcal{B}| \cdot |\mathcal{N}| \cdot b_{j,k}$ dimensions). To address the aforementioned problem, we employ DRL to solve the decision-making problem of a large size of state-action space. Meanwhile, in the paradigm of distributed machine learning, FL can be exploited to efficiently facilitate the collaboration between devices and reduce the network bandwidth consumption for transmitting training data.

4.4 Hybrid Federated Deep Reinforcement Learning for Device Association

4.4.1 Framework of HDRL

We propose a collaborative hybrid federated deep reinforcement learning, called HDRL, by incorporating the DRL into the FL framework. Figure 4.2 shows the architecture of HDRL, which is composed of DRL running on individual devices, and two levels of model aggregation based on DRL: horizontal weights aggregation (called hDRL) and vertical access feature aggregation (called vDRL).

4.4.1.1 DRL on Smart Devices

In FL, smart devices train their local models based on the local data, which should be kept where it is generated. On the one hand, modern smart devices (i.e., smartphones) have fast processors (e.g., GPUs) and artificial intelligence (AI) chips to accelerate training and reduce energy consumption (McMahan et al. [2016]). On the other hand, the state space of MDP for device association) is a discrete space with $|\mathcal{B}| \cdot |\mathcal{N}| + 1$ dimensions, and the action space is a discrete space with $|\mathcal{B}| \cdot |\mathcal{N}| \cdot b_{j,k}$ dimensions. Therefore, we employ the discrete-action

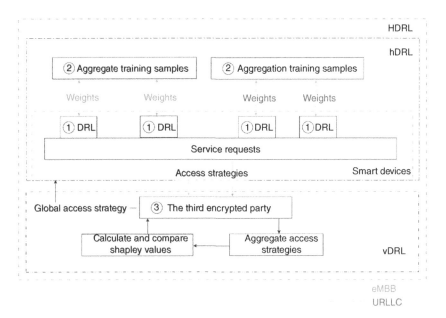

Figure 4.2 The hybrid federated deep reinforcement learning architecture-based device association.

DRL algorithm, double deep Q-Network (DDQN), to train the local model on individual smart devices. DDQN can address MDP with large state-action space by introducing the experience pool, improve the stability of the training results by introducing the target network, and decouple the selection from the evaluation to reduce the correlation between data (Van Hasselt et al. [2016]).

4.4.1.2 Horizontal Model Aggregation (hDRL) Level

As different smart devices may generate local data with different patterns based on the usage of the devices, no device has a representative sample of the popular distribution in general. However, the same type services from various devices strongly correlated, as they not only have similar data features (i.e., the service type mark) but also compete for radio and computing resources in the same/similar slices. Therefore, for the different services of the same type, we propose horizontal aggregation to integrate the similar data samples to train a global access model by adopting an iterative approach that requires a number of model update iterations, where each model update iteration is called a communication round. A communication round consists of five steps, including initialization of DDQN parameters, local model training, local model transmission, global model update, and global model transmission. In each communication round, we aim to update model through the cooperation between BSs and smart devices (i.e., using DRL for updating the local model and aggregating training samples for updating the global model). As a result, an individual device can share the updates of parameters with other devices.

hDRL is performed in two steps: (i) DRL for training and updating local model; and (ii) horizontal weights aggregation for aggregating training samples of the services with similar data features. Specifically, In the first step, based on the global model formed at the BS, all devices first update the local models at the beginning of each communication round, and then continue to train local models by using DRL with the aim to approach optimal parameters that minimize the loss function. In the second step, individual devices send their local models to the corresponding BSs at the end of each communication round. Upon receiving all local models of the trained devices, BSs form the global model and then send the updated global model back to individual devices.

4.4.1.3 Vertical Model Aggregation (vDRL) Level

Since the radio access network (RAN) always supports multiple types of services, horizontal aggregation for aggregating the similar data samples may be not optimal. Therefore, to further promote the collaboration between devices, we aggregate the local access features to form a global access feature. Furthermore, compared with aggregating the data of the same type services, directly aggregating the data of services of different types brings much more communication cost, as the parameters are more frequently transmitted and updated in each communication round (Yang et al. [2019]). Therefore, to reduce communication cost in vDRL, based on the aggregated global access feature, we introduce Shapley values (Wang et al. [2019a]). Shapley values are the average marginal contribution of a specific feature across all possible feature combinations, through which we can compare the importance of global access feature,

$$\phi_f = \frac{1}{M} \sum_{m=1}^{M} \left(f(x_{+i}^m) - f(x_{-i}^m) \right), \tag{4.4}$$

where M denotes the number of iterations. $f(x_{+i}^m)$ denotes the prediction for instance x. x_{-i}^m represents identical to x_{+i}^m, except that x_{+i}^m is different. Thus, we can derive the global optimal access selection by selecting the maximal Shapley value. Note the access feature refers to the access selection and its corresponding estimated value (i.e., the target value for local access selection and the Shapley value for global access selection).

In vDRL, both the local and global access selection can be represented by a 0–1 matrix (i.e., local 0–1 matrix, global 0–1 matrix). Specifically, the global 0–1 matrix consists of the row vectors of the local 0–1 matrices, where the row vectors of the global 0–1 matrix denote access selections of a specific device. Moreover, a local 0–1 matrix is composed of possible local access selection, where row vectors of a local 0–1 matrix denote specific access selections of this device. Therefore, we can update the local and global access selection by changing the row vectors of the global 0–1 matrix. Furthermore, vDRL includes three steps: (i) Aggregate access features. All devices send their own local access features (0–1 matrix and the corresponding target values) to the trusted third encrypted party every v communication rounds. Thus, different global access selection (i.e., global 0–1 matrix) can be derived by selecting different row vectors of the local 0–1 matrices; (ii) calculate and compare Shapley values. Based on the formed global 0–1 matrices, we can calculate the Shapley values and obtain a global optimal 0–1 matrix (global optimal access selection) with the maximal Shapley value by comparing these Shapley values; and (iii) store and update the global access selection. The third encrypted party stores the global optimal access selection and Shapley value until it is replaced by a better one. Here the third encrypted party is a logical entity used to aggregate the common devices of different parties without exposing their respective data. Currently, there are no standards or researchers explicitly specify which entity can play the role of the third encrypted party in a mobile network. In our views, a mobile edge computing (MEC)/cloud encrypted server or a secure computing node in core network (CN)/RAN can serve as the third encrypted party, as the aggregation on this entity needs certain computing resources.

4.4.2 Algorithm of Horizontal Model Aggregation

4.4.2.1 DDQN for Training Local Model

At the beginning of a communication round, smart devices receive the global model from BSs to update their local models (i.e., weight θ) if communication round $r \neq 1$. Otherwise the devices will update their local model directly with initial weights (which are set to zero). We assume that each communication round is composed of τ time slots. During each time slot, each device performs local training once. Therefore, after completing the local model update, the devices train their local models independently with DDQN during τ time slots. DDQN evaluates the greedy policy according to the Q-network with weights θ and estimates state-action value $Q(\cdot)$, according to the target network $\hat{Q}$ with weights $\hat{\theta}$ (Van Hasselt et al. [2016]). The update in DDQN is similar to that in DQN, but the target is replaced by

$$y_t^i = r_{t+1} + \gamma Q(s_{t+1}^i, \text{argmax}_{a_t^i} Q(s_{t+1}^i, a_t^i; \theta_t^i); \hat{\theta}_t^i), \tag{4.5}$$

where $\text{argmax}_{a_t^i} Q(s_{t+1}^i, a_t^i; \theta_t^i)$ represents an ϵ-greedy policy used to select access or handoff actions, and θ_t^i denotes the weight vector of Q-network for device d_i.

For a specific device d_i, if d_i satisfies the access condition and takes access or handoff action a_t^i at the beginning of time slot t, we can obtain the corresponding state-action value, as follows:

$$Q(s_t^i, a_t^i) = \mathbb{E}\left[\sum_{k=t}^{T} \gamma^k r_t | s_t^i, a_t^i\right], \tag{4.6}$$

where $\gamma \in [0,1]$ represents the discount factor which denotes the discounted impact of the future reward. The objective of DDQN is to minimize the gap between the estimated $Q(\cdot)$ and the target value. Therefore, DDQN running on d_i can be trained by minimizing the loss function, expressed by

$$L(\theta_t^i) = \mathbb{E}[(y_t^i - Q(s_t^i, a_t^i; \theta_t^i))^2]. \tag{4.7}$$

Moreover, when DDQN approximates the value function using the neural network, it indeed updates the parameter value θ_t^i by using the gradient descent method. Therefore, the update algorithm in DDQN can be given by

$$\theta_{t+1}^i = \theta_t^i + \alpha \left[y_t^i - Q\left(s_t^i, a_t^i; \theta_t^i\right)\right] \nabla Q\left(s_t^i, a_t^i; \theta_t^i\right). \tag{4.8}$$

After training local data for τ time slots, device d_i sends the local model θ_t^i to the BSs where the global model is formed.

4.4.2.2 Update Models

Once receiving all local models from individual devices, BSs begins updating the global model according to,

$$g_r(t) = \frac{\sum_{i=1}^{u_x} K_i \theta_t^i}{K}, \forall 1 \leq t \leq T, \tag{4.9}$$

where K_i represents the amount of training data of d_i, $K = \sum_{i=1}^{u_x} K_i$ represents the total amount of training data of the devices with service of type x, u_x represents the number of devices that have the same service type x, and r represents the rth communication round of hDRL. After updating the global model in the rth communication round, BSs sends the global mode $g_r(t)$ to all devices with the same type services to update the local DDQN models based on (4.10).

$$\theta_{t+1}^i = g_r(t) - \frac{\lambda}{K_i} \sum_{i=1}^{u} \nabla L(\theta_t^i), \forall i \in \mathcal{D}, 1 \leq t \leq T, \tag{4.10}$$

where λ and $L(\theta_t^i)$ represent the learning rate and the loss function of DDQN, respectively. After updating the local models, the devices begin training their local models. The horizontal model aggregation algorithm is presented as Algorithm 4.1, where the complexity of horizontally FL framework is $\mathcal{O}(R(u + |\mathcal{B}|))$ because each communication round includes the computation of BS aggregation and local model updating, where R, u, and $|\mathcal{B}|$ are the number of communication rounds, smart devices, and BSs.

4.4.3 Algorithm of Vertical Model Aggregation

The aforementioned horizontal model aggregation is applied for the same type services that have similar data samples. As multiple types of services are taken into account in this work, vertical model aggregation can be exploited for further improving the network performance, by aggregating local access features incurred from different types of services. As the data on individual devices is private and not visible to other devices, we use 0–1 matrix to represent the local and global BS-NS access selection, where we can update global access feature by transforming these 0–1 matrices. In this work, according to Wang et al. [2019a], the estimated global target value of a global access selection is given by

$$\varphi_f = \sum_{i=1}^{u} y_t^i - \mathbb{E}\left[\sum_{i=1}^{u} y_t^i\right], \tag{4.11}$$

where $f \subseteq \mathcal{X}$ denotes a specific global access scheme (0–1 matrix), each row vector of f denotes a local access selection. Moreover, y_t^i represents the target value in Eq. (4.5). For example, we assume there are two devices (i.e., d_1 and d_2) sending

---

Algorithm 4.1 Algorithm of Horizon Model Aggregation

Input: $s^i, a^i, \alpha, \gamma, C, R, K_i, u_x, x, \lambda, \tau$
output: Access scheme π^i, target value y_t^i.

1:    Initialize experience relay pool $D_x^i, \forall i \in D$;
2:    Initialize the global weights g_0;
3:    **for** communication round $r = 1, 2, \dots, R$ **do**
4:      **if** $r == 1$ **then**
5:        Initialize θ_0^i;
6:      **else**
7:        ▷ Update local model
8:        **for** $i = 1, 2, \dots, u_x$ **do**
9:          $\theta_0^i = g_{r-1}(t) - \frac{\lambda}{K_i} \sum_{i=1}^{u_x} \nabla L(\theta_t^i)$.
10:       **end for**
11:      **end if**
12:      ▷ Local model training
13:      Let $\hat{\theta}_0^i = \theta_0^i$, initialize target action-value function $\hat{Q}(\cdot)$ according to the parameter $\hat{\theta}_0^i$;
14:      **for** $t = 1$ to τ **do**
15:        Receive the initial observed state $s_1^1, s_1^2, \dots, s_1^{u_x}$;
16:        **if** $t \leq |D_x^i|$ **then**
17:          Randomly select $a_t^1, a_t^2, \dots$;
18:        **else**
19:          Select $a_t^i = \text{argmax}_a Q(\cdot)$ using ϵ-greedy policy;
20:          Execute action a_t^i, obtain r_t^i and s_{t+1}^i;
21:          Store $(s_t^i, a_t^i, r_t^i, s_{t+1}^i)$ into $D_x^i, \forall i \in D$;
22:          Randomly select a sample $\left(s_j^i, a_j^i, r_j^i, s_{j+1}^i \right)$ from the experience relay pool $D^i, \forall i \in D$;
23:          Calculate y_t^i according to equation (4);
24:          Perform a gradient descent step;
25:          Update the parameter $\theta_t^i, \forall i \in D$;
26:          Every C slots reset $\hat{Q} = Q$;
27:        **end if**
28:      **end for**
29:      ▷ Update global model
30:      **for** $i = 1, 2, \dots, u_x$ **do**
31:        $g_r(t) = \frac{\sum_{i=1}^{u} K_i \theta_t^i}{K}$.
32:      **end for**
33:    **end for**
34: Obtain access scheme π^i, target value y_t^i.

---

two service requests in the overlapping area of multiple BSs (i.e., BS 1, BS 2), based on which $\mathcal{X}$ can be given by

$$\mathcal{X} = \left\{ \begin{bmatrix} 1 & 0 \\ 0 & 1 \end{bmatrix}, \begin{bmatrix} 0 & 1 \\ 1 & 0 \end{bmatrix}, \begin{bmatrix} 1 & 0 \\ 1 & 0 \end{bmatrix}, \begin{bmatrix} 0 & 1 \\ 0 & 1 \end{bmatrix} \right\},$$

where $\mathcal{X}$ represents the set of possible global schemes. The elements of $\mathcal{X}$ represent global access selection schemes, which is composed of row vectors of matrix $\mathbf{A}$ and $\mathbf{H}$. Here $\mathbf{A}$ and $\mathbf{H}$ are as follows:

$$\mathbf{A} = \mathbf{H} = \begin{bmatrix} 1 & 0 \\ 0 & 1 \end{bmatrix}.$$

In this case, $\mathbf{A}$ and $\mathbf{H}$ denote the access selection scheme of device 1 and device 2, respectively. For example, in $\mathbf{A}$, the first row [10] means that device 1 accesses to BS 1 and the second row [01] means that device 1 accesses to BS 2. Moreover, the sum of each row of $\mathbf{A}$ or $\mathbf{H}$ is 1, which represents that a device can only access to one BS.

In Wang et al. [2019a], the authors proposed an Monte–Carlo sampling, where the Shapley value is given by

$$\phi_f = \frac{1}{M} \sum_{m=1}^{M} \left(\varphi_{+f} - \varphi_{-f} \right), \tag{4.12}$$

where M represents the number of access feature updates in vDRL. ϕ_f represents the Shapley value for an access selection scheme f, which represents the average marginal contribution of f across all possible access feature combinations $\mathcal{X}$. For example, in $\mathcal{X}$ abovementioned, if $f = \mathcal{X}\{1\}$, we can get the $+f = \mathcal{X}\{1\}$ and $-f$ is randomly selected in $\{\mathcal{X}\{2\}, \mathcal{X}\{3\}, \mathcal{X}\{4\}\}$. Thus, we can derive the Shapley value ϕ_f of access selection scheme f through (4.12) and derive the global optimal 0–1 matrix by comparing these Shapley values. Therefore, when the devices send their service requests, the third encrypted party will send the ith row vector of f to devices, where the ith row vector of f denotes the local access selection of d_i.

4.4.4 HDRL Algorithm for Device Association

Next we elaborate the model training process of HDRL scheme. Figure 4.3 gives an illustrative example of HDRL process, where two types of services are considered. In this process, a communication round is composed of τ time slots. Note only the first and last time slots in a communication round are relevant to the global parameter aggregation. Between BSs and devices, the parameters (i.e., global model $g_r(t)$ and local models θ_t^i) are transmitted and updated for the same type services. At the first time slot of a communication round, the global model is sent to individual

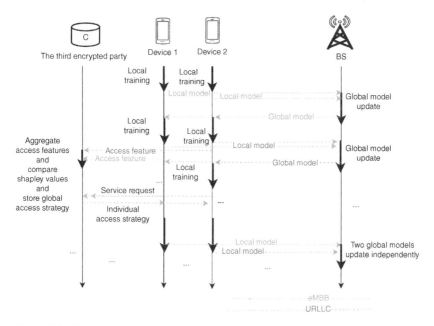

Figure 4.3 The process of HDRL.

devices if communication round $r \neq 1$. Otherwise the devices updates their local models directly with initial weights zero. During a communication round, based on the received global model or initial weights, the devices update local weights and train their own data with DDQN. At the last time slot of each communication round, according to the service type, the devices send their local models to BSs to update corresponding global model. Note that devices and BSs store all models of all the service types.

The global access feature aggregation is performed at the last time slot every v communication rounds. Between the third encrypted party and devices, access features are transmitted and aggregated for making a global BS-NS access selection scheme. Based on the aggregated global access feature, we can calculate the corresponding Shapley values and derive a global optimal NS-BS access selection scheme with the maximal Shapley value. Then the third encrypted party stores the global access selection scheme until it is replaced by a better one. Note that if devices send their service requests simultaneously, the third encrypted party sends the corresponding local access selection scheme to devices. Otherwise the smart devices can make access decision according to their own local model. Based on the training results of Algorithm 4.1, the HDRL algorithm is presented as Algorithm 4.2.

Algorithm 4.2 HDRL

Input: M, access selection scheme π_i, and target value y_t^i from **Algorithm 4.1**, iterations.

output: Shapley value ϕ_f, global access selection scheme.

  1: Initialize the maximum Shapley value $\phi_{max} = 0$;

  2: Initialize access selection scheme for all devices $f_0 = \varnothing$;

  3: **for** $m = 1, 2, \ldots, M$ **do**

  4:    Get π_i, θ_i, and y_t^i of the devices with different service categories in the same overlapping converges of multiple BSs;

  5:    Get $\mathcal{X}$ through π_i;

  6:    **for** $i = 1, 2, \ldots, |\mathcal{X}|$ **do**

  7:      $f = \mathcal{X}\{i\}$;

  8:      Initial the sets of $-f$;

  9:      Calculate φ_f;

10:      **for** iterations= $1, 2, \ldots$ **do**

11:        Choose $-f$ in $\{\mathcal{X} - \mathcal{X}\{i\}\}$;

12:        **if** $-f \subseteq F$ **then**

13:          Continue.

14:        **else**

15:          Calculate φ_{-f}.

16:        **end if**

17:      **end for**

18:      Calculate ϕ_f

19:      **if** $\phi_{max} \leq \phi_f$ **then**

20:        $\phi_{max} = \phi_f$;

21:        $f_0 = f$.

22:      **end if**

23:    **end for**

24: **end for**

25: Obtain Shapley value ϕ_f and the optimal global access selection scheme $f = f_0$.

4.4.5 Convergence Analysis

HDRL can be regarded as fully distributed DRL if neither horizontal aggregation nor vertical aggregation is performed (i.e., $r = R = v = 0$). Also, HDRL can be regarded as centralized DRL if we perform global aggregation after every local update (i.e., $\tau = v = 1$) if the data samples and features are visuable for the centralized controller, and the communication cost can be ignored (Wang et al. [2019b]). Furthermore, we use an auxiliary parameter vector $\mathbf{v}_t^r$, which follows a centralized gradient descent, given by

$$\mathbf{v}_{t+1}^r = \mathbf{v}_t^r - \eta \nabla L(\mathbf{v}_t^r), 1 \le t \le \tau, \forall r \in R. \tag{4.13}$$

According to Yang et al. [2019], Wang et al. [2019b], the global parameters $g_r(t)$ should be much close to $\mathbf{v}_t^r$ when $\tau = v = 1$. Formally, we have an upper bound on the difference between $L(g_r(t))$ and $L(\mathbf{v}_t^r)$ within $[t - (r - 1)\tau, t]$, as follows:

$$|L(g_r(t)) - L(\mathbf{v}_t^r)| \le h(\tau, r). \tag{4.14}$$

We have $h(\tau, r) = 0$ if fully distributed DRL or centralized DRL is performed. However, the fully distributed DRL, $r = R = 0$, is always the worst solution compared with centralized DRL and our proposed HDRL, as fully distributed DRL only considers independent training/access decision. Moreover, it is always optimal when setting $\tau = 1$ and $v = 1$ if we have unlimited resource budget and ignore the privacy issue, as centralized DRL jointly trains a global model for all services. Theoretically, the performance of model training in HDRL should be between that of the fully distributed DDQN and centralized DDQN. From the studies in Wang et al. [2019b], $h(\tau, r)$ is affected by data distribution, r, and τ. Thus, when given data distribution, τ and r, we can derive the upper bound of the divergence (i.e., $h(\tau, r)$) between loss function and the global loss function. Moreover, as we use a non-linear sigmoid function in neural networks, the loss function in this work is nonconvex. Therefore, we can obtain $h(\tau, r)$ and $[L(\mathbf{v}_t^r) - h(\tau, r), L(\mathbf{v}_t^r) + h(\tau, r)]$ through training. Intuitively, the frequency of performing global weights aggregation (i.e., τ) should be carefully specified, as the communication cost with a large number of communication rounds cannot be ignored. Numerical results in the Section 4.5.2 will illustrate this.

4.5 Numerical Results

In this section, we evaluate the performance of our proposed HDRL scheme through simulation experiments. We employ three reference device association (DA) schemes as comparison reference:

(1) *Greedy algorithm for DA (GDA)*: In this scheme, each device selects NS-BS to access which provides the maximal available bandwidth. Moreover, GDA aims to find the maximal communication efficiency based on instantaneous network conditions, instead of considering long-term optimal communication efficiency.

(2) *Centralized DDQN for DA (CDA)*: In this scheme, all devices transmit data to a controller for centralized training in DDQN. Then the controller makes access decisions for all devices. The performance of CDA can be viewed as

the upper bound of DDQN where the cost for transferring training data is igonored.

(3) *Distributed DDQN without model aggregation for DA (DDA):* In this scheme, individual devices train their own data through DDQN and make access decision independently, where no model aggregation of FL is used. Moreover, the reward function in CDA and DDA remains the same as that in HDRL except that the cost of communication round is zero.

(4) *Received signal strength (RSS)-based BS-NS access selection for device association (RDA):* In this scheme, devices first select the BS with highest RSS and then select the slice that can provide the maximum communication efficiency on this BS.

4.5.1 Simulation Settings

We consider a network scenario consisting of four BSs that are randomly distributed in a square area of 1060×1060 m^2 (Zhao et al. [2020]). Moreover, five end-to-end slices are randomly associated with the four BSs in the network. In the simulation, the maximal transmit power and the noise power of BSs are set to 47 and -174 dBm/Hz, respectively, (Zhao et al. [2020], Galeana-Zapién and Ferrús [2010]). The path loss for BSs is modeled as $L(d) = 34 + 40 \log(d)$ (Zhao et al. [2020], Galeana-Zapién and Ferrús [2010]). The wireless bandwidth of each BS is set to 20 MHz. For a specific BS k, the wireless bandwidth is randomly allocated to all NSs deployed at BS k. In other words, $b_{j,k}$ is randomly chosen from [0,20] MHz with the constraint $\sum_{j=1}^{5} b_{j,k} \leq 20$ MHz (Zhao et al. [2020]). In addition, devices are randomly distributed within the square area with different transmission rate and delay requirements. We assume that three types of services (i.e., eMBB, mMTC, and URLLC) are supported. The service type is characterized by transmission rate and delay, where the required transmission rate $r_{i,t}^{j,k}$ is randomly set within [5,30] Mbps (Chowdhury et al. [2020]) and the delay in CN is randomly set within [1,10] ms (Chowdhury et al. [2020]). For each device, we consider a three-layer fully connected neural network. Specifically, the input layer is composed of 12 neurons, representing the input of access condition and access/handoff action. The hidden layer is composed of 25 neurons, where the activation function is set to sigmoid function. To avoid correlation between action values and target values, we copy the weights of Q-network θ to the weights of target network $\hat{\theta}$ every 5 training steps (Xu et al. [2018]). The memory size for each service type is set to 1000, the batch size is set to 64, and the discount factor is set to 0.99 (Xu et al. [2018]). In addition, both learning rate for training and the learning rate for updating local model are set to 0.001.

4.5.2 Numerical Results and Discussions

First, we examine the relationship between the total long-term reward and the frequency of performing global aggregation. In this simulation experiment, the communication cost for one communication round, $\alpha^c_{i,k}$, is a constant. Figure 4.4 illustrates the total long-term reward with respect to the number of communication rounds. From Figure 4.4, we can see that the total long-term reward increases in the beginning and then decreases with the number of communication rounds. There is an optimal number of rounds, say 14, which leads to the maximal total reward when the number of trainings in a communication round (i.e., τ) is set to 2000. Furthermore, we observe that the number of trainings in a communication round (i.e., τ) affects the optimal number of communication rounds. The reasons are (i) the local model changes with the number of trainings in a communication round (i.e., τ) and (ii) the communication cost increases with the number of communication rounds.

Then, we verify the convergence property of our proposed HDRL scheme by depicting its learning curve (the curve of weights versus the total number of trainings). We set $\tau = 2000$ and randomly select three corresponding local models (i.e., $\theta^1, \theta^2, \theta^3$) on three different devices. As shown in Figure 4.5, HDRL converges with the total number of trainings increasing. Furthermore, Figure 4.6 shows the partial convergence curve of Figure 4.5 within $[5r, 50r]$. From Figure 4.6, we can see that the three corresponding weights from different smart devices coincide when they tend to be stable, which further illustrates the effectiveness of training a global

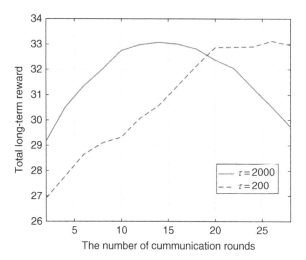

Figure 4.4 The relationship between the number of communication rounds and the total long-term reward.

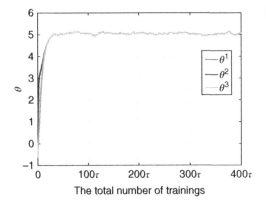

Figure 4.5 Convergence of HDRL.

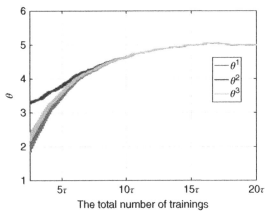

Figure 4.6 Partial convergence curve of Figure 4.5 within [5r, 50r].

model with multiple independent smart devices. Combining Figures 4.4–4.6, we set $r = 2000$ and $R = 14$ to evaluate the performance of HDRL in the following experiments.

Then, we compare the total long-term reward of five schemes (i.e., HDRL, CDA, DDA, GDA, and RDA) when the number of devices is 35 (i.e., $u = 35$). Figure 4.7 shows the total long-term reward of the five schemes. From Figure 4.7, we can see that HDRL and CDA achieve higher long-term reward than other three schemes. The reason is that HDRL and CDA explore the global optimal access selection schemes, while other three schemes including DDA, GDA, and RDA focus on the local access decision. Moreover, HDRL aggregates the same type services on BSs, reducing the correlation between the training data from different devices when compared with CDA. Therefore, the total reward of HDRL is higher than that of CDA.

Next, we examine the network throughput of the five schemes. Figure 4.8 shows the network throughput with respect to the number of devices. From

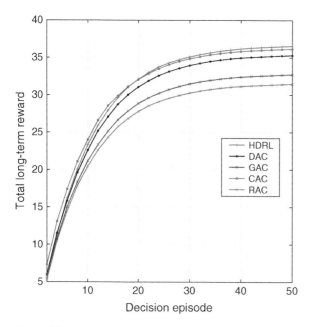

Figure 4.7 The performance of the total long-term reward.

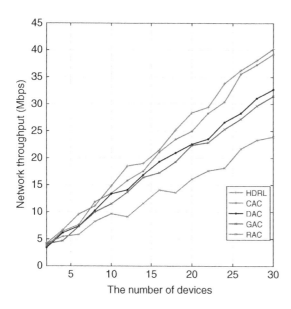

Figure 4.8 Comparison of network throughput as a function of the number of devices in four schemes.

Figure 4.8, we can see that HDRL always outperforms other four schemes on network throughput. The reason is that HDRL integrates the similar data samples into a global model before aggregating the access features. Moreover, the access feature aggregation in HDRL considers the global optimal access selection scheme. In comparison, the duplicate data samples in CDA for centralized training increase the correlation of data, resulting in overfitting easily. Moreover, DDA, GDA, and RDA focus on individual devices access condition without global aspects. In addition, GDA and RDA make access decisions based on instantaneous conditions, instead of considering long-term optimization objectives.

Next, we compare handoff cost of the five schemes. Figure 4.9 shows the comparison of handoff cost of the five schemes. From Figure 4.9, we can see that the handoff cost of HDRL and DDA is always higher than that of CDA, while HDRL incurs the highest handoff cost. The reason is that smart devices train their own data independently in HDRL and DDA schemes. Although HDRL employs two levels of aggregation, training on smart devices independently is not affected. Furthermore, combining Figures 4.8 and 4.9, we can see that a trade-off exists between network throughput and handoff cost, where higher network throughput implies more handoff cost. Thus, we compare the communication efficiency by combining network throughput and handoff cost to further evaluate the performance of the five schemes.

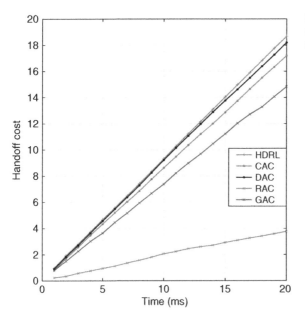

Figure 4.9 Comparison of handoff cost of four schemes.

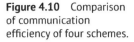

Figure 4.10 Comparison of communication efficiency of four schemes.

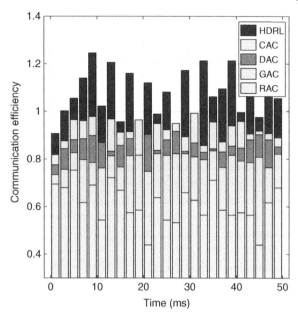

Finally, we compare the communication efficiency of the five schemes, shown in Figure 4.10. From Figure 4.10, we can see that HDRL always outperforms other four schemes on communication efficiency. The reason is that HDRL not only considers the optimal global access selection scheme but also integrates the similar data samples. Numerical results show that HDRL achieves higher communication efficiency by about 14.19%, 20.80%, 26.60%, and 36.36% on average compared with CDA, DDA, GDA, and RDA, respectively.

4.6 Conclusion

In this chapter, with the aim to improve network throughput while reducing handoff cost, we have modeled the device association problem for RAN slicing as an MDP model and developed a novel HDRL scheme to solve it by exploiting hybrid FL based on DRL. In HDRL, two levels of model aggregation based on DRL are employed to promote the collaboration between smart devices while enforcing the privacy and security of local data. Numerical results show that our proposed HDRL scheme can achieve a significant performance gain in terms of network throughput and communication efficiency when compared with some known state-of-the-art algorithms.

Acknowledgment

The work was supported by the National Key Research and Development Program of China (Grant No. 2020YFB1806804).

References

Mingzhe Chen, Zhaohui Yang, Walid Saad, Changchuan Yin, H. Vincent Poor, and Shuguang Cui. A joint learning and communications framework for federated learning over wireless networks. *arXiv preprint arXiv:1909.07972*, 2019.

Mostafa Zaman Chowdhury, M. D. Shanjalal, Shakil Ahmed, and Yeong Min Jang. 6G wireless communication systems: Applications, requirements, technologies, challenges, and research directions. *IEEE Open Journal of the Communications Society*, 1(1):957–975, 2020.

Ericsson AB. 5G Systems –enabling the transformation of industry and society. *White Paper*, (January):14, 2017. URL https://www.ericsson.com/res/docs/whitepapers/wp-5g-systems.pdf.

ETSI. TS 136 331 - V15.3.0 - LTE; Evolved Universal Terrestrial Radio Access (E-UTRA); Radio Resource Control (RRC); Protocol specification. *3GPP TS 36.331 version 13.7.1 Release 13*, pages 1:1–1:649, 2018. URL https://portal.etsi.org/TB/ETSIDeliverableStatus.aspx.

Hiram Galeana-Zapién and Ramon Ferrús. Design and evaluation of a backhaul-aware base station assignment algorithm for OFDMA-based cellular networks. *IEEE Transactions on Wireless Communications*, 9(10):3226–3237, 2010.

S.-B. Lee, Sayantan Choudhury, Ahmad Khoshnevis, Shugong Xu, and Songwu Lu. Downlink MIMO with frequency-domain packet scheduling for 3GPP LTE. In *Proceeding of IEEE INFOCOM 2009*, pages 1269–1277. IEEE, 2009.

Wei Yang Bryan Lim, Nguyen Cong Luong, Dinh Thai Hoang, Yutao Jiao, Ying-Chang Liang, Qiang Yang, Dusit Niyato, and Chunyan Miao. Federated learning in mobile edge networks: A comprehensive survey. *IEEE Communications Surveys & Tutorials*, 22(3):2031–2063, 2020.

Y. Mansouri, A. N. Toosi, and R. Buyya. Cost optimization for dynamic replication and migration of data in cloud data centers. *IEEE Transactions on Cloud Computing*, 7(3):705–718, 2019.

H. Brendan McMahan, Eider Moore, Daniel Ramage, Seth Hampson, and Blaise Aguera y Arcas. Communication-efficient learning of deep networks from decentralized data. *arXiv preprint arXiv:1602.05629*, 2016.

Abdelkader Sbihi. A best first search exact algorithm for the multiple-choice multidimensional knapsack problem. *Journal of Combinatorial Optimization*, 13(4):337–351, 2007.

Y. Sun, G. Feng, L. Zhang, M. Yan, S. Qin, and M. A. Imran. User access control and bandwidth allocation for slice-based 5G-and-beyond radio access networks. In *Proceeding of IEEE International Conference on Communications (ICC)*, pages 1–6, 2019a.

Yao Sun, Gang Feng, Lei Zhang, Paulo Valente Klaine, Muhammad Ali Iinran, and Ying-Chang Liang. Distributed learning based handoff mechanism for radio access network slicing with data sharing. In *Proceeding of IEEE International Conference on Communications (ICC)*, pages 1–6. IEEE, 2019b.

N. H. Tran, W. Bao, A. Zomaya, M. N. H. Nguyen, and C. S. Hong. Federated learning over wireless networks: Optimization model design and analysis. In *Proceeding of IEEE INFOCOM - Conference on Computer Communications*, pages 1387–1395, 2019.

Hado Van Hasselt, Arthur Guez, and David Silver. Deep reinforcement learning with double Q-Learning. In *Proceeding of 30th AAAI Conference on Artificial Intelligence, AAAI 2016*, pages 2094–2100, 2016.

S. Wang, R. Urgaonkar, M. Zafer, T. He, K. Chan, and K. K. Leung. Dynamic service migration in mobile edge-clouds. In *Proceeding of 2015 IFIP Networking Conference (IFIP Networking)*, pages 1–9, 2015.

Guan Wang, Charlie Xiaoqian Dang, and Ziye Zhou. Measure contribution of participants in federated learning. *arXiv preprint arXiv:1909.08525*, 2019a.

Shiqiang Wang, Tiffany Tuor, Theodoros Salonidis, Kin K. Leung, Christian Makaya, Ting He, and Kevin Chan. Adaptive federated learning in resource constrained edge computing systems. *IEEE Journal on Selected Areas in Communications*, 37(6):1205–1221, 2019b.

Zhiyuan Xu, Jian Tang, Jingsong Meng, Weiyi Zhang, Yanzhi Wang, Chi Harold Liu, and Dejun Yang. Experience-driven networking: A deep reinforcement learning based approach. In *Proceeding of IEEE INFOCOM - Conference on Computer Communications*, pages 1871–1879, 2018.

Qiang Yang, Yang Liu, Tianjian Chen, and Yongxin Tong. Federated machine learning: Concept and applications. *ACM Transactions on Intelligent Systems and Technology (TIST)*, 10(2):1–19, 2019.

Shunliang Zhang. An overview of network slicing for 5G. *IEEE Wireless Communications*, 26(3):111–117, 2019.

L. Zhang, J. Tan, Y. Liang, G. Feng, and D. Niyato. Deep reinforcement learning-based modulation and coding scheme selection in cognitive heterogeneous networks. *IEEE Transactions on Wireless Communications*, 18(6):3281–3294, 2019.

Guanqun Zhao, Shuang Qin, Gang Feng, and Yao Sun. Network slice selection in softwarization-based mobile networks. *Transactions on Emerging Telecommunications Technologies*, 31(1), 2020. ISSN 21613915. doi: 10.1002/ett.3617

5

Deep Federated Learning Based on Knowledge Distillation and Differential Privacy

Hui Lin[1,2], Feng Yu[1,2], and Xiaoding Wang[1,2]

[1] College of Computer and Cyber Security, Fujian Normal University, Fuzhou, Fujian Province, China
[2] Engineering Research Center of Cyber Security and Education Informatization, Fujian Province University, Fuzhou, Fujian Province, China

5.1 Introduction

The Internet of Things (IoT), which is also referred to as a sensor network, connects various objects to the internet through sensing equipment such as radio frequency identification, infrared sensors, global positioning systems, and laser scanners. This connection facilitates information exchange and communication, allowing for intelligent identification, positioning, tracking, monitoring, and management. The deployment and application of various sensors are essential for the widespread implementation of IoT. Different applications require different sensors, such as those used in smart industry, smart security, smart home, smart transportation, and smart medical care. As a result, IoT sensor technology plays a crucial role in economic development and promotes social progress.

Currently, most intelligent services provided by the IoT require outsourcing user data to service providers for analysis and processing, which may result in sensitive information leakage (Al-Rubaie and Chang [2019]). With the increasing awareness of user privacy protection and the promulgation of relevant laws and regulations, traditional machine learning-based data analysis services can no longer satisfy users' privacy protection needs. Although existing cryptographic technology can address some privacy leakage issues, both symmetric and asymmetric encryption systems have the risk of key leakage, and high-cost encryption chips cannot be widely used in terminal devices. To address the problem of user data privacy leakage in related service scenarios, Google proposed federated learning (FL) technology (Konečný et al. [2016a,b]).

Federated Learning for Future Intelligent Wireless Networks, First Edition.
Edited by Yao Sun, Chaoqun You, Gang Feng, and Lei Zhang.
© 2024 The Institute of Electrical and Electronics Engineers, Inc. Published 2024 by John Wiley & Sons, Inc.

Federated learning is a distributed machine learning approach that ensures privacy preservation and can generate secure, accurate, and robust data models without analyzing the real data of users. In an intelligent service scenario based on federated learning, the service provider invites various participants to provide data models by releasing federated learning tasks. The aggregation server then combines all data models to generate a reliable global model that provides related services. The reliability of the global model is crucial, as it provides secure and stable services for IoT applications. To compress models, knowledge distillation (Zhang et al. [2022]) is a common technique. Unlike model compression methods such as pruning and quantization, knowledge distillation trains a small, lightweight model using the supervision information of a larger model with better performance to achieve higher accuracy. Therefore, combining knowledge distillation with federated learning can help protect data privacy.

It is worth noting that cryptography technology can also be utilized to achieve data privacy protection. However, the use of cryptography often requires a trusted third party to generate a key for data encryption, which can be challenging to achieve in the context of IoT. In comparison, federated learning does not require a trusted third party and is easier to deploy. Even if the federated learning server is not trusted, adding noise to the model can still achieve data privacy protection, making federated learning more advantageous for data sharing.

The problem addressed in this chapter is how to enable data sharing while ensuring privacy protection. To solve this problem, we propose a data fusion architecture based on knowledge distillation-based federated learning (KDFL), as shown in Figure 5.1. The architecture comprises three layers: the perception layer, the data

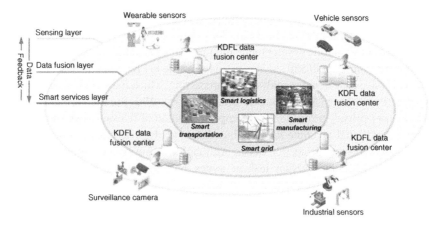

Figure 5.1 Knowledge distillation-based federated learning data fusion architecture. Source: Connect world/Adobe Stock; zapp2photo/Adobe Stock; gopixa/Adobe Stock; jeson/Adobe Stock.

fusion layer, and the intelligent service layer. The perception layer acquires perception data from various sensors such as wearable sensors, vehicle-mounted sensors, surveillance cameras, and industrial sensors and sends the data to the data fusion layer. In the data fusion layer, each KDFL data fusion center is responsible for intelligent fusion processing of perception data and provides the fusion data to the intelligent service layer. The intelligent service layer provides technical support for various IoT intelligent services like intelligent transportation, smart grid, intelligent manufacturing, and intelligent logistics. The perception layer provides the necessary data to the intelligent service layer via the data fusion layer, and the intelligent service layer provides feedback information to improve the quality of intelligent services.

1. Our proposed approach for privacy-preserving data fusion involves a federated learning algorithm based on knowledge distillation. This strategy integrates differential privacy noise into both the local model training process and the federated training process, which ultimately results in a certain amount of model accuracy loss. However, this trade-off leads to the protection of differential privacy of the model.
2. The experimental outcomes demonstrate that our approach not only achieves high-precision IoT data fusion but also offers superior privacy protection.

5.2 Related Work

Federated learning for data fusion is an effective method for IoT to provide intelligent services, and the reliability of the federated learning global model is crucial for ensuring service quality. Many scholars from around the world have conducted research on how to ensure the reliability of the federated learning global model under various requirements and have produced numerous outstanding research results.

The reliability of the global model is a crucial aspect for federated learning task publishers. To ensure global model reliability, researchers have developed methods for detecting anomalous models in the models to be aggregated. Cao et al. [2019] mapped the local models into a graph using Euclidean distance and solved the Maximum Clique problem in the graph to select the local model for aggregation, achieving the detection of anomalous models in federated learning. Zhao et al. [2022] generated a dataset for auditing the local model using a trained generative adversarial network and used the prediction and evaluation results of the local model in the dataset as a criterion for judging whether it was an abnormal model, thus achieving the detection of abnormal models. Zhao et al. [2020] proposed a proxy-based anomaly model detection mechanism, selecting participants with relatively stable performance in federated learning to perform anomaly model detection. Tolpegin et al. [2020] extracted abnormal model features by performing dimensionality reduction and principal component analysis on the local model, enabling abnormal model detection in the process. Liu et al. [2021b] proposed a federated learning scheme called PEFL to mitigate poisoning attacks under privacy enhancement. In Ma et al. [2021], an asynchronous update paradigm for real-time identification of client network parameters was proposed, adopting a linear fusion method based on sequential filtering. It asynchronously fused the parameters of the federated center with consideration of communication delay and established a linear filtering-based client real-time recognition method that obtained new labeled samples at unequal intervals, expecting better performance on the client side.

The primary demand of federated learning participants is the protection of their private training data from being leaked during the model training and aggregation process. Existing research work focuses on striking a balance between privacy protection and model reliability, as both cannot be simultaneously prioritized. The research aims to reduce the loss of global model reliability while meeting the privacy protection needs of participants. The authors of Truex et al. [2020] proposed LDP-Fed, a privacy-preserving federated learning scheme where federated learning participants can protect their model's privacy using personalized local differential privacy technology to prevent deep information leakage in the local model. In Hu et al. [2020], differential privacy technology was introduced in federated learning by Hu et al. They utilized the uncertainty brought by the heterogeneity of IoT devices to perform differential privacy on the model, which helped to reduce the risk of privacy leakage. The authors of Ibitoye et al. [2022] introduced differential privacy technology and self-normalization technology in federated learning. They added a differential privacy noise layer and SELU security training layer during model training to provide privacy protection for uploaded models. The authors of Kumar et al. [2020] proposed a blockchain-based federated learning training strategy that utilizes differential privacy and homomorphic encryption technology to ensure the privacy of participants during local model transmission and aggregation. Zhang and Luo [2020] generated training data for local training using a GAN model. They proposed a new loss function that made the generated training data have indistinguishable visual features from the original data, which helped protect the privacy of the training data of the participants. Liu et al. presented in Liu et al. [2021a] a method that uses the sparse features of the feature map in the network model to represent the data used by participants for local training. This technique enabled the privacy protection of the real data. Xu et al. [2021] proposed FedV, an efficient and privacy-preserving vertical federated learning framework. FedV implemented a two-stage noninteractive secure federated aggregation method by introducing functional encryption, which helped to achieve privacy protection of the real data of the participants. Lin et al. proposed a secure joint learning mechanism in Lin et al. [2021] based on variational autoencoders to resist inference attacks. In this approach, participants reconstructed the original data through variational autoencoders and trained local models based on this reconstruction, which helped protect the data privacy. The authors of Lu et al. [2022] proposed a heterogeneous model fusion federated learning mechanism wherein each node trains learning models of different scales according to its computing power. The parameter server receives the training gradient of each node, corrects it using the repetition matrix, updates the corresponding region of the global model according to the mapping matrix, and assigns the compressed model to the respective node.

Several studies have demonstrated the advantages of KDFL. However, the manual annotation data required for training deep learning models is typically collected and stored at different sites due to privacy concerns and transmission costs, thereby making it challenging to utilize distributed data. To overcome these obstacles for object recognition tasks in real-world IoT scenarios, Wang et al. [2021] propose a heterogeneous brainstorming method that optimizes temperature parameters on distributed datasets and trains heterogeneous models.

A joint distillation learning system is proposed for multitask time series classification in Xing et al. [2022]. The system comprises a central server and multiple mobile users, each of whom may run different tasks. For each user, the system employs knowledge distillation to transfer knowledge from the hidden layer of teachers to the hidden layer of students. The teachers and students have the same network structure, and the weight of the hidden layer of student models is regularly uploaded to the server.

In Sattler et al. [2022], the authors analyzed the impact of active distillation data storage, soft label quantification and incremental coding technology, studies federal distillation from the perspective of communication efficiency, and proposes an efficient combined distillation method, compression combined distillation, based on the insights collected from this analysis.

The authors of Sattler et al. [2022] analyzed the impact of active distillation data storage, soft label quantification, and incremental coding technology on communication efficiency in federal distillation. Based on the insights garnered from this analysis, they propose an efficient combined distillation method, compression combined distillation.

The literature in Taya et al. [2022] is concerned with the convergence of machine learning models in the function space. The authors proposed an algorithm for directly updating the function in the function space, which can converge to the optimal solution. The convergence of the proposed algorithm in the function space is analyzed, and it is shown that spectral theory can be applied to function space similarly to numerical vectors. Furthermore, a consensus-based multihop combined distillation method is developed for neural networks to achieve functional aggregation between adjacent devices without parameter averaging.

The above literature provides a large number of excellent algorithms for data fusion in the IoT. However, how to combine the privacy protection of local data with the privacy protection of the learning process to further strengthen the privacy protection of knowledge distillation-based federated learning is still a problem worthy of research.

5.3   System Model

When striving for privacy-preserving data fusion in IoT, it is necessary to take into account the following three components:

- *Sensor (data provider)*: The sensor collects the sensed data and transmits it to the data fusion center either through wired or wireless means for aggregation.
- *Federated learning (FL) based data fusion center*: The data fusion center utilizes local sensor data for model training, which allows the local model to encapsulate the information of the respective local data. Moreover, it is imperative to add differential privacy noise during the local model training process to ensure the differential privacy protection of the local model.
- *Knowledge distillation-based federated learning server*: This server aggregates the local models' logits and adds differential privacy noise to further improve the differential privacy protection capability of the model.

5.3.1   Security Model

We are considering various scenarios where privacy breaches could occur. The first scenario involves IoT smart service providers who may exploit private information of objects that are being monitored by the sensors, thereby exposing their privacy. Federated learning is a possible solution to mitigate this risk. However, there exists a possibility of privacy leakage during the federated learning process when the aggregation server is also interested in the privacy of the objects. This risk can be reduced through the implementation of differential privacy protection in local model training. Another scenario relates to malicious attackers who attempt to retrieve private information about perceptual objects from the model through inference attacks. Adding differential privacy protection to the model can effectively prevent such attacks.

5.4 The Implementation Details of the Proposed Strategy

Local data fusion involves the training of a deep neural network model using sensor data from the local environment. A deep neural network deploys multiple neurons across multiple levels and by adjusting the connection weights between them via layer-by-layer training, the input feature data can undergo multiple nonlinear transformations. The final product of this process is stable features that are useful for subsequent problem analysis. In order to evaluate the difference between the predicted value and the actual value in a deep neural network algorithm, a loss function is used. The goal of deep neural network algorithm training is to minimize this loss function. For complex networks, this is typically achieved through stochastic gradient descent, which involves randomly selecting batches of training samples during each iteration, calculating the partial derivative of the loss function, and updating the weight coefficients along the negative gradient direction toward the local minimum. To ensure privacy, we employ the differential privacy stochastic gradient algorithm. During each iteration, we begin by calculating the gradient of a randomly generated batch of samples and then perform gradient clipping. In order to protect the privacy of the sample data, the clipped gradient is updated using the additional Gaussian noise method of (Dwork and Roth [2014]) that involves adding random noise to the mean value of the gradient sum. In summary, in federated distillation, each user acts like a student and considers the output of the average model of all other users as their teacher's output. To achieve this, each user stores the average logit vector of each tag and uploads these local average logits to the server. During local training, it is important to add Gaussian noise. To obtain the average logit vector of each tag, the server averages the local average logit vector uploaded by all users while also adding Gaussian noise. This process is repeated for all labels. The average logit vectors of all labels are then downloaded to each device for distillation loss calculation. For the current training sample, the teacher's output is selected as the average logit vector that matches its label.

5.4.1 Security Analysis

The approach presented in this chapter is effective at thwarting privacy leakage attacks. First, the data fusion center adds differential privacy noise to the gradient of the model update during local model training to ensure that the local model is protected by differential privacy once training is complete. Second, after the logits are aggregated to the server, additional differential privacy noise is applied to further enhance the model's differential privacy protection. This process is repeated until federated learning convergence is achieved. With differential privacy's post-processing properties, the entire federated learning process is safeguarded with differential privacy protection. Because the model is shielded by differential privacy, it becomes more challenging for attackers to conduct known plaintext attacks or ciphertext only attacks and also increases the difficulty of attackers attempting to recover user data through reasoning attacks.

5.5 Performance Evaluation

5.5.1 Experimental Environment

To assess the effectiveness of the proposed approach, we conducted an experiment on a computer equipped with an i7 6.4 GHz processor, 32 GB of memory, and running a 64-bit Windows 7 operating system. The federated learning was established using the Python-based deep learning framework Tensorflow 2.2.0.

Through the experiment, both the local model and the global model employed CNN with 2 convolutional layers (1×10, kernelsize $= 5$; 10×20, kernelsize $= 5$), dropout layers, and 2 fully connected layers (320×50; 50×10). The Mnist dataset and the Fashion Mnist dataset were utilized in the experiment. The Mnist dataset, widely known for handwritten digit recognition, is commonly utilized to evaluate image classification performance in computer vision. Comprising 70,000 grayscale images with a 28×28 resolution, this dataset encompasses 10 digit classes – from digit 0 to digit 9. Out of these, 60,000 images are employed for model training while another 10,000 images are used for validation. The Fashion Mnist dataset, an extended variant of Mnist, contains 70,000 grayscale images including a training set of 60,000 images and a test set of 10,000 images. Each image is a 28×28 grayscale image depicting different types of t-shirts, dresses, boots, and so on.

In the experiments, we fix other hyperparameters, adjust $\epsilon \in (10, 20, 30)$ for multiple experiments. The rest of parameters are as follows: $Dp\_\delta = 1e - 5$, $Dp\_\epsilon \in \{10, 20, 30\}$, Epochs $= 100$, Num_users $= 10$, Frac $= 0.5$, Local_ep $= 1$, Local_bs $= 32$, Learning rate $= 0.01$, Lr_decay $= 0.995$, and Momentum $= 0.5$, $Dp\_clip = 10$.

In this experiment, we evaluate the performance of three strategies, FedAvg (McMahan et al. [2017]), client level DP (denoted by noise-added local model training), and round level DP (denoted by noise-added model training), in terms of model accuracy.

5.5.2 Experimental Results

Observed from Figure 5.2, the above six figures are divided into upper and lower parts. When the privacy budget is the same, the test accuracy of round level dp is

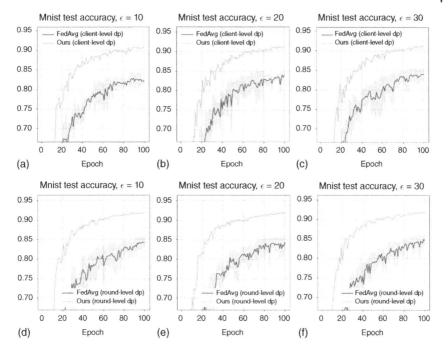

Figure 5.2 Accuracy on Mnist with different differential privacy protection (a) $\epsilon = 10$, (b) $\epsilon = 20$, (c) $\epsilon = 30$, (d) $\epsilon = 10$, (e) $\epsilon = 20$, and (f) $\epsilon = 30$.

slightly higher than that of client level dp. To some extent, it shows that adding noise to the local model has a greater impact than adding noise to the aggregated global model, which means that its model test accuracy is lower after training. Divide the above six figures into three parts: left, middle, and right. We can see that when the privacy budget increases, we can roughly see that the test accuracy jitters more and more fiercely in the early stage, which indicates that the more noise is added, the more contributions to the decline of model availability. For the same type of DP, there is little difference in the final model test accuracy. In the above figure, our method is not only much more accurate than FedAvg but also has a faster convergence speed. This shows that the application of knowledge distillation to traditional federal learning methods can greatly improve the quality of the model, and thus improve the accuracy of the test.

Observed from Figure 5.3, by dividing the above six figures into upper and lower parts, we can see that the maximum test accuracy of round level dp is higher than that of client level dp. When the privacy budget is 10, the difference between the two maximum accuracy rates is obvious, while in the other two privacy budget experiments, there is almost no difference. We analyze that this is because EMnist dataset is more complex than Mnist dataset, and its model test accuracy is not high in the experiment without adding noise. Therefore, for the same type of DP, when

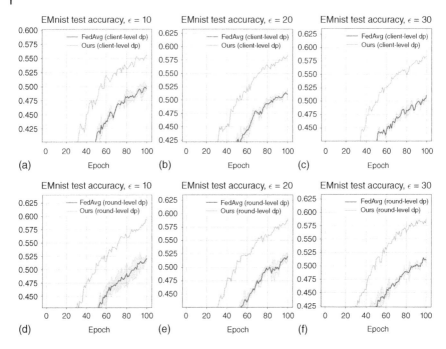

Figure 5.3 Accuracy on EMnist with different differential privacy protection (a) $\epsilon = 10$, (b) $\epsilon = 20$, (c) $\epsilon = 30$, (d) $\epsilon = 10$, (e) $\epsilon = 20$, and (f) $\epsilon = 30$.

the noise volume increases, the impact on its model quality is not obvious. Similarly, our method is not only much more accurate than FedAvg but also has faster convergence speed. This shows that the application of knowledge distillation to traditional federal learning methods can greatly improve the quality of the model, and thus improve the accuracy of the test.

5.6 Conclusions

The merging of data collected by various sensors in the IoT represents an urgent problem requiring resolution. During the process of data fusion, it is possible for the privacy of objects collected by sensors to be compromised, highlighting the need for both data fusion and privacy protection. To address this, we propose a deep federated learning algorithm based on knowledge distillation and differential privacy for data fusion. Our approach incorporates Gaussian noise at various stages of KDFL to ensure privacy protection throughout the data fusion process. The experimental results indicate that this strategy provides superior privacy protection while still achieving precise IoT data fusion.

Bibliography

Mohammad Al-Rubaie and J. Morris Chang. Privacy-preserving machine learning: Threats and solutions. *IEEE Security & Privacy*, 17(2):49–58, 2019.

Di Cao, Shan Chang, Zhijian Lin, Guohua Liu, and Donghong Sun. Understanding distributed poisoning attack in federated learning. In *2019 IEEE 25th International Conference on Parallel and Distributed Systems (ICPADS)*, pages 233–239, 2019. doi: https://doi.org/10.1109/ICPADS47876.2019.00042.

Cynthia Dwork and Aaron Roth. The algorithmic foundations of differential privacy. *Foundations and Trends® in Theoretical Computer Science*, 9(3–4):211–407, 2014.

Rui Hu, Yuanxiong Guo, Hongning Li, Qingqi Pei, and Yanmin Gong. Personalized federated learning with differential privacy. *IEEE Internet of Things Journal*, 7(10):9530–9539, 2020.

Olakunle Ibitoye, M. Omair Shafiq, and Ashraf Matrawy. Differentially private self-normalizing neural networks for adversarial robustness in federated learning. *Computers & Security*, 116:102631, 2022.

Jakub Konečný, H. Brendan McMahan, Daniel Ramage, and Peter Richtárik. Federated optimization: Distributed machine learning for on-device intelligence. *arXiv preprint arXiv:1610.02527*, 2016a.

Jakub Konečný, H. Brendan McMahan, Felix X. Yu, Peter Richtárik, Ananda Theertha Suresh, and Dave Bacon. Federated learning: Strategies for improving communication efficiency. *arXiv preprint arXiv:1610.05492*, 2016b.

Swaraj Kumar, Sandipan Dutta, Shaurya Chatturvedi, and M. P. S. Bhatia. Strategies for enhancing training and privacy in blockchain enabled federated learning. In *2020 IEEE Sixth International Conference on Multimedia Big Data (BigMM)*, pages 333–340. IEEE, 2020.

Hui Lin, Wenxin Liu, and Xiaoding Wang. A secure federated learning mechanism for data privacy protection. In *2021 20th International Conference on Ubiquitous Computing and Communications (IUCC/CIT/DSCI/SmartCNS)*, pages 25–31, 2021. doi: https://doi.org/10.1109/IUCC-CIT-DSCI-SmartCNS55181.2021.00019.

Bingyan Liu, Yao Guo, and Xiangqun Chen. PFA: Privacy-preserving federated adaptation for effective model personalization. In *Proceedings of the Web Conference 2021*, pages 923–934, 2021a.

Xiaoyuan Liu, Hongwei Li, Guowen Xu, Zongqi Chen, Xiaoming Huang, and Rongxing Lu. Privacy-enhanced federated learning against poisoning adversaries. *IEEE Transactions on Information Forensics and Security*, 16:4574–4588, 2021b.

Xiaofeng Lu, Yuying Liao, Chao Liu, Pietro Lio, and Pan Hui. Heterogeneous model fusion federated learning mechanism based on model mapping. *IEEE Internet of Things Journal*, 9(8):6058–6068, 2022. doi: 10.1109/JIOT.2021.3110908.

Xue Ma, Chenglin Wen, and Tao Wen. An asynchronous and real-time update paradigm of federated learning for fault diagnosis. *IEEE Transactions on Industrial Informatics*, 17(12):8531–8540, 2021. doi: 10.1109/TII.2021.3063482.

Brendan McMahan, Eider Moore, Daniel Ramage, Seth Hampson, and Blaise Aguera y Arcas. Communication-efficient learning of deep networks from decentralized data. In *Artificial Intelligence and Statistics*, pages 1273–1282. PMLR, 2017.

Felix Sattler, Arturo Marban, Roman Rischke, and Wojciech Samek. CFD: Communication-efficient federated distillation via soft-label quantization and delta coding. *IEEE Transactions on Network Science and Engineering*, 9(4):2025–2038, 2022. doi: 10.1109/TNSE.2021.3081748.

Akihito Taya, Takayuki Nishio, Masahiro Morikura, and Koji Yamamoto. Decentralized and model-free federated learning: Consensus-based distillation in function space. *IEEE Transactions on Signal and Information Processing over Networks*, 8:799–814, 2022. doi: 10.1109/TSIPN.2022.3205549.

Vale Tolpegin, Stacey Truex, Mehmet Emre Gursoy, and Ling Liu. Data poisoning attacks against federated learning systems. In *European Symposium on Research in Computer Security*, pages 480–501. Springer, 2020.

Stacey Truex, Ling Liu, Ka-Ho Chow, Mehmet Emre Gursoy, and Wenqi Wei. LDP-Fed: Federated learning with local differential privacy. In *Proceedings of the Third ACM International Workshop on Edge Systems, Analytics and Networking*, pages 61–66, 2020.

Chengjia Wang, Guang Yang, Giorgos Papanastasiou, Heye Zhang, Joel J. P. C. Rodrigues, and Victor Hugo C. de Albuquerque. Industrial cyber-physical systems-based cloud IoT edge for federated heterogeneous distillation. *IEEE Transactions on Industrial Informatics*, 17(8):5511–5521, 2021. doi: https://doi.org/10.1109/TII.2020.3007407.

Huanlai Xing, Zhiwen Xiao, Rong Qu, Zonghai Zhu, and Bowen Zhao. An efficient federated distillation learning system for multitask time series classification. *IEEE Transactions on Instrumentation and Measurement*, 71:1–12, 2022. doi: https://doi.org/10.1109/TIM.2022.3201203.

Runhua Xu, Nathalie Baracaldo, Yi Zhou, Ali Anwar, James Joshi, and Heiko Ludwig. FedV: Privacy-preserving federated learning over vertically partitioned data. In *Proceedings of the 14th ACM Workshop on Artificial Intelligence and Security*, pages 181–192, 2021.

Xianglong Zhang and Xinjian Luo. Exploiting defenses against GAN-based feature inference attacks in federated learning. *arXiv preprint arXiv:2004.12571*, 2020.

Linfeng Zhang, Chenglong Bao, and Kaisheng Ma. Self-distillation: Towards efficient and compact neural networks. *IEEE Transactions on Pattern Analysis and Machine Intelligence*, 44(8):4388–4403, 2022. doi: https://doi.org/10.1109/TPAMI.2021.3067100.

Lingchen Zhao, Shengshan Hu, Qian Wang, Jianlin Jiang, Chao Shen, Xiangyang Luo, and Pengfei Hu. Shielding collaborative learning: Mitigating poisoning attacks through client-side detection. *IEEE Transactions on Dependable and Secure Computing*, 18(5):2029–2041, 2020.

Ying Zhao, Junjun Chen, Jiale Zhang, Di Wu, Michael Blumenstein, and Shui Yu. Detecting and mitigating poisoning attacks in federated learning using generative adversarial networks. *Concurrency and Computation: Practice and Experience*, 34(7):e5906, 2022.

6

Federated Learning-Based Beam Management in Dense Millimeter Wave Communication Systems

Qing Xue[1,2] and Liu Yang[1]

[1] School of Communications and Information Engineering, Chongqing University of Posts and Telecommunications, Chongqing, China
[2] State Key Laboratory of Internet of Things for Smart City, University of Macau, Macao SAR, China

6.1 Introduction

The past few years have witnessed the explosive growth of wireless data traffic, and this growth trend will continue (Cisco [2020]) due to the fast development of mobile multimedia applications and Internet of Things. With beamforming and massive multiple-input multiple-output (MIMO) (Feng et al. [2019], Wu et al. [2020]), millimeter wave (mmWave) communication has been widely acknowledged as a promising means to meet the projected requirements by dint of the abundant spectrum resources. However, mmWave communications face two critical challenges compared with the traditional microwave communications. One is the limited coverage caused by severe propagation path loss. An effective technique to address this problem is ultradense network (UDN) (Kamel et al. [2016]), where various small base stations (SBSs) are densely deployed and thus the distance between the user and SBS becomes closer. The other is that the directional mmWave link is very sensitive to mobility. An efficient method for enabling reliable transmissions and enhanced data rates is to employ multiconnectivity (Wolf et al. [2019]), which enables a user to connect to multiple SBSs simultaneously. In recent years, intelligent reflecting surface (IRS) (Liang et al. [2019], Gong et al. [2021], Zhang et al. [2021], Xu et al. [2021]) has also become a powerful means to overcome this difficulty. Meanwhile, to achieve universal coverage and improve overall capacity, heterogeneous networks (HetNets) have been treated as one of the promising candidates, where SBSs are deployed underlying a core macro network. Inspired by these research results, this chapter considers ultradense mmWave HetNets supporting multi-connectivity and focuses on the beam management issue since it affects the system performance principally.

Although each mmWave SBS (mSBS) may only support limited beams simultaneously due to hardware constraints, the number of transmitting and/or receiving beams could be extremely large for an ultradense mmWave system. This leads to a more complex and critical beam management problem than that in conventional nondense mmWave systems. Currently, beam management usually refers to fine alignment of the transmitting and receiving beams to perform a variety of control tasks including initial access for idle users and beam tracking for connected users (Giordani et al. [2019b]). In this chapter, we manage beams in a systematic manner instead of beam-by-beam basis to improve beam utilization and reduce interbeam interference. In other words, we are committed to dynamically controlling the beam directions at the mSBS side (i.e., mSBS beam configuration) based on periodically sensing instantaneous user distributions.

Moreover, machine learning (ML) is adopted to realize intelligent and proactive beam management. Although the existing ML-based beam management mechanisms bring a number of benefits, they also expose some potential risks, most typically, in security and privacy protection. It is mainly because that traditional ML algorithms usually require data collection and processing by a central controller/server, but the data for model training may be privacy sensitive in nature. This problem becomes a bottleneck of large-scale implementation of the centralized ML in practical applications. In addition, the overhead of the centralized data aggregation and processing is usually significant. These reasons have led to increased interest in a new ML model namely federated learning (FL) (Niknam et al. [2020]). In FL, participants train a shared model collaboratively by exploiting their local data and computation capability. In this way, we only need to update the local model without sending the data to the central server. Thereby, FL can be exploited to train ML models in a distributed way while preserving user privacy. These natures of FL motivate our work in this chapter, which is the first time to exploit FL for designing mmWave beam management scheme in the open literature.

6.1.1 Prior Work

There are several papers that have provided an overview of beam management for mmWave in 5G and beyond, e.g., (Giordani et al. [2019a, 2019b], Onggosanusi et al. [2018], Li et al. [2020]). Beam management procedures for handling mobility can be categorized into beam sweeping, beam measurement and reporting, beam determination, beam maintenance, and beam failure recovery (Li et al. [2020]). To date, most of the investigations in beam management tackle the problem by resorting to beam training, sparse channel estimation, and location aided beamforming (Arajo and de Almeida [2019]). Among them, beam training is the most popular. In beam training, both ends of a link search through the available beam set in either an exhaustive or iterative mode until a good link is established. In particular, a fixed sector-level beam training is specified in IEEE 802.11ad/ay for initial access [Ghasempour et al., 2017]. For mobile users, to ensure the accurate acquisition of the channel state information (CSI), the beam training procedure should be performed frequently. But this process faces a number of problems including high complexity, significant training overhead, and access delays. Motivated by this, we investigated the challenges and potential solutions of downlink beamspace SU-/MU-MIMO including multibeam training, cooperative beam tracking, and multibeam power allocation in Xue et al. [2017] and Xue et al. [2019]. To solve the problem of co-channel interference in dense mSBS scenarios, the authors in Feng et al. [2017] proposed a large-scale CSI-based interference coordination approach. Moreover, the priori-aided beam training termed beam/channel tracking is of crucial importance to improve the training efficiency especially in dynamic environments (Zhang et al. [2016, 2019], Jayaprakasam et al. [2017]). Nevertheless, beam training/tracking is generally used to ensure the quality of service (QoS) requirements of a specific user. Furthermore, the assumption of mmWave channel modeling required by channel tracking techniques for most practical scenarios is too stringent to meet (Zhang et al. [2020]).

Nowadays, ML has attracted significant attention to optimize wireless communication systems, owning to its ability in creating smart systems that can take sequential decisions and make accurate predictions. Prior work has established that ML is a good tool for beam management in mmWave communication systems, especially in mobile applications or dynamic environments (Sun et al. [2018], Zhang et al. [2020], Alkhateeb et al. [2018], Zhou et al. [2019], Moon et al. [2020], Xue et al. [2021]). For example, the authors in Alkhateeb et al. [2018] proposed a deep-learning-based coordinated beamforming algorithm to reduce the training overhead. The authors in Zhou et al. [2019] proposed a deep neural network (DNN)-based beam management and interference coordination algorithm to reduce the interference and improve the sum rate of dense mmWave network. A deep Q-network based user-centric association scheme is designed in our previous work (Xue et al. [2021]) to provide reliable connectivity and high achievable data rate for ultradense mmWave networks. Furthermore, some work as in Wang et al. [2019] and Liu et al. [2020] uses FL framework to address their corresponding research issues. Intuitively, FL can improve the performance of the adaptive beam management scheme in terms of security and privacy protection, but the relevant topic is still in the exploratory research stage.

6.1.2 Contributions

In this chapter, we first employ FL to realize intelligent and proactive beam configuration on the mSBS side, named beam management based on FL (BMFL), and then adopt a μWave-assisted multiple association (Xue et al. [2021]) to ensure the QoS of users in ultradense mmWave networks (UDmmNs). The key contributions can be summarized as follows.

- Different from the conventional beam management schemes based on beam training and/or tracking (i.e., the optimal pairing of transmitting and receiving beams), we present a systematic beam management scheme that pays more attention to the global performance of UDmmN.
- Considering that the capacity of each mSBS to support simultaneous beams is limited, instead of static beam deployment as in traditional scenarios, we perform dynamic beam configuration by periodically sensing instantaneous user distributions to improve beam utilization.
- Different from existing work, we address the issue of data privacy in BMFL by adopting an FL framework to avoid any exchange of user private information (such as location, trajectory, and behavior). To the best of the authors' knowledge, it is the first attempt to apply federated DRL to beam management of UDmmN.
- According to the coverage of mSBS and the frequency of user participation in training, a data cleaning method is used in the BMFL algorithm to further strengthen the privacy protection of specific users while improving the learning convergence speed.

The remainder of this chapter is organized as follows. System model of the UDmmN is described in Section 6.2. Systematic beam management problem is formulated in Section 6.3. In Section 6.4, algorithm of BMFL is presented. Performance of BMFL is evaluated in Section 6.5. Finally, Section 6.6 concludes the paper.

6.2 System Model

A two-tier heterogeneous UDmmN is illustrated as Figure 6.1, where ultra-dense mSBSs are deployed randomly under the coverage of one macro BS (MBS) operating on conventional microwave band. The MBS and mSBSs can communicate and exchange control information via traditional backhaul X2 interfaces. Meanwhile, multiconnectivity is employed to improve the robustness of mmWave communications. All mSBSs are denoted by $B = \{1, \ldots, B\}$, and the users moving randomly within the UDmmN are denoted by $U = \{1, \ldots, U\}$, where $B = |B|$ and $U = |U|$. We assume that all mSBSs share the total mmWave bandwidth W_{mm}, which mainly takes into account that the interference caused by the short distance and directivity of mmWave transmission is usually not large.

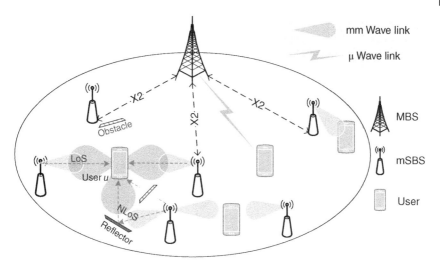

Figure 6.1 An illustration of ultradense mmWave network.

By adopting beamforming technique, we assume that mSBS b ($b \in B$) can form up to M_b transmit beams simultaneously. In order to compensate the high propagation loss of mmWave signals, the beamwidth is generally narrow. Consequently, the beams may not cover the entire range of the small cell. For ease of illustration, we divide the small cell b ($b \in B$) into S_b transmit sectors (or beam directions), where $0 < M_b \leq S_b$. We assume that the beams covering different sectors are mutually orthogonal in space, and each beam can serve multiple users within its coverage, for example, in a time division multiplexing manner. Different from traditional beam management, which is used to maximize the quality of a single link through beam training and tracking, the beam management in this chapter aims to maximize the system-level performance through the beam configuration on the mSBS side. In order to improve beam utilization, beam management leveraging double-deep Q-network (DDQN) (van Hasselt et al. [2016]) under an FL framework is used in our work. Meanwhile, as shown in Figure 6.2, beam management is assumed to be performed in a synchronous time-slotted fashion (Shen and van der Schaar [2017]) and the beams are static during each time slot. A time slot here is defined as a beam adjustment interval, the duration of which is generally related to user mobility and sector range. Within each slot, there are three main operations. First, mSBSs calculate the accumulated network performance over previous slots at the beginning of a slot and then manage beams by adopting FL-based algorithm. Second, users choose suitable mSBSs to associate with. Third, the selected beam/mSBS is used for data transmission.

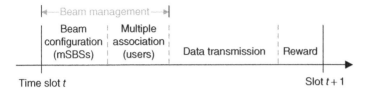

Figure 6.2 Beam management operations in time slot t.

In addition to the MBS, user u $(u \in \mathcal{U})$ in UDmmN may receive data from several mSBSs surrounding it and can associate with up to $B_u^{\max}$ of them. Denoting by $x_{u,b}(t) = \{0, 1\}$ the binary association indicator variable for user u $(u \in \mathcal{U})$ and mSBS b $(b \in B)$, we have $x_{u,b}(t) = 1$ if user u is associated with mSBS b at time t, otherwise $x_{u,b}(t) = 0$. Let $B_u(t)$ $(B_u(t) \subseteq B)$ denote the set of mSBSs connected to user u at time t, the number of the associated mSBSs is $B_u(t) = |B_u(t)| = \sum_{b \in B} x_{u,b}(t) \leq B_u^{\max}$. We assume that user u is always associated with the MBS to ensure seamless communication. To ensure the QoS requirements, the mSBS association for the user is carried out after the beam configuration. Moreover, the propagation models in this chapter are as follows.

For microwave transmission with distance d_{mbs}, the channel gain is (Zang et al. [2019])

$$h\left(d_{\text{mbs}}\right) = \kappa + 10\rho\log_{10}\left(d_{\text{mbs}}\right), \tag{6.1}$$

where κ is the path loss factor (in dB) per meter and ρ is the path loss exponent. Denoting by p_u^0, p_N^0, and I_u the transmission power of MBS allocated to user u, the received noise power and the cochannel interference power, respectively, the corresponding signal-to-interference-plus-noise ratio (SINR) for user u can be expressed as

$$SINR_{u,\text{mbs}} = \frac{p_u^0 \cdot h\left(d_{\text{mbs}}\right)}{p_N^0 + I_u}. \tag{6.2}$$

For mmWave transmission with distance $d_{u,b}$, the path loss can be modeled as (Akdeniz et al. [2014])

$$PL\left(d_{u,b}\right) = \alpha + 10\beta\log_{10}\left(d_{u,b}\right) + \xi \text{ [dB]}, \quad \xi \sim N\left(0, \sigma^2\right), \tag{6.3}$$

where α and β are the least square fits of floating intercept and slope over the measured distances (30–200 m), and σ^2 is the lognormal shadowing variance. The values of α, β, and σ^2 are different for line-of-sight (LoS) and Non-LoS (NLoS) cases. Let $p_{u,b}$ denote the allocated transmit power of mSBS b to user u, $G_{u,b}^T$ $(G_{u,b}^R)$ denote the transmit (receive) antenna gain, and $CF_{u,b}$ denote the small-scale channel fading, respectively, the received power at user u can be written as $p_{u,b}^R = \frac{p_{u,b} \cdot G_{u,b}^T \cdot G_{u,b}^R}{PL(d_{u,b}) \cdot CF_{u,b}}$. Considering that this work focuses on the beam management at the mSBS side

Table 6.1 Summary of key notations.

Notation	Description
B	Set of total mSBSs in the UDmmN
$\mathcal{V}$	Set of total users in the UDmmN
W_{mm}	Total mmWave bandwidth
W_{mbs}	Total available bandwidth of the MBS
$B_u(t)$	Set of the serving mSBSs of user u at time t
$x_{u,b}(t)$	Binary association indicator variable
M_b	Number of beams of mSBS b
S_b	Number of transmit sectors of mSBS b
B_u^{max}	Maximum number of mSBSs associated with user u
$B_u(t)$	Number of the serving mSBSs of user u at time t
$N_b(t)$	Number of users served by mSBS b at time t
d	Distance between MBS/mSBS and user
$h(d)$	Microwave channel gain
$PL(d)$	MmWave path loss model
$G\left(\varphi_{u,b}^T\right)$	Transmit antenna gain between user u and mSBS b
$G\left(\varphi_{u,b}^R\right)$	Receive antenna gain between user u and mSBS b
p_u^0	Transmit power of the MBS to user u
$p_{u,b}$	Transmit power of mSBS b to user u
χ	SINR threshold

to maximize user coverage, the small-scale channel fading is assumed to have little impact on the network-level beam adjustment. Hence, denoting p_N the noise power, the observed SINR can be expressed as

$$SINR_{u,b} = \frac{p_{u,b} \cdot G_{u,b}^T \cdot G_{u,b}^R / PL\left(d_{u,b}\right)}{p_N + \sum_{k \in B, k \neq b} \frac{p_{u,k} \cdot G_{u,k}^T \cdot G_{u,k}^R}{PL(d_{u,k})}}, \tag{6.4}$$

where the right part of the denominator represents the total interference power.

A summary of key notations is presented in Table 6.1.

6.3 Problem Formulation and Analysis

In this chapter, we first formulate the beam management problem in UDmmN as a long-term optimization and then discuss the tractability of this optimization problem. For the case that user u is associated with multiple mSBSs, the achievable rate of the user at time t should be the sum of the data rate received from all the associated mSBSs. The sum rate can be given as

$$r_u(t) = \begin{cases} \sum_{b\in B_u(t)} W_{u,b} \log_2\left(1 + SINR_{u,b}(t)\right), & \text{if } B_u(t) \neq \emptyset, \\ \frac{W_{mbs}}{N_{mbs}(t)} \log_2\left(1 + SINR_{u,mbs}(t)\right), & \text{if } B_u(t) = \emptyset, \end{cases} \tag{6.5}$$

where $W_{u,b}$ and W_{mbs} are the allocated mmWave bandwidth of mSBS b to user u and the total available bandwidth of the MBS, respectively, $N_{mbs}(t)$ is the number of users served by the MBS at time t, and $SINR_{u,b}(t)$ $(SINR_{u,mbs}(t))$ is the obtained SINR of user u from mSBS b (the MBS) at time t. Hence, the system throughput is

$$R(t) = \sum_{u\in U} r_u(t). \tag{6.6}$$

In order to improve beam utilization and reduce interbeam interference, we optimize the beam configuration of mSBSs, i.e., determine which sectors should be covered at time t based on periodically sensing instantaneous user distributions. Denote the optimization variable $\pi_b(t)$ as the set of sectors covered by mSBS b at time t, and the beam management policy for the whole system at time t is denoted by $\pi(t) = \left[\pi_1(t), \pi_2(t), \ldots, \pi_{|B|}(t)\right]$. Taking a suitable policy can let more sectors be covered by mSBSs and thus improve the system throughput. To this end, the beam management problem can be formulated as follows with the objective of maximizing the long-term system throughput.

$$P1.1 : \max_{\pi(t)} \lim_{T\to\infty} \mathbb{E}\left[\frac{1}{T}\sum_t R(t)\right] \tag{6.7}$$

$$\text{s.t. } 0 \leq |\pi_b(t)| \leq M_b, \quad \forall b \in B, t \in \mathcal{T}, \tag{6.7a}$$

$$SINR_{u,b}(t) \geq \chi, \quad \forall (u,b) \in \left\{(u,b)|x_{u,b}(t) = 1\right\}, t \in \mathcal{T}, \tag{6.7b}$$

$$0 \leq \sum_{b\in B} x_{u,b}(t) \leq B_u^{max}, \quad \forall u \in U, t \in \mathcal{T}, \tag{6.7c}$$

$$x_{u,b}(t) = \{0,1\}, \quad \forall u \in U, b \in B, t \in \mathcal{T}, \tag{6.7d}$$

where $\mathbb{E}[\cdot]$ is the expectation of the variable, $\mathcal{T}$ with cardinality T is the set of time slots for adjusting beam management policy, χ is the SINR threshold that users can correctly receive and decode the information, and $\hat{r}_u$ is the minimum requirement on data rate of user u. In problem P1.1, constraint (6.7a) ensures that the maximum number of beams for mSBS b is M_b. Constraint (6.7b) can ensure that the established link quality meets the requirements. Equations (6.7c) and (6.7d) are the constraints on user association, where the number of associated mSBSs for user u cannot exceed the access capability B_u^{max}.

Examining problem P1.1, we realize that it is hard to solve by using the traditional optimization methods. The rational behind is that the long-term optimization objective with unknown user movement behavior is formulated. Thus, the network environment (including user locations, channel quality, network resources, etc.) of future time slot cannot be obtained or even mathematically modeled at the beginning. An efficient and promising way to solve P1.1 is to resort to ML algorithms, especially reinforcement learning algorithms. However, as mentioned earlier that the raw data in terms of user locations is quite private and should be carefully protected rather than being exchanged among multiple mSBSs like that in most reinforcement learning algorithms. To this effect, FL, which requires the exchanges of learning model rather than raw data, is next adopted to derive the optimal beam management policy of P1.1.

6.4 FL-Based Beam Management in UDmmN

In this chapter, we focus on beam management for UDmmN. Considering the problem has a high state and action dimensions, we exploit DDQN using neural networks to estimate the value function, which improves the learning accuracy with a small compromise on the learning convergence speed. Meanwhile, if all the user information of the mSBSs is sent to the MBS for centralized data aggregation and processing, there are several great challenges in terms of computation energy consumption, computational latency, learning time, and privacy protection. Therefore, a decentralized learning framework (i.e., FL) is used, where training datasets are distributed over mSBSs so as to reduce the scale of the learning problem. The user location information is regarded as sensitive information in the system and should not be shared with those mSBSs that are not associated with the user and can be protected under FL.

In order to maximize the long-term throughput while enforcing the protection of user location privacy in UDmmN, a novel beam management scheme is proposed for mSBS beam configuration based on FL, called BMFL. Specifically, we first formulate the beam management problem as a Markov decision process (MDP) model and then propose BMFL based on the MDP model by exploiting federated DRL.

6.4.1 Markov Decision Process Model

An MDP process $\mathcal{M} = (\mathcal{S}, \mathcal{A}, \mathcal{P}, \mathcal{R})$ is composed of a four-tuple, where $\mathcal{S}$ and $\mathcal{A}$ represent the space of state and action, respectively, $\mathcal{P}$ is the transition probability from the current state S_t to the next state S_{t+1}, and $\mathcal{R}$ represents the reward function. In our problem, each mSBS makes a decision (action) on beam directions periodically to maximize the long-term throughput, which may lead to changes in the network state. The state, action, transition probability, and reward are defined as follows:

- *State*: The set of all network states for mSBSs at time t is denoted by S_t, where the system state is described by the current operating beam sectors and the serving users of each mSBS. For mSBS b, the state is $s_t^b = \{\mathcal{U}_b, \pi_b(t), S_t^\mathcal{K}\} \in S_t$, where $\mathcal{U}_b$ is the set of serving users, $\pi_b(t)$ is the corresponding beam sectors occupied by these users, and $S_t^\mathcal{K} = \{\pi_k(t)\}_{k \in B, k \neq b}$.
- *Action*: Let $\mathcal{A}_t = \{a_t^b\}$ be the set of actions for all mSBSs at time t, where $a_t^b = \pi_b(t)$ be the action of mSBS b, which means that mSBS b serves users in $\mathcal{U}_b$ with covered beams in $\pi_b(t)$ at time t. Note that an mSBS is an agent which trains local model independently.
- *State transition probability*: Let the transition probability of mSBS b be $P_b = Pr(s_{t+1}^b | s_t^b)$, which indicates the probability that the network state of mSBS b changes from s_t^b to s_{t+1}^b.
- *Reward*: In order to maximize the long-term system throughput, we define the reward as $\mathcal{R}_t = R(t)$, where $R(t)$ is the optimization objective of P1.1.

In the MDP for beam management, the state for an mSBS consists of three elements, i.e., the set of serving users, the occupied beam sectors, and the available sectors of all other mSBSs. Therefore, the state-space dimensions should be the combination of the number of users and the number of sectors of all mSBSs, i.e., $|\mathcal{U}_b| \cdot \prod_{b \in B} M_b$. Similarly, for a local agent (i.e., an mSBS), the action space dimensions can be given by M_b. Please note that a distributed learning scheme is exploited, so the action space is calculated for each mSBS separately.

6.4.2 FL-Based Beam Management

In this section, we propose the BMFL in UDmmN. The BMFL consists of two steps, as shown in Figure 6.3, the data cleaning and the model training (including the local model updating, local model training, and the global model aggregating (Liu et al. [2020])). Specifically, to reduce the computing resources occupied by training, an mSBS will first clean the training data, i.e., choose training users (participants) according to the frequency of participating training and the distance between users and this mSBS. Then, DRL under an FL framework is introduced into the model training to enhance user location protection while coping with large state-action space issues. The data cleaning and model training are described later.

6.4.2.1 Data Cleaning

In order to ensure the quality and diversity of training data, the mSBS should perform data cleaning at the beginning of each communication round (i.e., each model update iteration). Figure 6.4 shows the process of a communication round,

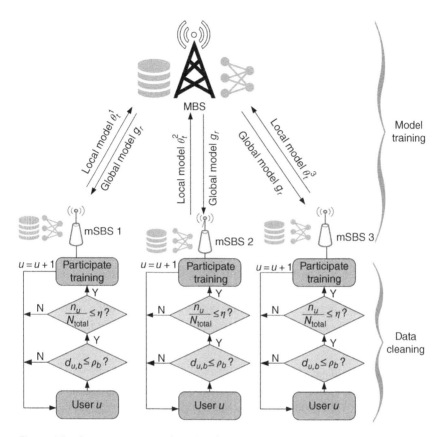

Figure 6.3 Beam management based on FL (BMFL) in UDmmN.

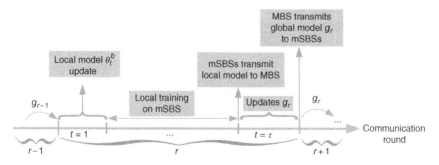

Figure 6.4 The process of a communication round.

which mainly includes five steps: the local parameter initializing, local model training, local model transmission, global model updating, and the global model transmission. At the beginning of each communication round, mSBS will make a selection of users participating in training.

Even if each mSBS can obtain all the users' location information in its coverage range, it is unrealistic for the mSBS to choose all the serving users to participate in the local training as (i) the computing resources occupied for training on a specific mSBS may be inadequate and (ii) the location of some certain users needs to be protected. To solve the above issue and increase the diversity of samples as much as possible, mSBS performs data cleaning according to the following two parameters. One is the coverage of the mSBS. Users not covered will not be selected to participate in the training. The other is the frequency of participating in training. If some users have not participated in the local model training for a certain period of time, mSBS will choose them as the participants in the training for the next global model updating to ensure the diversity of samples. For user u, denoting by n_u, the number of participating in training and N_{total} the total training times of the relevant mSBS so far, the frequency of participating in training is $\frac{n_u}{N_{\text{total}}}$. Therefore, the users will be chosen to train the local model on the mSBS once both conditions $d_{u,b} \leq \rho_b$ and $\frac{n_u}{N_{\text{total}}} \leq \eta$ are met, where ρ_b is the coverage radius threshold of mSBS b and η is the frequency threshold of participating in training. Here we assume $\mathcal{U}_b^*$ $(\mathcal{U}_b^* \subseteq \mathcal{U}_b)$ represents the set of the users participating in the local model training on mSBS b.

6.4.2.2 Model Training
Once the data is cleaned, mSBS begins to train the local model including the local model updating, local model training, and the global model aggregating, which are shown as follows.

Local Model Updating We assume that each communication round consists of τ time slots. The local beam management model on mSBS b at time t and the global model at communication round r are denoted by θ_t^b and g_r, respectively. Each mSBS performs the local training once during each time slot. At the beginning of communication round r, mSBS b will receive the global model g_r from the MBS to update θ_t^b according to

$$\theta_{t+1}^b = g_r - \frac{\lambda}{K_b} \sum_{b=1}^{|B|} \nabla L(\theta_t^b), \quad 1 \leq t \leq T, \tag{6.8}$$

where λ is the step size, K_b is the total amount of training data of mSBS b, and $L(\theta_t^b)$ is the loss function which will be given in the next part.

Local Model Training For a specific mSBS, once all training data is cleaned and the local model is updated, the mSBS begins to train the local beam management model based on the location information of participants within its coverage range. As mentioned earlier, a large number of mSBSs and users will result in a large state space and action space. Therefore, during each communication round, a discrete-action DRL algorithm, DDQN, is employed to train the local beam management model on individual mSBSs. The DDQN can tackle the issue of large state/action space by introducing the experience pool and decoupling the selection from the evaluation to reduce the correlation among data. The DDQN evaluates the greedy policy according to the Q-network with weight θ and estimates the state-action value Q according to the target network $\hat{Q}$ with weight $\hat{\theta}$. The update in DDQN is the same as that in deep Q-network, but the target is replaced by

$$y_t^b = R_{t+1} + \gamma Q(s_{t+1}^b, \arg\max_{a_t^b} Q(s_{t+1}^b, a_t^b; \theta_t^b); \hat{\theta}_t^b), \tag{6.9}$$

where $\pi = \arg\max_{a_t^b} Q(s_{t+1}^b, a_t^b; \theta_t^b)$ is an ϵ-greedy policy used to manage beam sectors, θ_t^b and $\hat{\theta}_t^b$ are the weight vectors of Q-network and $\hat{Q}$-network for mSBS b respectively, and $\gamma \in [0, 1]$ is the discount factor representing the discounted impact of future reward. If mSBS b is in state s_t^b with action a_t^b at time slot t, the corresponding state-action value is given by $Q(s_t^b, a_t^b) = \mathbb{E}[\sum_{k=t}^{T} \gamma^k \mathcal{R}_t | s_t^b, a_t^b]$. As the objective of DDQN is to minimize the gap between Q and $\hat{Q}$, the DDQN running on each mSBS can be trained by minimizing the following loss function

$$L(\theta_t^b) = \mathbb{E}[(y_t^b - Q(s_t^b, a_t^b; \theta_t^b))^2]. \tag{6.10}$$

Moreover, when DDQN is used to approximate the value function using the neural network, a gradient descent method is employed to update the parameter value θ_t^b, i.e.,

$$\theta_{t+1}^b = \theta_t^b + \alpha \left[y_t^b - Q\left(s_t^b, a_t^b; \theta_t^b\right) \right] \nabla Q\left(s_t^b, a_t^b; \theta_t^b\right), \tag{6.11}$$

where α is a scalar step size. After training local data for τ time slots, mSBS b ($b \in B$) will send its model parameter θ_t^b to the MBS to update the global model.

Global Model Aggregating When all local models (i.e., θ_t^b for $\forall b \in B$) are received at the end of communication round r, the MBS updates the global model by

$$g_r = \frac{\sum_{b \in B} K_b \theta_t^b}{K}, \quad t = \tau, \tag{6.12}$$

where $K = \sum_{b \in B} K_b$ is the total amount of training data. After updating the global model g_r, the MBS broadcasts g_r to all the mSBSs to update their local models.

Therefore, the workflow for the proposed BMFL is described as follows. Each mSBS (local agent) conducts local training for a DNN to predict Q value, which is used for guiding action decisions (which sectors should be covered) in the reinforcement framework. After each round of the local training, the weights in the

---

Algorithm 6.1 Beam management based on FL

Input: $\mathcal{U}$, $\mathcal{B}$, $\mathcal{U}_b^* = \varnothing$, s^b, a^b, η, ρ_b, γ, C, K_b, λ, τ

output: Beam sectors π_b.

1: Initialize experience relay pool D_b, $\forall b \in \mathcal{B}$;
2: Initialize the global weights g_0;
3: **for** communication round $r = 1, 2, \ldots, J$ **do**
4:   ▷Data cleaning
5:   **for** $b = 1, 2, \ldots, |\mathcal{B}|$ **do**
6:     **for** $u = 1, 2, \ldots, |\mathcal{U}|$ **do**
7:       **if** $d_{u,b} \leq \rho_b$ and $\frac{n_u}{N_{\text{total}}} \leq \eta$ **then**
8:         $\mathcal{U}_b^* = \{\mathcal{U}_b^*, u\}; u = u + 1$;
9:       **else**
10:         $\mathcal{U}_b^* = \{\mathcal{U}_b^*\}; u = u + 1$;
11:       **end if**
12:     **end for**
13:   **end for**
14:   Collect data from $\mathcal{U}_b^*$ for mSBS b, $\forall b \in \mathcal{B}$;
15:   ▷ Update local model
16:   **if** $r == 1$ **then**
17:     Initialize θ_0^b, $\forall b \in \mathcal{B}$;
18:   **else**
19:     $\theta_0^b = g_{r-1} - \frac{\lambda}{K_b} \sum_{b=1}^{|\mathcal{B}|} \nabla L(\theta_t^b), \forall b \in \mathcal{B}$.
20:   **end if**
21:   ▷ Train local model
22:   Let $\hat{\theta}_0^b = \theta_0^b$, initialize target action-value function $\hat{Q}(\cdot)$ according to $\hat{\theta}_0^b$;
23:   **for** $t = 1$ to τ **do**
24:     Receive the initial state $s_t^1, s_t^2, \ldots, s_t^{|\mathcal{B}|}$;
25:     Select $a_t^b = \text{argmax}_a Q(\cdot)$ using ϵ-greedy policy;
26:     Execute action a_t^b;
27:     Obtain $\mathcal{R}_t^b$ and s_{t+1}^b; Store $(s_t^b, a_t^b, \mathcal{R}_t^b, s_{t+1}^b)$ into D_b, $\forall b \in \mathcal{B}$;
28:     Randomly select a sample $\left(s_j^b, a_j^b, \mathcal{R}_j^b, s_{j+1}^b\right)$ from D_b, $1 \leq j \leq t$;
29:     Calculate y_t^b according to Eq. (6.9);
30:     Perform a gradient descent step on $(y_t^b - Q(s_t^b, a_t^b; \theta_t^b))^2$;
31:     Update the parameter θ_t^b according to Eq. (6.11);
32:     Reset $\hat{Q} = Q$ every C steps;
33:   **end for**
34:   ▷ Update global model
35:   $g_r = \frac{\sum_{b=1}^{|\mathcal{B}|} K_b \theta_\tau^b}{K}$.
36: **end for**
37: Obtain beam sectors π_b.

---

neural network from all local agents should be aggregated by the MBS based on Eq. (6.12) to update the global model, which then is shared with all the local agents to guide them in obtaining a more accurate DNN as per the rule in Eq. (6.8). The BMFL algorithm for beam management is presented as Algorithm 6.1. As each communication round includes the computation of mSBS data cleaning and local model updating, the computational complexity of the proposed algorithm is given by $\mathcal{O}_{\text{BMFL}} = \mathcal{O}(J \cdot (B \cdot U + \tau))$, where J denotes the number of communication rounds. In fact, for BMFL, the local updating step can be regarded as fully distributed DRL, as individual mSBSs train a local learning model based on local dataset without data interaction or aggregation (i.e., each mSBS performs gradient descent to adjust the local model parameter to minimize the loss function defined on its own dataset). The global aggregation step can be regarded as centralized DRL if the global aggregation is performed after every local update and the data samples and features are available for the aggregator (i.e., the MBS). However, the BMFL algorithm has greater advantages in privacy protection than traditional centralized and distributed DRL algorithms, due to the data cleaning and non-raw-data aggregation.

6.4.3 User Association in UDmmN

We propose a μWave-assisted user association in UDmmN, as shown in Figure 6.5, where the information exchange is realized with the aid of the μ Wave link. The details are as follows.

6.4.3.1 Downlink Measurements

In the association process, since mSBS b ($\forall b \in \mathcal{B}$) has no information about idle users (or the users associated with other mSBSs) in either beam steering directions or the signal attenuation, the mSBS explores its antenna elements to form a sweeping beam to the users, while users operate in an omni receive pattern mode or quasi-omni mode (i.e., closely approximating the omni mode) to listen for the association frames broadcast by the mSBS. Note that mSBS b only sweeps the transmit sectors in $\pi_b(t)$.

6.4.3.2 User Perception

After receiving the association frames, user u ($u \in \mathcal{U}$) can know the mSBSs that can be associated and the corresponding beam indexes. By measuring the received signal power, the user determines the set of the candidate serving mSBSs at time t, denoted by $\mathcal{B}_u^{\text{can}}(t)$ ($\mathcal{B}_u^{\text{can}}(t) \subseteq \mathcal{B}$), and sends $\mathcal{B}_u^{\text{can}}(t)$ to the MBS. The perceived signal power of user u from beam i of mSBS b ($b \in \mathcal{B}_u^{\text{can}}(t)$, $i \in \pi_{b,u}(t) \subseteq \pi_b(t)$) satisfies $\zeta_{u,b,i}(t) \geq \varsigma$, where ς is a given threshold of the power, and $\pi_{b,u}(t)$ is the set of the candidate transmit sectors of mSBS b for user u.

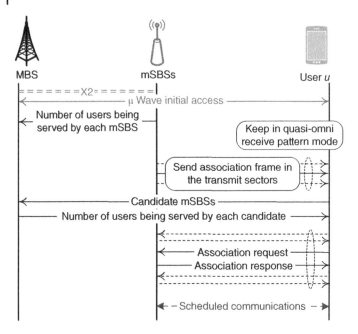

Figure 6.5 Procedure of μWave-assisted mmWave association.

6.4.3.3 Multiple Association

If user u is an idle user, i.e., $B_u(t-1) = \emptyset$, the association process can be outlined as follows.

- **repeat**
- Request to associate with mSBS b^* $(b^* \in B_u^{\text{can}}(t))$ that meets $\zeta_{u,b^*,i^*}(t) = \max_{\forall b \in B_u^{\text{can}}(t), \forall i \in \pi_{b,u}(t)} \zeta_{u,b,i}(t)$;
- Record mSBS b^* into $B_u(t)$ if $x_{u,b}(t) = 1$;
- Remove mSBS b^* out of $B_u^{\text{can}}(t)$;
- **until** $B_u(t) = |B_u(t)| = B_u^{\text{max}}$ or $B_u^{\text{can}}(t) = \emptyset$

For user u of which $B_u(t-1) \neq \emptyset$, some handover steps may be required before the above association. Moreover, we may jointly consider the perceived signal power $\zeta_{u,b,i}(t)$ and the user load balancing between mSBSs when dealing with user association. For instance, user u can choose the optimal candidate serving mSBS by calculating the variable $y_{u,b,i}(t)$, which is defined as $y_{u,b,i}(t) = k_1 \frac{\zeta_{u,b,i}(t)}{\varsigma} + k_2 \frac{U}{\sum_{u \in U} x_{u,b}(t)}$, where k_1 and k_2 are the proportion factors, $0 < k_1, k_2 < 1$, and $k_1 + k_2 = 1$. The higher the value of $y_{u,b,i}$, the higher the priority of mSBS b selected by user u. This method may overcome the problem of unbalanced mSBS loads, which in turn affects the network fairness and may result in overly frequent handovers between the adjacent mSBSs.

Meanwhile, as a coarse-grained beam training between user u ($u \in \mathcal{U}$) and mSBS b ($b \in \mathcal{B}_u(t)$) has been carried out during the above downlink measurements and user perception, user u can be served by beam i^* of mSBS b^* for the follow-up mmWave communications. For users with high QoS requirements, they may use directional beams to receive signals from the serving mSBSs, and thus a fine-grained beam training will be further required before the data transmission.

6.5 Performance Evaluation

In the simulations, a square area with the size $100\,m \times 100\,m$ is considered, where an MBS is located at the center, multiple mSBSs are distributed within the macro cell uniformly, and multiple users are randomly distributed in the cell. To make the results convincing, we compare the performance of the proposed algorithm with other schemes under the same user distribution.

Table 6.2 summarizes the detailed simulation parameters. For the coverage of each mSBS, we uniformly divide it into eight sectors (i.e., each sector is with the coverage of $45°$). Each mSBS generates three beams covering different sectors. Each user can be associated with up to three mSBSs. Assuming that the bandwidth of the MBS is evenly allocated to the serving users and there is no cochannel interference between users, we have $W_{u,\text{mbs}} = \frac{W_{\text{mbs}}}{N_{\text{mbs}}}$. Meanwhile, the mmWave bandwidth for user u is assumed to be $W_{u,b} = W_{\text{mm}}$. As both the mSBSs and users in mmWave network usually transmit/receive signals with directional beams pointing in different angular directions (spatially orthogonal to each other), the interference between simultaneous mmWave links will only be caused by beam sidelobes, which is usually very small and negligible. Furthermore, the co-beam interference (the interference between the users served by the same mSBS transmit beam) may be eliminated by appropriate spatial precoding. Therefore, the SINR can be approximated by the SNR for mmWaves, which is quite different than that in sub-6 GHz networks.

In the learning part settings, a fully connected neural network is considered for each mSBS. We mainly use the nn.Module, nn.Sequential, and nn.Linear of PyTorch to build a four-layer neural network of which the structure is given in Table 6.3, where in_dim $= |\mathcal{U}| + 2M_b + 1$, n_hidden_1 $= 40$, n_hidden_2 $= 60$, n_hidden_3 $= 40$, and out_dim $= 1$. Specifically, the two parameters in nn.Linear() are the size of each input sample and the size of each output sample, respectively. Meanwhile, the size of experience replay pool for each mSBS is set to 400, the batch-size is set to 36, and the learning rate is set to 0.1. All the other settings such as reward, action, and state keep consistent with those in our modeling part.

We first evaluate the convergence performance of the proposed BMFL algorithm in terms of the average loss function value, as shown in Figure 6.6. From this

Table 6.2   Key simulation parameters.

Parameter	Value
Carrier frequency of the MBS	$f_{c1} = 2.1\,\text{GHz}$
Carrier frequency of mSBSs	$f_{c2} = 28\,\text{GHz}$
Bandwidth of the MBS	$W_{\text{mbs}} = 100\,\text{MHz}$
mmWave bandwidth	$W_{\text{mm}} = 2\,\text{GHz}$
Number of transmit sectors of mSBS b	$S_b = 8$
Number of beams of mSBS b	$M_b = 3$
Maximum number of mSBSs serving user u	$B_u^{\text{max}} = 3$
Transmit power of the MBS	$p_u^0 = 50\,\text{dBm}$
Transmit power of mSBSs	$p_{u,b} = 37\,\text{dBm}$
Parameters of microwave channel gain	$\kappa = 38.8, \rho = 2$
Parameters of mmWave path loss	$\alpha = 61.3, \beta = 2.1, \xi \sim N(0,4)$
mmWave transmit antenna gain	$G_{u,b}^T = 12\,\text{dB}$
mmWave receive antenna gain	$G_{u,b}^R = 10\,\text{dB}$
Number of layers in neural network	4
Target network update interval step	4
Discount factor	0.8
Learning rate for training	0.1
Replay memory size	400
Minibatch size	36

Table 6.3   Structure of the neural network.

Layer	Generate
1	nn.Sequential(nn.Linear(in_dim, n_hidden_1), nn.ReLU())
2	nn.Sequential(nn.Linear(n_hidden_1, n_hidden_2), nn.ReLU())
3	nn.Sequential(nn.Linear(n_hidden_2, n_hidden_3), nn.ReLU())
4	nn.Sequential(nn.Linear(n_hidden_3, out_dim))

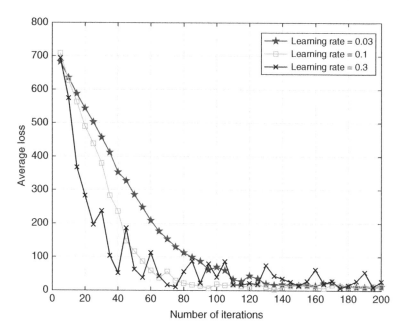

Figure 6.6 Convergence of the BMFL in UDmmN.

figure, we find that all curves under the three typical learning rates, 0.03, 0.1, and 0.3, reach the convergence after a certain number of iterations. These convergence results clearly demonstrate the effectiveness and rationality of BMFL. Meanwhile, observed from the figure, the average loss value of the BMFL algorithm presents dramatical decreasing in the beginning stages. It is because that the gradient descent approach is employed by the algorithm to train the DDQN-based optimization framework, thereby the loss function converges faster at the beginning while becoming gentle near the minimum point. In addition to the learning rate, the convergence performance of BMFL is also related to some other hyperparameters such as the size of replay memory and minibatches. The parameters in Table 6.2 are chosen to make a trade-off between the communication performance and the computational complexity according to the simulation results that are not shown here. These settings may not be optimal, but they can make our BMFL algorithm achieve a good convergence performance, which will stimulate its practical application.

We then compare the BMFL with the following four beam management schemes in terms of user coverage and network throughput.

(1) *Brute-force search (BFS)*: Find the optimal beam coverage by searching all the possible beam sectors. This algorithm can reach the optimal beam management solution but has extremely high computational complexity.

(2) *Evenly deployed beam (EDB)*: Deploy the beams in a uniform manner. In EDB, only one beam direction for each mSBS needs to be optimized, and the direction of other beams can be determined by the rule of uniform deployment.

(3) *Beam management based on distributed learning (BMDL)*: Individual mSBSs train their own data through DDQN and make decision on beam configuration independently, where no data aggregation of FL is used.

(4) *Beam management based on centralized learning (BMCL)*: All mSBSs transmit data to a controller (i.e., the MBS) for centralized training in DDQN. Then the MBS makes global decision on beam configuration for all mSBSs.

6.5.1 Comparison with BFS and EDB

In this section, we compare the performance of BMFL with that of the two traditional schemes, BFS and EDB.

Compared with BFS and EDB, the user coverage performance of BMFL versus SNR threshold χ and user density Den_{user} is evaluated as shown in Figure 6.7. As multiple association is considered for users, the user coverage here is defined

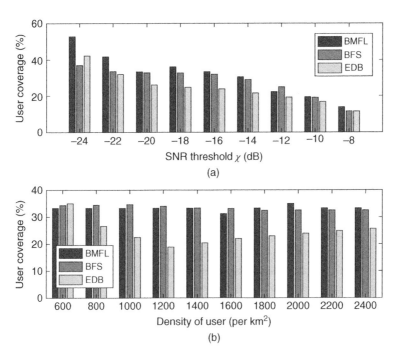

Figure 6.7 Comparisons of user coverage for the BMFL, BFS, and EDB, versus (a) SNR threshold, given that $B = |\mathcal{B}| = 3$ and $U = |\mathcal{U}| = 12$ and (b) user density, given that $B = |\mathcal{B}| = 6$ and $\chi = -20\,\mathrm{dB}$.

as $Cov_{\text{user}} = \left(\sum_{\forall b \in B} N_b\right) / \left[U \cdot \min\left(B, B_u^{\max}\right)\right]$. As a simple example, when $U = 1$, $B = 4$, and $B_u^{\max} = 3$, we have $Cov_{\text{user}} = \frac{1+0+1+0}{1\times3} = 66.67\%$ if the user is associated with mSBS 1 and mSBS 3 simultaneously. Meanwhile, the user density is defined as $Den_{\text{user}} = U/Area$, where $U = |\mathcal{U}|$ is the number of users in a certain area at time t. For example, we have $Den_{\text{user}} = 20/(0.1 \times 0.1) = 2000$ per km^2 when we set $U = 20$. By analyzing the results in Figure 6.7, we can see that the performance of the proposed scheme in user coverage is generally consistent with the optimal scheme BFS, but better than EDB. Also worth noting is that for all the three schemes, the value of Den_{user} in the results is not very high. This is mainly because we divide the service range of mSBS b ($\forall b \in B$) into eight sectors and only three operating beams, i.e., $S_b = 8$ and $M_b = 3$.

Furthermore, the network throughput performance of BMFL is compared with BFS and EDB, as shown in Figure 6.8. Similar to the definition of user coverage, the mSBS density is defined as $Den_{\text{mSBS}} = B/Area$, where $B = |\mathcal{B}|$ is the number of mSBSs in a certain area at time t. The mSBS density is set to 600 per km^2 in this simulation. As expected, the BFS beam management algorithm achieves the highest throughput as all the potential solutions have been searched and tested. Importantly, we find that BMFL achieves the second highest network throughput

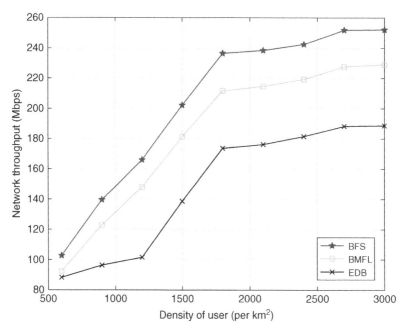

Figure 6.8 Comparisons of network throughput for the BMFL, BFS, and EDB, given that $B = |\mathcal{B}| = 6$ and $\chi = -20$ dB.

with relatively small difference of that in BFS but much higher than that of EDB. For example, when $Den_{user} = 3000$ per km^2, the network throughput of the three schemes is $R_{BFS} \approx 252$ Mbps, $R_{BMFL} \approx 229$ Mbps, and $R_{EDB} \approx 188$ Mbps, respectively. These results further demonstrate the performance gain of the proposed BMFL algorithm. Meanwhile, although the performance of the proposed scheme in terms of user coverage is comparable to that of the optimal scheme BFS, it is slightly weak in terms of network throughput, which may be due to the difference of users they serve. We believe that when there is enough training data, the performance of BMFL in terms of network throughput can also reach the level of the BFS.

6.5.2 Comparison with BMDL and BMCL

In this section, we compare the performance of BMFL with that of the two adaptive schemes, BMDL and BMCL. For BMDL and BMCL, the setting of simulation parameters (neural network, user distribution, etc.) is consistent with the proposed BMFL.

Figure 6.9 shows the performance comparison in terms of the user coverage. When the SINR threshold χ is fixed, we see that the proposed BMFL can achieve

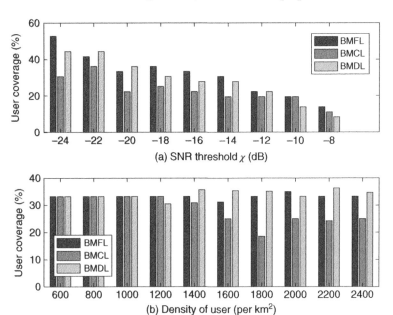

Figure 6.9 Comparisons of user coverage for the BMFL, BMCL, and BMDL, versus (a) SNR threshold, given that $B = |\mathcal{B}| = 3$ and $U = |\mathcal{U}| = 12$ and (b) user density, given that $B = |\mathcal{B}| = 6$ and $\chi = -20$ dB.

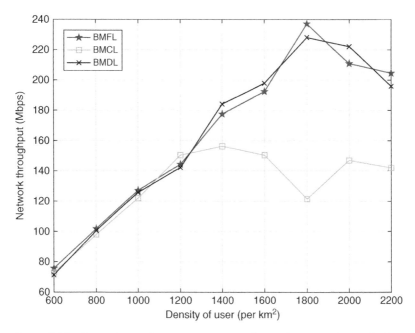

Figure 6.10 Comparisons of network throughput for the BMFL, BMCL, and BMDL, given that $B = |\mathcal{B}| = 6$ and $\chi = -20\,$dB.

a higher user coverage than BMDL and BMCL. For relatively low user density, e.g., when $Den_{user} < 1200$ per km^2, the user coverage of these three schemes is almost the same. With the increase of Den_{user}, the user coverage performance of BMFL is not much different from that of BMDL, but better than that of BMCL. The variation of the network throughput with the user density is similar to that of the user coverage, as shown in Figure 6.10. For example, when $Den_{user} = 800$ per km^2, the network throughput of the three schemes has little difference, where $R_{BMFL} \approx R_{BMCL} \approx R_{BMDL} \approx 100\,$Mbps. For a high Den_{user}, the throughput of BMFL and BMDL is usually close, but higher than that of BMCL. Compared with BMDL, the proposed BMFL also adopts a distributed learning architecture. However, it has an advantage that the training model accuracy and learning convergence speed can be improved through the cooperation of multiagent (e.g., mSBSs). Due to the small number of mSBSs in the simulation, this advantage is not obvious, so the performance of the two is similar. For BMCL, it may be that the algorithm converges to a suboptimal solution, which leads to the low results. The essence of ML algorithms determines that the optimal result cannot be guaranteed. It is an iterative updating process.

6.6 Conclusions

Due to the directional transmission and dense network deployment, the complexity of beam management problem in mmWave communication systems becomes a real challenge. To address the complex and dynamic beam control issue, in this chapter, we proposed a federated DRL-based adaptive beam management algorithm, BMFL. In BMFL, individual mSBSs train a local ML model based on the cleaned local dataset and then send the model features to the MBS for aggregation. Meanwhile, we employed DDQN to train the local model on mSBSs under an FL framework. Due to the data cleaning and nonraw-data aggregation, the proposed BMFL algorithm has great advantages in privacy protection and wireless resource conservation (e.g., transmit power and bandwidth). Simulation results have shown that the BMFL provides a better trade-off between computational complexity and network throughput. Moreover, the performance of the proposed scheme in user coverage is generally comparable to that of the optimal scheme BFS, which is also verified by the simulations. In general, this work can be seen as a pioneer of using FL to solve the systematic beam management problem under UDN scenarios.

Bibliography

M. R. Akdeniz, Y. Liu, M. K. Samimi, S. Sun, S. Rangan, T. S. Rappaport, and E. Erkip. Millimeter wave channel modeling and cellular capacity evaluation. *IEEE Journal on Selected Areas in Communications*, 32(6):1164–1179, 2014. doi: 10.1109/JSAC.2014.2328154

A. Alkhateeb, S. Alex, P. Varkey, Y. Li, Q. Qu, and D. Tujkovic. Deep learning coordinated beamforming for highly-mobile millimeter wave systems. *IEEE Access*, 6:37328–37348, 2018. doi: 10.1109/ACCESS.2018.2850226

D. C. Arajo and A. L. F. de Almeida. Beam management solution using Q-learning framework. In *2019 IEEE 8th International Workshop on Computational Advances in Multi-Sensor Adaptive Processing (CAMSAP)*, pages 594–598, 2019. doi: 10.1109/CAMSAP45676.2019.9022645

Cisco. Cisco annual internet report (2018–2023). *White paper*, March 2020.

Wei Feng, Yanmin Wang, Dengsheng Lin, Ning Ge, Jianhua Lu, and Shaoqian Li. When mmWave communications meet network densification: A scalable interference coordination perspective. *IEEE Journal on Selected Areas in Communications*, 35(7):1459–1471, 2017. doi: 10.1109/JSAC.2017.2698898

Junjuan Feng, Shaodan Ma, Sonia Aïssa, and Minghua Xia. Two-way massive MIMO relaying systems with non-ideal transceivers: Joint power and hardware scaling. *IEEE Transactions on Communications*, 67(12):8273–8289, 2019. doi: 10.1109/TCOMM.2019.2940634

Y. Ghasempour, C. R. C. M. da Silva, C. Cordeiro, and E. W. Knightly. IEEE 802.11ay: Next-generation 60 GHz communication for 100 Gb/s Wi-Fi. *IEEE Communications Magazine*, 55(12):186–192, 2017. doi: 10.1109/MCOM.2017. 1700393

M. Giordani, M. Polese, A. Roy, D. Castor, and M. Zorzi. Standalone and non-standalone beam management for 3GPP NR at mmWaves. *IEEE Communications Magazine*, 57(4):123–129, 2019a. doi: 10.1109/MCOM. 2019.1800384

M. Giordani, M. Polese, A. Roy, D. Castor, and M. Zorzi. A tutorial on beam management for 3GPP NR at mmWave frequencies. *IEEE Communications Surveys & Tutorials*, 21(1):173–196, 2019b. doi: 10.1109/COMST.2018.2869411

Shiqi Gong, Chengwen Xing, Xin Zhao, Shaodan Ma, and Jianping An. Unified IRS-aided MIMO transceiver designs via majorization theory. *IEEE Transactions on Signal Processing*, 69:3016–3032, 2021. doi: 10.1109/TSP. 2021.3078571

Hado van Hasselt, Arthur Guez, and David Silver. Deep reinforcement learning with double Q-learning. In *AAAI'16: Proceedings of the Thirtieth AAAI Conference on Artificial Intelligence*, pages 2094–2100, 2016.

S. Jayaprakasam, X. Ma, J. W. Choi, and S. Kim. Robust beam-tracking for mmwave mobile communications. *IEEE Communications Letters*, 21(12):2654–2657, 2017. doi: 10.1109/LCOMM.2017.2748938

M. Kamel, W. Hamouda, and A. Youssef. Ultra-dense networks: A survey. *IEEE Communications Surveys &Tutorials*, 18(4):2522–2545, 2016. doi: 10.1109/COMST. 2016.2571730

Y. R. Li, B. Gao, X. Zhang, and K. Huang. Beam management in millimeter-wave communications for 5G and beyond. *IEEE Access*, 8:13282–13293, 2020. doi: 10.1109/ACCESS.2019.2963514

Ying-Chang Liang, Ruizhe Long, Qianqian Zhang, Jie Chen, Hei Victor Cheng, and Huayan Guo. Large intelligent surface/antennas (LISA): Making reflective radios smart. *Journal of Communications and Information Networks*, 4(2):40–50, 2019. doi: 10.23919/JCIN.2019.8917871

Yi-Jing Liu, Gang Feng, Yao Sun, Shuang Qin, and Ying-Chang Liang. Device association for RAN slicing based on hybrid federated deep reinforcement learning. *IEEE Transactions on Vehicular Technology*, 69(12):15731–15745, 2020. doi: 10.1109/TVT.2020.3033035

S. Moon, H. Kim, and I. Hwang. Deep learning-based channel estimation and tracking for millimeter-wave vehicular communications. *Journal of Communications and Networks*, 22(3):177–184, 2020. doi: 10.1109/JCN.2020.000012

S. Niknam, H. S. Dhillon, and J. H. Reed. Federated learning for wireless communications: Motivation, opportunities, and challenges. *IEEE Communications Magazine*, 58(6):46–51, 2020. doi: 10.1109/MCOM.001.1900461

E. Onggosanusi, M. S. Rahman, L. Guo, Y. Kwak, H. Noh, Y. Kim, S. Faxer, M. Harrison, M. Frenne, S. Grant, et al. Modular and high-resolution channel state information and beam management for 5G new radio. *IEEE Communications Magazine*, 56(3):48–55, 2018. doi: 10.1109/MCOM.2018.1700761

C. Shen and M. van der Schaar. A learning approach to frequent handover mitigations in 3GPP mobility protocols. In *2017 IEEE Wireless Communications and Networking Conference (WCNC)*, pages 1–6, 2017. doi: 10.1109/WCNC. 2017.7925950

Yao Sun, Gang Feng, Shuang Qin, Ying-Chang Liang, and Tak-Shing Peter Yum. The SMART handoff policy for millimeter wave heterogeneous cellular networks. *IEEE Transactions on Mobile Computing*, 17(6):1456–1468, 2018. doi: 10.1109/TMC. 2017.2762668

Shiqiang Wang, Tiffany Tuor, Theodoros Salonidis, Kin K. Leung, Christian Makaya, Ting He, and Kevin Chan. Adaptive federated learning in resource constrained edge computing systems. *IEEE Journal on Selected Areas in Communications*, 37(6):1205–1221, 2019. doi: 10.1109/JSAC.2019.2904348

A. Wolf, P. Schulz, M. Dörpinghaus, J. C. S. Santos Filho, and G. Fettweis. How reliable and capable is multi-connectivity? *IEEE Transactions on Communications*, 67(2):1506–1520, 2019. doi: 10.1109/TCOMM.2018.2873648

Xianda Wu, Shaodan Ma, and Xi Yang. Tensor-based low-complexity channel estimation for mmWave massive MIMO-OTFS systems. *Journal of Communications and Information Networks*, 5(3):324–334, 2020. doi: 10.23919/JCIN.2020.9200896

Kaizhe Xu, Jun Zhang, Xi Yang, Shaodan Ma, and Guanghua Yang. On the sum-rate of RIS-assisted MIMO multiple-access channels over spatially correlated rician fading. *IEEE Transactions on Communications*, 69(12):8228–8241, 2021. doi: 10.1109/TCOMM.2021.3111022

Qing Xue, Xuming Fang, and Cheng-Xiang Wang. Beamspace SU-MIMO for future millimeter wave wireless communications. *IEEE Journal on Selected Areas in Communications*, 35(7):1564–1575, 2017. doi: 10.1109/JSAC.2017.2699085

Q. Xue, X. Fang, M. Xiao, S. Mumtaz, and J. Rodriguez. Beam management for millimeter-wave beamspace MU-MIMO systems. *IEEE Transactions on Communications*, 67(1):205–217, 2019. doi: 10.1109/TCOMM.2018.2867487

Qing Xue, Yao Sun, Jian Wang, Gang Feng, Li Yan, and Shaodan Ma. User-centric association in ultra-dense mmWave networks via deep reinforcement learning. *IEEE Communications Letters*, 25(11):3594–3598, 2021. doi: 10.1109/LCOMM. 2021.3108013

S. Zang, W. Bao, P. L. Yeoh, B. Vucetic, and Y. Li. Managing vertical handovers in millimeter wave heterogeneous networks. *IEEE Transactions on Communications*, 67(2):1629–1644, 2019. doi: 10.1109/TCOMM.2018.2877326

C. Zhang, D. Guo, and P. Fan. Tracking angles of departure and arrival in a mobile millimeter wave channel. In *2016 IEEE International Conference on Communications (ICC)*, pages 1–6, 2016. doi: 10.1109/ICC.2016.7510902

D. Zhang, A. Li, M. Shirvanimoghaddam, P. Cheng, Y. Li, and B. Vucetic. Codebook-based training beam sequence design for millimeter-wave tracking systems. *IEEE Transactions on Wireless Communications*, 18(11):5333–5349, 2019. doi: 10.1109/TWC.2019.2935731

J. Zhang, Y. Huang, Y. Zhou, and X. You. Beam alignment and tracking for millimeter wave communications via bandit learning. *IEEE Transactions on Communications*, 68(9):5519–5533, 2020. doi: 10.1109/TCOMM.2020.2988256

Jun Zhang, Jie Liu, Shaodan Ma, Chao-Kai Wen, and Shi Jin. Large system achievable rate analysis of RIS-assisted MIMO wireless communication with statistical CSIT. *IEEE Transactions on Wireless Communications*, 20(9):5572–5585, 2021. doi: 10.1109/TWC.2021.3068494

P. Zhou, X. Fang, X. Wang, Y. Long, R. He, and X. Han. Deep learning-based beam management and interference coordination in dense mmWave networks. *IEEE Transactions on Vehicular Technology*, 68(1):592–603, 2019. doi: 10.1109/TVT.2018.2882635

7

Blockchain-Empowered Federated Learning Approach for An Intelligent and Reliable D2D Caching Scheme

Runze Cheng[1], Yao Sun[1], Yijing Liu[2], Le Xia[1], Daquan Feng[3], and Muhammad Imran[1]

[1] *James Watt School of Engineering, University of Glasgow, Glasgow, UK*
[2] *Shenzhen Key Laboratory of Digital Creative Technology, Shenzhen University, Shenzhen, China*
[3] *National Key Lab on Communications, University of Electronic Science and Technology of China, Chengdu, China*

7.1 Introduction

Content caching is becoming significantly essential in next-generation wireless networks due to the growing demands of information requests for user equipments (UEs). This predownload method can mitigate the pressure of backhaul links during peak times and reduce the latency of fetching content for clients (Amer et al. [2018], Feng et al. [2013, 2015], Zhuang et al. [2019]). Meanwhile, device-to-device (D2D) communication technology has been proposed recently, which enables multiple direct transmissions between pairs of nearby devices in cellular networks (Ji et al. [2015]), thus dramatically improving the spectrum efficiency. Borrowing the D2D communication technology, a promising and attractive trend is to allow UEs to play an active role as caching servers, thus establishing caching-enabled D2D networks.

In the caching-enabled D2D networks, a UE can obtain the required contents (e.g., popular videos) from other neighbor UEs via D2D links, in addition to base station (BS)s or the server in the core network. In this way, the network resource (wireless bandwidth and power, etc.) utilization could be greatly improved, and service provisioning quality, especially download latency for UEs, could be enhanced. Furthermore, while the traditional content distribution network demands high costs for massive users requiring the same content, the D2D caching is becoming more cost efficient in wireless networks with high user density (Chen et al. [2016b]). Additionally, with recent advancements in

Federated Learning for Future Intelligent Wireless Networks, First Edition.
Edited by Yao Sun, Chaoqun You, Gang Feng, and Lei Zhang.

UE battery, storage, and GPU/CPU models, mobile terminals can now utilize machine learning techniques to develop a fast and accurate caching policy. Recently, extensive studies have been conducted in the design of intelligent caching schemes based on deep reinforcement learning (DRL). The DRL-based caching schemes use neural networks to learn the optimal strategy, thus solving complex multiagent caching problems, when there is sufficient training data (Sun et al. [2020], Ye et al. [2021]).

Although traditional DRL-based schemes are capable of addressing the accuracy of decision-making, some challenges remain unresolved. First, users are reluctant in raw data sharing when there is a risk of privacy leakage, hence it may cause a lack of training data. Furthermore, users tend to be self-interested, which means fewer users willing to participate in the D2D caching network when there is no direct reward or obvious benefit. In order to attract more users to participate in mobile D2D networks and stay active in the caching systems, it is thereby crucial to establish a privacy-preserved and secure caching scheme.

To address the difficulty of designing a privacy-preserved caching scheme, federated learning (FL) emerges as a promising alternative for intelligent decision-making in the wireless system without sharing private raw data (Konečný et al. [2016]). As FL only requires the exchange of weight or gradient of the local model, the local training data is merely kept in user local storage, and the terminal and personal privacy can be carefully protected. Furthermore, the FL can perform efficient machine learning among distributed multiagent by sharing training models, even when there is insufficient local training data (Bonawitz et al. [2019]).

In spite of these superiorities, one critical challenge faced is the reliability of model updates in the FL process, especially in untrusted D2D networks with faults and malicious attacks. For example, once some D2D nodes do not send the updates or send fault and fake updates, the accuracy and reliability of information exchanges cannot be guaranteed, leading to a heavy degradation of learning performance. Fortunately, a technical solution to jointly maintain reliable databases through decentralization and trustlessness, i.e., blockchain, can be introduced here. Blockchain, being a distributed technology, is employed in numerous cryptocurrencies and has proven to be a reliable method for tamper-proof information (Cao et al. [2020]). Theoretically, blockchain has the potential to provide immutable and persistent data records, (Sun et al. [2019]), and serve as an information verification and storage tool in FL. Additionally, as a by-product of blockchain, users' willingness to participate in cache sharing and model training can be improved.

In this study, we propose a privacy-preserved and secure D2D caching scheme by exploiting DRL-based FL under a double-layer blockchain architecture. Simulations are conducted to demonstrate the convergence of blockchain-based deep reinforcement FL (BDRFL) and verify the performance gain compared with several traditional learning-based caching schemes and a heuristic algorithm-based scheme. Here, the main contributions of our work are listed as follows:

- We formulate the D2D caching problem as a multiagent Markov decision process (MDP) problem and propose a novel, privacy-preserved, and secure caching scheme named BDRFL.
- We develop an FL with the double-stage cluster-based framework to establish a reliable learning scheme. This FL method allows users to train models in a distributed way without raw data exchange. In addition, we determine the update method of FL parameters for local models, area models, and the global model.
- We exploit a double-layer blockchain architecture to underpin the above FL. Specifically, multisubchains based on the Raft consensus mechanism are used to store area models and stimulate users to participate in D2D caching. Meanwhile, a mainchain with the practical Byzantine fault tolerance (PBFT) consensus mechanism verifies area models to resist Byzantine failures, thus ensuring the accuracy of the global model.

The remainder of this chapter is structured as follows. Section 7.2 provides an overview of some related works. Section 7.3 describes the system model in detail, while Section 7.4 presents the problem formulation and BDRFL-based scheme. In Section 7.5, we show the blockchain consensus mechanism and the FL model's updating details. After that, we also conduct extensive simulations to further evaluate our scheme in Section 7.6 with some discussions. Finally, we conclude our chapter in Section 7.7.

7.2 Related Work

7.2.1 Learning-Based D2D Caching Schemes

In recent year, numerous researches exploit learning methods to design effective cache schemes with the aim of assisting traffic offloading and reducing transmission delay. In Jiang et al. [2019], the authors emphasize the influence of content popularity distribution and formulate the D2D caching problem as a multiagent multiarmed bandit learning problem and solve it via a centralized Q-learning. The authors in Chen and Yang [2018] analyze the synthesis of user preferences as well as content popularity. They predict the behavior of each user and update the caching strategy of Q-learning. To tackle the challenge of large action space, a deep Q network (DQN)-based caching scheme optimization method is proposed in Chen et al. [2016a] by learning the preferences of the users and analyzing the similarity between the users and their neighbors. Similarly, the authors of Li et al. [2019] use long short-term memory (LSTM) to learn user behaviors and DQN to update the state value function.

The majority of existing schemes do not consider privacy and security, while only a few related works are privacy friendly of user information. The authors in Kumar et al. [2018] propose a distributed Q-learning resource reservation framework based on multi D2D controllers. As users only share their privacy with the trusted neighbors, the risk of privacy leakage can be mitigated to some extent. A novel weighted distributed DQN is proposed for edge caching replacement optimization in Li et al. [2020a], where each BS trains its own DQN model based on the local data. In addition, they use a reward-based adaptive boosting manner to update the global model. Unfortunately, once failures or attacks (like a malicious node sends fake updates) happen, the caching system is easily crashing, if no consensus mechanism adopted in the caching scheme.

7.2.2 Blockchain-Enabled D2D Caching Schemes

A few recent works propose incorporating blockchain into FL to ensure D2D caching performance in attack or failure scenarios. The majority of relevant works focus on how to establish a blockchain-based platform to incentivize users in caching networks without taking into account the benefits blockchain provides in terms of enhancing security. The partial PBFT (pPBFT)-based blockchains are proposed in the smart contract-based cache delivery markets by the authors in Zhang et al. [2020a,b]. In these caching networks, UEs get rewards from the cache provider for content delivery. Meanwhile, blockchains serve only as distributed ledgers for recording the content service list. In Liu et al. [2018], the authors develop a multiaccess edge computing (MEC)-enabled blockchain framework, where the UEs take on the role of the miners and resort to the nearby edge nodes for performing the computation-intensive proof of work (PoW) puzzle and content caching. There is one work (Lu et al. [2019]) that improves the performance of caching policy by using blockchain to resist failures, where FL- and PBFT-based blockchains are combined thus the data and local model can be verified, retrieved, and shared securely. Furthermore, users can get rewards from the blockchain to improve their willingness of sharing content.

However, because of the massive user number, it is exceptionally complex to use the pPBFT or PBFT consensus in a large-scale caching sharing network. Additionally, proof-based consensus with high computing and storage consumption is hard to run at the side of mobile equipment. Therefore, when designing an efficient and privacy-preserved FL caching scheme, a permissioned blockchain consensus that can incentivize UEs in large-scale networks is necessary.

7.3 System Model

7.3.1 D2D Network

We consider that a D2D network consisting of a central base station (BS) and multiple UEs with caching capability. Each UE can transmit cached contents to its neighbors via D2D links (Jiang et al. [2019]). Moreover, we assume that the BS can retrieve any content items from the core network, and all the required contents can be obtained from the BS. Let $\mathcal{U} = \{1, 2, \dots, U\}$ be the set of UEs covered by the BS. We denote $C_{\mathcal{U}} = \{c_1, c_2, \dots, c_K\}$ as the set of storage capacity of UEs. Moreover, for a specific UE u, it requests content from the item library $\mathcal{F} = \{1, 2, \dots, F\}$ with the size of each item $\hat{C}_{\mathcal{F}} = \{\hat{c}_1, \hat{c}_2, \dots, \hat{c}_F\}$. When UE u requests for item f, it first broadcasts this request to the nearby UEs. The UEs with the required item f in their storage send a reply. Then, the UE u chooses the nearest UE (item holder) to obtain this content. If no UE replies to the request, BS should respond to the request.

We do not consider the UE mobility in this work, thus the locations of UEs are unchangeable. Denote the distance between UE u and UE v as $d_{u,v}$. UE u can connect to UE v only if $d_{u,v} < \hat{d}$, where $\hat{d}$ represents the D2D communication range threshold.

7.3.2 Content Caching Schemes

In this work, we consider three possible caching models as shown in Figure 7.1 including self-caching, D2D caching, and BS transmission.

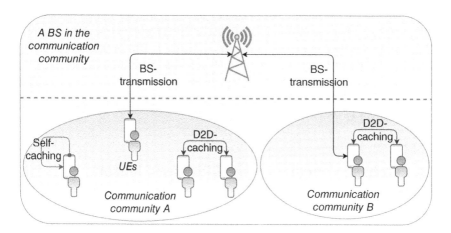

Figure 7.1 Three different caching schemes in the communication community, (1) self-caching, (2) D2D caching, and (3) BS transmission.

(1) *Self-caching*: When UE u requires content items, it will first check whether the exact cache has been cached in the local storage. The request will be satisfied immediately if the local cache hits.

(2) *D2D caching*: If the required content items are not cached at the local device, UE u will search the nearby devices that are located within the radius of $\hat{d}$ trying to get the required item. If all the UEs in this communication coverage do not have the requested content items, the request cannot be met. Note that a UE with needed caching can meet the requests of multiple UEs.

(3) *BS transmission*: BS transmission will be the last method for UE u to obtain the desired content when neither Self-caching nor D2D caching hits the requirement. The BS will receive the request of UE u and transmit items from the server in the core network.

7.3.3 Transmission Latency

There are two kinds of transmission links: (i) D2D link, (ii) BS-UE link. Therefore, we discuss the two transmission scenarios separately.

We assume that both the wireless bandwidth and transmit power are evenly allocated among the multiple serving devices. For D2D link, let $SNR_{u,v} = P_v \cdot G_{u,v}/\sigma_N^2$ represent signal-to-noise ratio (SNR), where P_v denotes the transmission power of UE v, $G_{u,v}$ represents the channel gain of UE v to u's D2D transmission link, and σ_N^2 is the Gaussian white noise power. Furthermore, for the channel gain of D2D transmission, we have $G_{u,v} = \varphi \cdot d_{u,v}^{-\varepsilon}$, where φ and ε denote the path loss constant and exponent of the D2D link, respectively. The transmission rate of a D2D pair UE u and v is $\omega_{u,v} = B_{u,v} \cdot \log_2(1 + SNR_{u,v})$, where available $B_{u,v}$ is the bandwidth of UE this D2D pair. Therefore, the latency for UE u to fetch item F from UE v is $\tau_{u,v}^f = \hat{c}_f/\omega_{u,v}$. For the BS-UE link, we use a similar way to calculate the SNR $SNR_{u,0}$ and transmission rate $\omega_{u,0}$ experienced by UE u. Unlike the D2D link, UE u cannot directly fetch item f from the BS. Before the BS transmits an item to UE, it first retrieves this item from the core network. We assume this retrieve delay as a constant ϵ. Hence, the transmission latency for u getting item f from BS is $\tau_{u,0}^f = \hat{c}_f/\omega_{u,0} + \epsilon$.

7.4 Problem Formulation and DRL-Based Model Training

In this section, to minimize the total transmission latency of all UEs in the D2D caching system, we model this cache scheme design as a multiagent MDP problem. Then, we propose a DRL-based local model training process.

7.4.1 Problem Formulation

Based on the system model described in Section 7.2, the caching scheme design should be formulated as a multiagent MDP problem as follows.

7.4.1.1 Action

In this system, UE can cache multiple content items in one period. Denote $\mathbf{A}_u = (a_{u,1}, a_{u,2}, \ldots, a_{u,F})$ as a set of actions for whether UE u caches content items or not. Specifically, $a_{u,f} \in \{0,1\}$, is a binary decision variable, where UE u caches item f if $a_{u,f} = 1$ and $a_{u,f} = 0$ otherwise. Note that the total size of cached content items cannot exceed the storage capacity c_u of UE u, i.e., $\sum_{f=1}^{F} \hat{c}_f \cdot a_{u,f} \leq c_u$.

7.4.1.2 State

The state of UE u at the time tth is denoted as $\mathbf{S}_u = (E_u, H_u)$. Here $E_u = (e_{u,1}, e_{u,2}, \ldots, e_{u,F})$ is the local content popularity, while $H_u = (h_{u,1}, h_{u,2}, \ldots, h_{u,F})$ is the local cache hitting rate. The local content popularity and local item hit-rate are determined by UE u and its neighbors. The request rate of content f is $e_{u,f}^t = \frac{n_{u,f}^t}{\sum_{f=1}^{F} n_{u,f}^t}$, where $n_{u,f}^t$ is the number of request for item f. Furthermore, $h_{u,f}^t = \frac{\dot{n}_{u,f}^t}{\hat{n}_u^t}$ is the local hit rate of item f, where $\dot{n}_{u,f}^t$ is the number of UEs fetch item f from UE u and $\hat{n}_u$ is the total number of UE u's neighbors.

7.4.1.3 Reward and Return

The transmission latency of UE u for caching content item f at time t is given by

$$\Gamma_{u,f}^t = (1 - a_{u,f}^t) \cdot \left[Z_{u,f}^t + \tau_{u0}^f \cdot \prod_{\mu \in \mathcal{N}(u)} (1 - a_{\mu,f}^t) \right], \tag{7.1}$$

where $\mathcal{N}(u)$ denotes the credible potential neighbors of u, μ is the UE with the μth lowest latency for sending content items to UE u, and $Z_{u,f}^t$ denotes the lowest latency of UE u to fetch the content item f from $\mathcal{N}(u)$, which is given by

$$Z_{u,f}^t = \sum_{u=1}^{|\mathcal{N}(u)|} \left(\tau_{u,\mu}^f \cdot \prod_{v=1}^{\mu-1} (1 - a_{v,f}^t) a_{\mu,f}^t \right), \tag{7.2}$$

where $\prod_{v=1}^{\mu-1} (1 - a_{v,f}^t) a_{\mu,f}^t$ is indicator function with the meaning that no UE can fetch content faster than μ. When μ is larger than the total number of neighbors, it means that no UE in $\mathcal{N}(u)$ caches the content item f. In addition, when UE u fetches item f, the transmission latency reduction can be calculated by

$$Y_{u,f}^t = \tau_{u0}^f - \Gamma_{u,f}^t. \tag{7.3}$$

If $a_{u,f}^t = 1$, UE u caches the content item f. In other words, the request should be satisfied immediately by the self-caching. If $\prod_{v=1}^{\mu-1}(1 - a_{v,f}^t)a_{\mu,f}^t = 1$, UE μ will deliver item f to UE u through the D2D link. Otherwise, UE u will fetch the content item from BS with $\prod_{v=1}^{|\mathcal{N}(u)|}(1 - a_{v,f}^t) = 1$. Therefore, based on Eq. (4), we have

$$
Y_{u,f}^t = \begin{cases} \tau_{u0}^f, & self-cache; \\ \tau_{u\mu}^f, & D2D-cache; \\ 0, & otherwise. \end{cases} \tag{7.4}
$$

The total reduction of the transmission latency for UE u in round t is defined as

$$
Y_u^t(a_{u,1}^t, \ldots, a_{u,f}^t) = \sum_{f=1}^{\beta} b_{u,f}^t \cdot Y_{u,f}^t
$$
$$
+ \sum_{f=1}^{\beta} \sum_{\mu \in \mathcal{N}(u)} a_{\mu,f}^t \cdot b_{u,f}^t \cdot Y_{u,f}^t, \tag{7.5}
$$

where $b_{u,f}^t \in \{0,1\}$ is a binary decision variable that indicates whether UE u request for content items f at time t.

Therefore, for all the participated UEs, the total reduction of the transmission latency for self-cache and D2D cache can be given by

$$
Y_t(\mathbf{a_1^t}, \ldots, \mathbf{a_u^t}) = \sum_{u=1}^{\kappa} \sum_{f=1}^{\beta} b_{u,f}^t \cdot Y_{u,f}^t
$$
$$
+ \sum_{u=1}^{\kappa} \sum_{f=1}^{\beta} \sum_{\mu \in \mathcal{N}(u)} a_{\mu,f}^t \cdot b_{u,f}^t \cdot Y_{i,f}^t. \tag{7.6}
$$

The total reward that can be cooperatively obtained by UEs is given as

$$
R_t(\mathbf{a_1^t}, \ldots, \mathbf{a_u^t}) = \frac{1}{|\mathcal{N}(u)|} Y_t(\mathbf{a_1^t}, \ldots, \mathbf{a_u^t})
$$
$$
= \frac{1}{|\mathcal{N}(u)|} \sum_{u=1}^{\kappa} \sum_{f=1}^{\beta} b_{u,f}^t \cdot Y_{u,f}^t
$$
$$
+ \frac{1}{|\mathcal{N}(u)|} \sum_{u=1}^{\kappa} \sum_{f=1}^{\beta} \sum_{\mu \in \mathcal{N}(u)} a_{\mu,f}^t \cdot b_{u,f}^t \cdot Y_{i,f}^t. \tag{7.7}
$$

In order to ensure the highest overall reward, some UEs may sacrifice their own storage space to meet the needs of others. Therefore, the content transmission latency of partial UEs could be relatively high in some cases. To avoid the problem, we set a constraint that the maximum average latency for each UE fetching an item cannot exceed the threshold $\tau_{\max}$.

The goal of the UEs is to cooperatively distribute content by selecting actions in a way that maximizes future returns (composed of the short-term rewards and long-term rewards). In this scheme, we assume that the future returns are discounted by a factor of γ per time step, $0 < \gamma < 1$.

Therefore, the future discounted return at time t is denoted as $\hat{R}_t$ and is shown as

$$\hat{R}_t = R_t + \gamma R_{t+1} + \gamma^2 R_{t+2} + \gamma^3 R_{t+3} + \cdots$$

$$= \sum_{i=0}^{m-1} \gamma^i R_{t+i} + \gamma^m \hat{R}_{t+m}. \tag{7.8}$$

7.4.2 DRL-Based Local Model Training

In this study, there are three kinds of models in our proposed BDRFL, which are the (i) local model, (ii) area model, and (iii) global model. As there are numerous areas in this D2D network, the parameter of the global model is updated by utilizing the area models. Each area model is updated by using multiple local models in the coverage area of the area leader. Meanwhile, each UE trains its local model using local data. In this part, we illustrate the DRL-based local model training process. The area model and the global model updates are elaborated in Section 7.5.

The basic idea behind the majority of value-based reinforcement learning (RL) is to estimate the action-value function. As actions $\{\mathbf{A}_{t+1}, \mathbf{A}_{t+2}, \ldots\}$ and state $\{S_{t+1}, S_{t+2}, \ldots\}$ are integrated out, only observations $\mathbf{A}_t = \mathbf{a}_t$ and $S_t = \mathbf{s}_t$ remain. The action-value function is as follows:

$$Q_\pi(\mathbf{s}_t, \mathbf{a}_t) = \mathrm{E}[\hat{R}_t | S_t = \mathbf{s}_t, \mathbf{A}_t = \mathbf{a}_t], \tag{7.9}$$

where Q_π is the return of taking action $\mathbf{a}_t$ in the current state $\mathbf{s}_t$, which is related to the policy function π. Moreover, we eliminate the influence of the strategy function on the choice of return action and maximize Q_π. The optimal action-value function is

$$Q^*(\mathbf{s}_t, \mathbf{a}_t) = \max_\pi Q_\pi(\mathbf{s}_t, \mathbf{a}_t), \tag{7.10}$$

where Q^* is independent of the strategy function, which represents the expected return of the best action $\mathbf{A}^\star$ in state S. Whatever policy function π is performed, the result of taking action $\mathbf{a}_t$ at state $\mathbf{s}_t$ cannot be better than $Q^*(\mathbf{s}_t, \mathbf{a}_t)$. Moreover, we approximate the optimal action-value function $Q^*(\mathbf{s}_t, \mathbf{a}_t)$ by $Q(\mathbf{s}_t, \mathbf{a}_t; \mathbf{w})$, where $\mathbf{w}$ is a neural network parameter. Note that all actions are scored by the optimal action-value function to select the best action.

Since our D2D caching problems have a large state space and action space, it is challenging for traditional RL methods to estimate action-value (Li et al. [2020b]).

The value-based DRL uses neural networks to approximate the action value, thus effectively tackling the large state space and action space. While the majority of value-based methods, including the original DQN and nature DQN, are prone to producing overoptimistic value estimations, Double DQN (DDQN) solves this issue by using two different value functions to decouple the selection and the evaluation (Van Hasselt et al. [2016]). Therefore, we use DDQN in this work to reduce the impact of nonuniform overestimation and yield more accurate value estimates. At time t, the state space, action space, reward, and the next state can be packed as a transition $(\mathbf{s}_t, \mathbf{a}_t, r_t, \mathbf{s}_{t+1})$. Moreover, we use a larger buffer to store these samples. Note that only recent certain transitions are stored in a replay buffer, and old transitions will be removed. Furthermore, we use the Stochastic gradient descent (SGD) scheme to randomly sample a transition from the buffer to compute the temporal difference (TD) error and then calculate the stochastic gradient. The expected future return $E[\hat{R}_t]$ parameterized by the weight $\mathbf{w}$ is denoted as $Q(\mathbf{s}_t, \mathbf{a}_t; \mathbf{w})$and is given by

$$Q(\mathbf{s}_t, \mathbf{a}_t; \mathbf{w}) \approx r_t + \gamma Q(\mathbf{s}_{t+1}, \mathbf{a}_{t+1}; \mathbf{w}). \tag{7.11}$$

In DDQN, the best action selection is given as

$$\mathbf{a}^* = \underset{x}{\operatorname{argmax}} \ Q(\mathbf{s}_{t+1}, \mathbf{a}; \mathbf{w}). \tag{7.12}$$

The evaluation in DDQN using TD target network is

$$y_t = r_t + \gamma \cdot \max_{\mathbf{a}} Q(\mathbf{s}_{t+1}, \mathbf{a}^*; \mathbf{w}^-), \tag{7.13}$$

where $\mathbf{w}^-$ is the parameter of the TD target network, and the structure of the TD target network is the same as that of the DQN. Moreover, γ helps to balance the short-term reward and long-term reward. DDQN aims to shorten the difference between $\mathbf{w}^-$ and $\mathbf{w}$ by minimizing the loss function. The loss function is given by

$$L_t(\mathbf{w}) = \frac{1}{2} \left[Q(\mathbf{s}_t, \mathbf{a}_t; \mathbf{w}) - y_t \right]^2 = \frac{\delta_t^2}{2}, \tag{7.14}$$

where δ_t is the target error.

We draw samples randomly from the pool of reply buffer. Then, we use the loss to update $\mathbf{w}$ as per the following rule:

$$
\begin{aligned}
\mathbf{w}_{u,t+1} &= \mathbf{w}_{u,t} - \alpha \cdot \mathbf{g}_{u,t} = \mathbf{w}_{u,t} - \alpha \cdot \frac{\partial \left(\delta_t^2 / 2 \right)}{\partial \mathbf{w}} \\
&= \mathbf{w}_{u,t} - \alpha \cdot \delta_t \cdot \frac{\partial Q(\mathbf{s}_t, \mathbf{a}_t; \mathbf{w})}{\partial \mathbf{w}},
\end{aligned} \tag{7.15}
$$

where $\mathbf{g}_t$ is the stochastic gradient and α denotes the learning rate.

Algorithm 7.1 DQN to implement the optimization of content delivery

---

1: Initialize the parameter of models $\mathbf{w}_u$, $u \in \mathcal{U}$, the target networks of UEs with random weights $\mathbf{w}_u^-$, reply memories $n_{b,u}$, capacity c_u and discount factor γ.
2: **for** epsiode $T = 1, \ldots, m$ **do**
3:      **for** $u = 1, \ldots, \kappa$ **do**
4:          Select an action $\mathbf{a}_{u,t}$ which is the best Q-value action or random action.
5:          Input the initial transmit power P_u, the channel gain G_u, the D2D range r_{d2d}, the distance from user u to its neighbors d_{uv}, namely $I_u = \{P_u, G_u, r_{d2d}, duv\}, v \in \mathcal{N}_u$.
6:          Request for content f.
7:          Fetch content item and calculate the reduction value of latency $Y_{u,t}$.
8:          **if** cache sharing stage end **then**
9:              Observe a new state $\mathbf{s}_u'$.
10:          **end if**
11:      **end for**
12:      $Y_t = \frac{1}{\kappa} \sum_{u=1}^{\kappa} Y_{u,t}$.
13:      Observe the reward r_t according to Y_t, $r_{u,t} = r_t$.
14:      **for** $u = 1, \ldots, \kappa$ **do**
15:          Store the transition $\{\mathbf{s}_u, \mathbf{a}_u, r_u, \mathbf{s}_u'\}$.
16:          Calculate the target loss according to (7.13), perform a gradient descent step on (7.14), thus updating $\mathbf{w}_u$
17:          Every j step update the $\mathbf{w}_u^-$ of target network.
18:      **end for**
19: **end for**

7.5 Privacy-Preserved and Secure BDRFL Caching Scheme Design

FL is a good alternative that enables UEs learning without any centralized server (Hosseinalipour et al. [2020]). Also, it is a solution for limiting raw data transfer and accelerating learning processes of UEs (Zhan et al. [2020]). However, various failures may occur in these distributed D2D caching systems. Basically, these failures can be divided into three classifications:

(1) *Crash failure*: The crash is termed a fail-stop, and it makes no further process. The crash failure is serious and comes without any signal.
(2) *Omission failure*: A node fails to do partial expected processes (like the sending process, and receiving process). Once an omission failure happens, the UE will report the error.

(3) *Byzantine failure*: A node exhibits arbitrary, erratic, and unexpected behaviors. Moreover, these behaviors may be malicious and disruptive.

These failures seriously interfere with the learning process. Fortunately, blockchain is capable of enabling FL coordination under failures by utilizing the consensus mechanism, and it has been proved validation of offering a reliable method for information tamperproof (Cao et al. [2019a–c, 2020]). Therefore, we propose a double-layer blockchain architecture in BDRFL to support an efficient D2D caching scheme, as shown in Figure 7.2, with the aim of enhancing data privacy while improving the transparency and credibility of D2D caching systems. The sublayer Raft-based blockchains help UEs in reaching consensus despite crash and omission failures in the large-scale sublayer network, whereas the small-scale mainlayer network's PBFT consensus can detect Byzantine failures. To achieve this secure and intelligent D2D caching framework, five steps are involved in our proposed BDRFL under the double-layer blockchain architecture, including the task and requirements publication, Appropriate UE selection, local model training, area model update and recording, and finally global model update and recording, as shown in the flow chart of Figure 7.3. In the following, let us illustrate these steps.

7.5.1 Task and Requirements Publication

The task publisher broadcasts FL tasks and determines the key requirements. Specifically, starting from the round 0, the publisher packages the initial weight parameters of global model $\mathbf{w}_g$ and the initial state to the first block, i.e., block 0. If UEs intend to join the task training and satisfy the data requirement, they need to send their participation requests, identity code, data resources information, and other relevant information (e.g., CPU/GPU model, storage, and battery).

7.5.2 Appropriate UE Selection

Appropriate UEs are chosen by the task publisher. The task publisher determines whether UEs could participate in cache sharing and model training based on the identity code and additional information such as computing power, storage size, and battery size. The approved UEs $\mathcal{U}$ participate in the initial round.

7.5.3 Local Model Training

Each area's approved UEs execute out the sublayer Raft consensus and elect an area leader from among their neighbors. UEs with the highest cache hit rate and computing power are the area leaders. The area leader first downloads the global

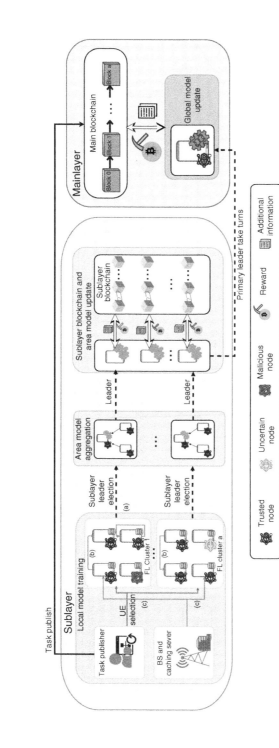

Figure 7.2 Blockchain and federated learning framework of the D2D caching system.

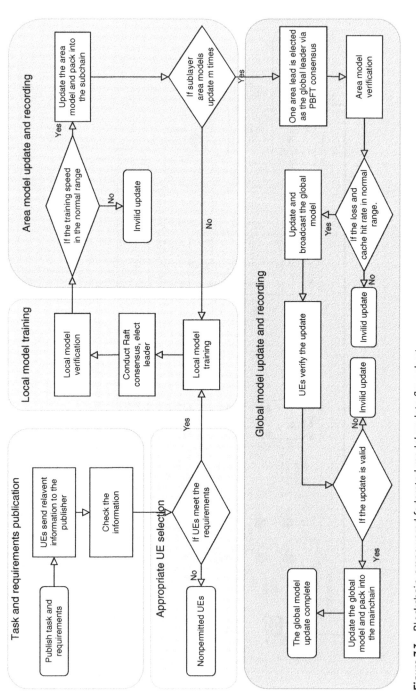

Figure 7.3 Blockchain-empowered federated model update flow chart.

model from the nearest mainchain block and broadcasts it to neighboring UEs. Subsequently, UEs $\mathcal{U}$ predict and cache the item that may request by themselves and their neighbors. In the cache sharing slot, UEs fetch required content in the order of self-caching, D2D caching then BS transmission. The UEs update the local cache popularity and local cache hit rate after all content requests from UEs are satisfied in the sharing slot. Next, the average latency reduction of users can be calculated, which is used to update the reward. At time t, every UE packs their state space, action space, reward, and the next state as a transition (s_t, a_t, r_t, s_{t+1}). Each UE has a larger buffer, which is used to store its own transitions. The replay buffer stores only recent certain samples, and old transitions are deleted. Each UE takes a random transaction from its buffer to compute the TD error and thus update the weights of its local model. Following the updating of all local models, the parameters of local models as well as some publicly available information such as training data size, training time, and local content popularity are sent to the area leaders.

7.5.4 Area Model Update and Recording

The sublayer area leaders are in charge of verifying local gradients and updating area models. These leaders are elected by UEs in each area by executing the Raft consensus. Each area leader is in charge of maintaining an area model.

Specifically, sublayer leaders receive the local models from UEs in their coverage area. Then the leaders check whether the training speed $v_{u,t}$ matches to the training time $\tau_{u,t}$ and training data size $\bar{c}_{u,t}$, thus determining the authenticity of a local model, $v_{u,t} = \frac{\bar{c}_{u,t}}{\tau_{u,t}}$. If the speed is within the normal training speed range $v_{\min} < v_{u,t} < v_{\max}$, the model will be certified as a reasonable update. Only those verified local models can be used to update the area models. The updated area model and additional information (such as the number of content distributions, training time, and training data size) in each area are recorded in the corresponding sublayer blockchain as reliable and tamper-proof transactions for future verification and updates. According to the number of content distributions for others, UEs get some rewards that can be an amount of virtual currency from the task publisher, which encourages more UEs to participate in this D2D caching system.

7.5.5 Global Model Update and Recording

There is one global leader each round takes charge of global model update. The sublayer leaders execute PBFT consensus and take turns to be the global leader. Once the global leader received a request for area model verification, it first checks the loss between the previous global model and each area model. Subsequently, the global leader tests the cache hit rate to evaluate the cache prediction performance of these area models. If both the loss $l_{a,r}$ and the hit-rate $h_{a,r}$ are in rational ranges, this area model can be utilized in the global model update. Once the global leader completes the model update, it broadcasts the update to area leaders, and area leaders check the update by testing the cache hit rate. Only over half of the leaders confirm the validity of the model, the updated model can be determined as the new global model. Then, the global leader packs the verified global model and additional information (e.g., area model gradient) into the mainchain. The global leader UE can get some rewards from the blockchain for package data and update blocks.

7.6 Consensus Mechanism and Federated Learning Model Update

In this section, we first investigate the consensus mechanism of the double-layer blockchain system, then discuss how the FL model is updated in this framework.

7.6.1 Double-Layer Blockchain Consensus Mechanism

In the D2D caching system, failures are classified into three types, i.e., crash failure, omission failure, and Byzantine failure. Among these failures, the Byzantine failure is the most difficult to detect and can have a significant impact on the security of the caching network. In this study, we consider the existence of Byzantine failures in order to construct a reliable and secure D2D caching scheme. However, most of the consensus mechanisms that can detect Byzantine failures are energy intensive and costly, like PoW and Proof of Stake (PoS). Limited by the low computing capability of user devices, these mechanisms cannot be directly applied to our scheme. Moreover, the numerous sublayer nodes also bring exponential growth of consensus complexity when using PBFT or BFT binomial. Therefore, a new blockchain framework and appropriate consensus mechanisms should be explored in this work. In this study, we proposed a double-layer blockchain empowered FL. The sublayer Raft-based blockchains only consider the crash and omission failures, while the PBFT-based mainchain resolves the Byzantine failure.

To allow more users to participate in D2D caching while ensuring model reliability, for the subchains, we require this algorithm has strong consistency and high consensus efficiency, regularly ensuring liveness, observing system status, and distributing machine data. In addition, this algorithm-based system should be independently and rationally maintained and managed. Furthermore, it should be simple to implement so that devices with limited computing power can quickly reach a consensus. Fortunately, these requirements are aligned with the characteristics of the Raft, prompting us to select it.

For the mainchain, we implement the PBFT consensus mechanism to prevent the global model from being severely affected by malicious nodes. After supernodes are selected from various areas, the number of nodes in the mainchain can be considerably reduced. Because there are only a few nodes involved in the consensus process, the computational complexity of employing the PBFT is reduced, which makes it suitable for the mainchain.

In this way, the subchains can resolve the crash and omission failures and help UEs reach consensus in a short slot, while the mainchain resists Byzantine failures caused by malicious UEs.

7.6.2 FL Area Model Update in Subchain Layer

There are three categories of nodes in Raft consensus, saying leader, candidate, and follower (Ongaro and Ousterhout [2014]). Each area can only select one leader UE that with the highest cache hit rate and strong computing power from all the UEs within this cluster region, and then the rest of UEs become follower nodes. Candidate nodes are followers at the intermediate state trying to become the next leader node.

The FL model update in sublayer is composed of two stages.

(1) *Leader election stage*: The leader UE in each cluster sends heartbeats to all followers. If followers cannot receive the heartbeats of the leader within the election timeout period, the leader election will be initiated. These followers transfer to the candidate state and update their term numbers. A follower first votes for itself and then sends RequestVoteRPC, which is a set of data about its information to request other nodes to vote for this follower. When it wins the majority of votes, it will become the new leader and regularly send heartbeats to all followers to maintain its rule.

(2) *FL model update stage*: The leader UE starts to receive information from other neighbor UEs, including local models and some extra information. Then, the leader checks whether the training speed matches to the training time and data size, thus determining the authenticity of a local model. Only verified local models can be utilized in area model updates. The area model update function is shown as follows:

$$\mathbf{w}_{u_a}^{t+1} = \mathbf{w}_{u_a}^t - \alpha \cdot \sum_{u \in \mathcal{N}(u_a)} \frac{\bar{c}_{n,u}}{\bar{C}_a} \cdot \mathbf{g}_{u,t}, \tag{7.16}$$

where $\mathbf{w}_{u_a}^t$ is the weight parameters of area model updated by leader UE u_a, $\bar{C}_a$ denotes the total size of training data samples from UEs in area a, $\bar{C}_a = \sum_{u \in \mathcal{N}(u)} \bar{c}_{n,u}$, and $\mathbf{g}_{u,t}$ is the gradient of the local model trained by UE u.

The updated area model and the relevant information are packed into a new block in the sublayer blockchain. These datasets are transparent so that can be checked and verified.

7.6.3 FL Global Model Update in Mainchain Layer

In the PBFT-based mainchain, the nodes are divided into two categories, (i) primary node and (ii) child node (Castro and Liskov [1999]). Only area leaders are involved in the mainlayer global model and mainchain update. The area leaders become the global leader in a Round-robin mode. That means there is merely one primary node each time, and other area leader UEs perform as the child nodes. All nodes are capable of communicating with each other. The ultimate goal is that all sublayer leaders reach a consensus on a principle of the minority obeying the majority.

Algorithm 7.2 Model update process in sublayer layer FL with blockchain

---

1: Initialize the parameter of global model $\mathbf{w}_g$, the local models of UEs with random weights $\mathbf{w}_u, u \in \mathcal{U}$, weights of area models $\mathbf{w}_{u_a}, u_a \in \mathcal{U}_a$, reply memories $n_{b,u}$, cache capacity c_u and discount factor γ.

2: **for** time T$=1, \ldots, m$ **do**

3:     **for** $u = 1, \ldots, \kappa$ **do**

4:         Download the global model.

5:         Select a joint action $\mathbf{a}_{u,t} = \max_{\mathbf{a}} Q\{\mathbf{s}_{u,t}, \mathbf{a}_u, \mathbf{w}_u\}$.

6:         Input the initial transmit power, the channel gain, the D2D range, the distance from user u to its neighbors d_{uv}, namely $\mathbf{S}_u = \{E_u, H_u\}$, $v \in \mathcal{N}(u)$.

7:         Observe the environment state s'_u.

8:         Get reward r_l.

9:         Update the parameters of the local model.

10:        Upload information and model to leader $u_a \in \mathcal{U}_a$.

11:     **end for**

12:     Implement Raft consensus and elect new leaders.

13:     **for** $u_a = 1, \ldots l$ **do**

14:         Validate updates.

15:         Perform model update step on (7.16).

16:         Pack updated model and additional information into the new block of sublayer blockchain.

17:     **end for**

18: **end for**

When the primary node receives an area model verification request, it first analyzes the loss between the previous global model and each area model and then evaluates the cache prediction performance of these models. To update the global model, only well-performed area models with normal loss and cache hit rate are used. Then, the primary node broadcasts the verification request to all the child nodes. Here should note that the maximum number of malicious nodes in PBFT that can be tolerated is $f = (n_a - 1)/2$, where n_a is the total number of sublayer leader UEs. Unlike the Raft, child UEs in PBFT have the right to question the reliability and rationality of the primary UE. After the primary UE analyzes all area models and updates parameters of the global model, the child UEs can test the update by using it to predict cache. Though check the cache hit rate, the child UEs determine whether the model update is valid, reasonable, and effective. The global model update has a longer update time slot, after area models update several times, the global model update once, it is capable of avoiding the waste of communication resources. The update function of global model is given by

$$\mathbf{w}_g^{r+1} = \sum_{u_a=1}^{l} \frac{\mathbf{w}_{u_a}^r}{l}, \tag{7.17}$$

where $\mathbf{w}_g^{r+1}$ is the weight parameters of global model at round $r + 1$.

Algorithm 7.3 Model update process in mainlayer FL with blockchain.

1: **for** Round=1, ..., r **do**
2:   **for** $u_a = 1, ..., l$ **do**
3:     Fetch area updates of every cluster.
4:     Verify update model and data.
5:   **end for**
6:   Implement PBFT consensus and an area leader UE become a new global leader u_g.
7:   **for** $u = u_g$ **do**
8:     Perform global model update step on (7.17).
9:     Pack updates and additional data to the mainchain.
10:   **end for**
11: **end for**

7.7 Simulation Results and Discussions

In this section, we evaluate the performance of our proposed BDRFL scheme through simulations.

We compare our proposed BDRFL caching scheme with the four following caching schemes:

(1) *Centralized DQN-based scheme*: This scheme collects and utilizes all users' information to train a DQN model by a central server (Li et al. [2019]).

(2) *Distributed DQN-based scheme*: Cluster leader UEs collect neighbor UEs' information and update models independently, each leader maintains a learning model (Kumar et al. [2018]).

(3) *FL-based scheme*: This scheme uses FL in model training, each UE maintains a local model. All the local models are utilized to update the global model (Li et al. [2020a]).

(4) *Zipf random*: UEs randomly cache content with an assumption that the content popularity obeys Zipf distribution.

We consider crash, omission, and Byzantine failures in this D2D caching network. All these failures are caused by attacks. In our simulations, a crash failure is that a UE halts all the activities (such as cache sharing and model training). A UE that stops updating the model or sending training data for 10 rounds is the omission failure. The Byzantine failure is that a malicious UE randomly generates fake model updates/training data.

7.7.1 Simulation Setting

We consider a cache-enabled D2D network scenario with a total of three clusters covering 5 UEs for each. All UEs have the same capacity with the size of $c_u = 1000$ MB. There are 10 types of contents with the same size. Moreover, we assume that UEs can communicate with each other in the same area via D2D

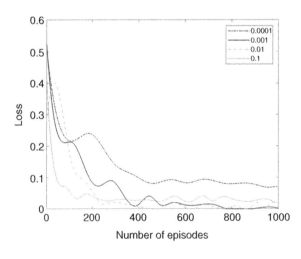

Figure 7.4 Training process of BDRFL under different learning rate.

links. The number of requested contents of a UE in a round is randomly generated within [1, 3]. The request pattern of a certain UE is modeled by the Zipf distribution. We set different content popularity parameters based on the heterogeneous preferences of UEs.

The details of DQN networks in this simulation are set as follows. The input layer neural number, the hidden layer neural number, and that of the output layer are 20, 10, and 10, respectively. We use *Tanh* as the activation function from the input layer to the hidden layer, and the activation function from the hidden layer to the output layer is *ReLU*. The size of the reply buffer is set to 100, and the number of transitions used to calculate the target loss is set to 5. The training process under different learning rate are shown as Fig. 7.4, and we fix the learning rate α as 0.001 with the influence factor $\gamma = 0.1$ in the other following simulation.

7.7.2 Numerical Results

Then, we compare the average reward of BDRFL when using three different DQN-based methods, which are the original DQN-based, nature DQN-based, and double DQN-based methods in the scenario without any failure, shown as Figure 7.5. From this figure, we find that the double DQN-based method always outperforms the other two methods in terms of reward convergence. This is because the original DQN and nature DQN always choose the action with the optimal value at the next decision slot to update Q function, which leads to an overestimation issue.

We first evaluate the average download latency reduction of fetching items for the five schemes. Figure 7.6a shows that the latency reduction of these five caching schemes under the scenario without any failure. The average latency reduction

Figure 7.5 Training process of BDRFL with different DQN-based methods.

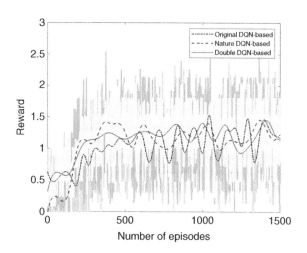

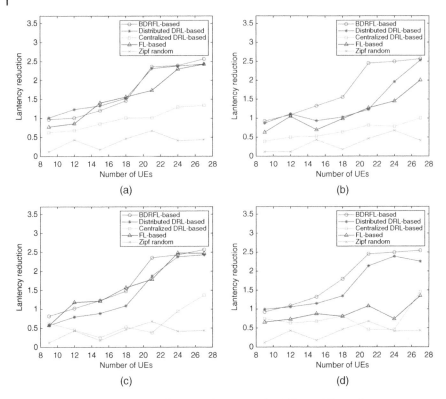

Figure 7.6 The reduced value of average latency under different UE numbers. (a) Comparisons of latency reduction without any failure, (b) comparisons of latency reduction under the crash failure, (c) comparisons of latency reduction under the omission failure, and (d) comparisons of latency reduction under the Byzantine failure.

of these five schemes all increase with the UE number. It is because the cache sharing network performs better under the higher user density. The latency reduction of the BDRFL-based caching scheme shows the fastest increase, while the FL-based scheme and distributed DRL-based scheme show similar performance. Meanwhile, applying the distributed DRL-based caching scheme only brings a bit more latency reduction than applying the worst schemes of Zipf random. The reason is the BDRFL, distributed DRL, and FL can achieve satisfactory model training and caching prediction without raw data sharing. However, the centralized DRL-based scheme requires global information which cannot be shared in this simulation.

In order to evaluate the performance under the scenario with a crash failure, we compare the reduced transmission latency of these five schemes while there is a UE accidentally fail-stop and cannot recover. From the Figure 7.6b, we find that our proposed BDRFL caching scheme always outperforms than other four schemes in terms of average latency reduction. That is because the impact of the crash failure is minimized via blockchain consensus by canceling the qualification of that fail-stop UE. Additionally, the crashed UE should be replaced by a qualified normal node. However, the other four schemes cannot detect and replace this failure node, thus the crashed UE does not cache and distribute content in the network.

Then, we assess the impact of the omission failure for the five caching schemes. Figure 7.6c shows the reduced latency, while a UE does not update the model or sharing the training data. From this figure, we find that the BDRFL caching scheme and the FL caching scheme achieve a similar latency reduction, which is larger than that of the other three schemes. It is because that FL helps the failure UE to replace the local bad-performance model with an updated global model via area leader broadcasting.

With the aim of evaluating the impact of the Byzantine failure, we compare the reduced transmission latency of the BDRFL scheme with the other four schemes in the scenario that a malicious node in the system randomly generates fake model updates/training data. Figure 7.6d shows that the latency reduction of the BDRFL scheme increases significantly with the number of UEs, which is almost the same as that in the nonfailure scenario, but much higher than that of the other schemes. The reason is that the introduced blockchain in BDRFL is capable to verify model updates and detect malicious UEs. Only well-performed models are used to update the global model; hence, BDRFL can ensure the valid global model update and achieve satisfied performance. Furthermore, the qualification of the malicious UE is canceled, and a new qualified UE is permitted to participate in, thus eliminating the impact of malicious UEs on the cache sharing network.

7.8 Conclusion

In this study, we have developed an intelligent and privacy-preserving caching scheme BDRFL in D2D network. BDRFL is based on a framework of FL underpinned by a double-layer blockchain system. We have illustrated the blockchain consensus of each layers and streamlined the process of BDRFL including FL model training as well as model data recording on blockchain. We have conducted simulations in scenarios with and without malicious attacks, where numerical demonstrated both the improvements of caching performance and the reliability of resisting attacks. In general, this work can be seen as a pioneer to explore the interplay of blockchain and FL, thus developing an intelligent and trusted caching scheme under an unreliable wireless network.

References

Ramy Amer, M. Majid Butt, Mehdi Bennis, and Nicola Marchetti. Inter-cluster cooperation for wireless D2D caching networks. *IEEE Transactions on Wireless Communications*, 17(9):6108–6121, 2018.

Keith Bonawitz, Hubert Eichner, Wolfgang Grieskamp, Dzmitry Huba, Alex Ingerman, Vladimir Ivanov, Chloe Kiddon, Jakub Konečný, Stefano Mazzocchi, H. Brendan McMahan, et al. Towards federated learning at scale: System design. *arXiv preprint arXiv:1902.01046*, 2019.

Bin Cao, Yixin Li, Lei Zhang, Long Zhang, Shahid Mumtaz, Zhenyu Zhou, and Mugen Peng. When Internet of Things meets blockchain: Challenges in distributed consensus. *IEEE Network*, 33(6):133–139, 2019a.

Bin Cao, Shichao Xia, Jiawei Han, and Yun Li. A distributed game methodology for crowdsensing in uncertain wireless scenario. *IEEE Transactions on Mobile Computing*, 19(1):15–28, 2019b.

Bin Cao, Long Zhang, Yun Li, Daquan Feng, and Wei Cao. Intelligent offloading in multi-access edge computing: A state-of-the-art review and framework. *IEEE Communications Magazine*, 57(3):56–62, 2019c.

Bin Cao, Zhenghui Zhang, Daquan Feng, Shengli Zhang, Lei Zhang, Mugen Peng, and Yun Li. Performance analysis and comparison of PoW, PoS and DAG based blockchains. *Digital Communications and Networks*, 6(4):480–485, 2020.

Miguel Castro and Barbara Liskov. Practical byzantine fault tolerance. In *OSDI*, volume 99, pages 173–186, 1999.

Binqiang Chen and Chenyang Yang. Caching policy for cache-enabled D2D communications by learning user preference. *IEEE Transactions on Communications*, 66(12):6586–6601, 2018.

Zheng Chen, Nikolaos Pappas, and Marios Kountouris. Probabilistic caching in wireless D2D networks: Cache hit optimal versus throughput optimal. *IEEE Communications Letters*, 21(3):584–587, 2016a.

Zhuoqun Chen, Yangyang Liu, Bo Zhou, and Meixia Tao. Caching incentive design in wireless D2D networks: A Stackelberg game approach. In *2016 IEEE International Conference on Communications (ICC)*, pages 1–6. IEEE, 2016b.

Daquan Feng, Lu Lu, Yi Yuan-Wu, Geoffrey Ye Li, Gang Feng, and Shaoqian Li. Device-to-device communications underlaying cellular networks. *IEEE Transactions on Communications*, 61(8):3541–3551, 2013.

Daquan Feng, Guanding Yu, Cong Xiong, Yi Yuan-Wu, Geoffrey Ye Li, Gang Feng, and Shaoqian Li. Mode switching for energy-efficient device-to-device communications in cellular networks. *IEEE Transactions on Wireless Communications*, 14(12):6993–7003, 2015.

Seyyedali Hosseinalipour, Sheikh Shams Azam, Christopher G. Brinton, Nicolo Michelusi, Vaneet Aggarwal, David J. Love, and Huaiyu Dai. Multi-stage hybrid federated learning over large-scale wireless fog networks. *arXiv preprint arXiv:2007.09511*, 2020.

Mingyue Ji, Giuseppe Caire, and Andreas F. Molisch. Fundamental limits of caching in wireless D2D networks. *IEEE Transactions on Information Theory*, 62(2):849–869, 2015.

Wei Jiang, Gang Feng, Shuang Qin, Tak Shing Peter Yum, and Guohong Cao. Multi-agent reinforcement learning for efficient content caching in mobile D2D networks. *IEEE Transactions on Wireless Communications*, 18(3):1610–1622, 2019.

Jakub Konečný, H. Brendan McMahan, Felix X. Yu, Peter Richtárik, Ananda Theertha Suresh, and Dave Bacon. Federated learning: Strategies for improving communication efficiency. *arXiv preprint arXiv:1610.05492*, 2016.

Naveen Kumar, Siba Narayan Swain, and C. Siva Ram Murthy. A novel distributed Q-learning based resource reservation framework for facilitating D2D content access requests in LTE-A networks. *IEEE Transactions on Network and Service Management*, 15(2):718–731, 2018.

Lixin Li, Yang Xu, Jiaying Yin, Wei Liang, Xu Li, Wei Chen, and Zhu Han. Deep reinforcement learning approaches for content caching in cache-enabled D2D networks. *IEEE Internet of Things Journal*, 7(1):544–557, 2019.

Ruibin Li, Yiwei Zhao, Chenyang Wang, Xiaofei Wang, Victor C. M. Leung, Xiuhua Li, and Tarik Taleb. Edge caching replacement optimization for D2D wireless networks via weighted distributed DQN. In *2020 IEEE Wireless Communications and Networking Conference (WCNC)*, pages 1–6. IEEE, 2020a.

Yixin Li, Bin Cao, Mugen Peng, Long Zhang, Lei Zhang, Daquan Feng, and Jihong Yu. Direct acyclic graph-based ledger for Internet of Things: Performance and security analysis. *IEEE/ACM Transactions on Networking*, 28(4):1643–1656, 2020b.

Mengting Liu, F. Richard Yu, Yinglei Teng, Victor C. M. Leung, and Mei Song. Computation offloading and content caching in wireless blockchain networks with mobile edge computing. *IEEE Transactions on Vehicular Technology*, 67(11):11008–11021, 2018.

Yunlong Lu, Xiaohong Huang, Yueyue Dai, Sabita Maharjan, and Yan Zhang. Blockchain and federated learning for privacy-preserved data sharing in industrial IoT. *IEEE Transactions on Industrial Informatics*, 16(6):4177–4186, 2019.

Diego Ongaro and John Ousterhout. In search of an understandable consensus algorithm. In *2014 USENIX Annual Technical Conference (USENIX ATC 14)*, pages 305–319, 2014.

Yao Sun, Lei Zhang, Gang Feng, Bowen Yang, Bin Cao, and Muhammad Ali Imran. Blockchain-enabled wireless Internet of Things: Performance analysis and optimal communication node deployment. *IEEE Internet of Things Journal*, 6(3):5791–5802, 2019.

Yao Sun, Wei Jiang, Gang Feng, Paulo Valente Klaine, Lei Zhang, Muhammad Ali Imran, and Ying-Chang Liang. Efficient handover mechanism for radio access network slicing by exploiting distributed learning. *IEEE Transactions on Network and Service Management*, 17(4):2620–2633, 2020.

Hado Van Hasselt, Arthur Guez, and David Silver. Deep reinforcement learning with double Q-learning. In *Proceedings of the AAAI Conference on Artificial Intelligence*, volume 30, 2016.

Qiang Ye, Weisen Shi, Kaige Qu, Hongli He, Weihua Zhuang, and X. Shen. Learning-based computing task offloading for autonomous driving: A load balancing perspective. In *Proceedings of the ICC*, volume 21, pages 1–6, 2021.

Yufeng Zhan, Peng Li, Zhihao Qu, Deze Zeng, and Song Guo. A learning-based incentive mechanism for federated learning. *IEEE Internet of Things Journal*, 7(7):6360–6368, 2020.

Ran Zhang, F. Richard Yu, Jiang Liu, Tao Huang, and Yunjie Liu. Deep reinforcement learning (DRL)-based device-to-device (D2D) caching with blockchain and mobile edge computing. *IEEE Transactions on Wireless Communications*, 19(10):6469–6485, 2020a.

Ran Zhang, F. Richard Yu, Jiang Liu, Renchao Xie, and Tao Huang. Blockchain-incentivized D2D and mobile edge caching: A deep reinforcement learning approach. *IEEE Network*, 34(4):150–157, 2020b.

Weihua Zhuang, Qiang Ye, Feng Lyu, Nan Cheng, and Ju Ren. SDN/NFV-empowered future IoV with enhanced communication, computing, and caching. *Proceedings of the IEEE*, 108(2):274–291, 2019.

8

Heterogeneity-Aware Dynamic Scheduling for Federated Edge Learning

Kun Guo[1], Zihan Chen[2], Howard H. Yang[3], and Tony Q. S. Quek[2]

[1]*School of Communications and Electronics Engineering, East China Normal University, Shanghai, China*
[2]*Information Systems Technology and Design Pillar, Singapore University of Technology and Design, Singapore, Singapore*
[3]*Zhejiang University/University of Illinois at Urbana-Champaign Institute, Zhejiang University, Haining, China*

8.1 Introduction

Extending cloud infrastructures to edge networks facilitates powerful edge computing capability (Hu et al. [Sep. 2015]) and inspires edge intelligence (Wang et al. [2020]). In future edge networks, new radio (NR) with the flexible numerology and mini-slot scheduling can be employed to satisfy the diverse user requirements through the spectrum flexibility and ultrashort response times (3GPP [2018]). An early area of interest in edge networks lies in industrial internet of things (IIoT) applications, where sensors are installed at the devices in the factory to monitor that everything is working as it should and to identify any potential issues before they occur.

To deliver on such requirements, edge intelligence draws attentions from industry and academia. There are two paradigms for edge learning. The one is centralized edge learning, where devices directly offload raw data to the edge server for global model training (Liu et al. [2021], Guo and Quek [2020]). However, due to privacy concerns, the other paradigm, referred to as federated learning (FL), is more appealing (McMahan et al. [2017], Li et al. [2019]). In this chapter, we thus pay attention on *federated edge learning* (FEEL).

In a nutshell, FEEL is a distributed learning paradigm, in which a group of devices train a global model together with the assistance of an edge server. As illustrated in Figure 8.1, the training of FEEL is generally accomplished by multiple communication rounds, each of which involves three procedures: (i) global model

Federated Learning for Future Intelligent Wireless Networks, First Edition.
Edited by Yao Sun, Chaoqun You, Gang Feng, and Lei Zhang.
© 2024 The Institute of Electrical and Electronics Engineers, Inc. Published 2024 by John Wiley & Sons, Inc.

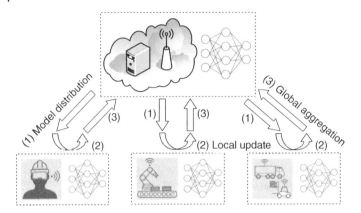

Figure 8.1 An illustration of FEEL with heterogeneous devices.

distribution from the edge server to participating devices; (ii) local model update at participating devices; and (iii) global aggregation for global model update. Considering the synchronous FEEL, global aggregation starts at the edge server after receiving local updates from all participating devices. That is, the duration of a round is dominated by the slowest participating devices, also known as stragglers. In this regard, the limited network resources impose restriction on the number of participating devices in each round. On one hand, the limited spectrum resources make it impossible to schedule all devices for model distribution and global aggregation. On the other hand, the energy-limited device cannot support its participation in all rounds. To guarantee the learning performance, it is thus of paramount importance to explore the spectrum flexibility for a dynamic scheduling policy so as to determine which devices to select in each round, with the consideration of straggler and limited device energy issues.

As for the scheduling policy design, the concern heterogeneity, including unbalanced data sizes across devices as well as diverse communication and computation capabilities at devices (Li et al. [2019], Cai et al. [2020]), brings along great challenges. First, it is not clear how the data sizes of scheduled devices affect the learning performance. Second, the unbalanced data sizes may incur nonindependent and identically distributed (IID) data across devices, which have an adverse effect on the learning performance. Third, the heterogeneity across devices makes the completion time of local updates asynchronous, which results in serious straggler issue. Lastly, heterogeneous devices have diverse energy consumption over time, which interferes with the current scheduling decision due to limited device energy. What is more, the energy consumption in the current round has direct impact on the available energy and scheduling decision in the future rounds.

To address the above challenges, we aim to devise a heterogeneity-aware dynamic scheduling policy in FEEL. The devised scheduling policy is expected to minimize the global loss, while simultaneously accounting for the device heterogeneity, straggler, and limited device energy issues. For the device heterogeneity, we put an emphasis on the unbalanced data sizes as well as diverse communication and computation capabilities, which is the specific perspective compared with the existing works. Our main contributions are summarized as follows:

- The device heterogeneity results in the asynchronous completion times of local model updates, which makes the typical multiplexing access scheme with reserved spectrum resources inefficient for global aggregation. To this end, we explore NR-supported spectrum flexibility and adopt the sequential transmission. In this way, the early completed local update is uploaded to the edge server using all spectral bandwidth, while the late completed local update is executing at the devices. Such a parallel communication and computation approach not only enhances the utilization of spectrum resources but also reduces the transmission delay of participating devices. Consequently, more feasible devices can participate in one communication round within the maximum allowable delay while consuming less energy. That is, our adopted sequential transmission can enlarge the feasible set of participating devices, under the constraints of maximum allowable delay and energy consumption.
- To select the best one from the feasible set of participating devices, we customize an optimization criterion with the consideration of unbalanced data sizes. In detail, we start with the convergence analysis of FEEL and then conclude that the more the data points participating in the model training in one communication round are, the better the learning performance is in terms of both the convergence rate and accuracy. On this basis, we propose the average scheduled data size maximization criterion to pick out the best set of participating devices, along with the global loss minimization.
- We formulate a heterogeneity-aware dynamic scheduling problem to maximize the average scheduled data size, constrained by maximum allowable delay and average energy level in each communication round. The formulated problem is a long-term optimization problem. By solving this problem with the Lyapunov optimization framework, we can make a decision on the set and order of scheduled devices in each round. The key technical challenge for the decision making is to optimally solve the one-round scheduling problem, which is a combinatorial optimization problem that requires one to simultaneously optimize the set and order of scheduled devices. To this end, we exploit its special properties: (i) the optimality of first coming and first scheduling (FCFS) policy and the monotonicity lying in its delay constraint. The proofs of these two properties provide the efficiency and novelty of the solution. Based on the solution of

one-round scheduling problem, we arrive at a *dynamic scheduling* algorithm (DISCO).

- Through theoretical analysis, we conclude that the learning performance and the feasibility of our proposed DISCO can be guaranteed under certain conditions. Besides, we present numerical and experimental results to demonstrate that the DISCO outperforms the baselines, in terms of the average scheduled data size, test accuracy, and training loss, respectively.

The rest of this chapter is organized as follows. In Section 8.2, we introduce the related works, which is followed by the system model and problem formulation in Sections 8.3 and 8.4, respectively. The dynamic scheduling algorithm is proposed and analyzed in Section 8.5. Then, we present the evaluation results in Section 8.6 and conclude this chapter in Section 8.7. Finally, the related theoretical proofs are given as supplementary materials in Section 8.A.

8.2 Related Works

In this section, we introduce the related works, with respect to the above-mentioned four challenges in heterogeneous FEEL.

In Yang et al. [2020], the convergence performance of three scheduling policies, i.e., random scheduling, round robin, and proportional fair, was analyzed from the perspective of transmission success probability of local updates. The similar analytical method was adopted in Chen et al. [2021]. On the other hand, the impact of device scheduling on the learning performance was investigated in Nguyen et al. [2021] and Ren et al. [2020] to highlight the importance of participating devices, regardless of unbalanced data sizes across devices. In this regard, the relation between the data sizes of scheduled devices and the learning performance still remains open.

In computer science community, heuristic federated learning algorithms were proposed to improve the learning performance, such as FedAvg, FedProx, and FedSS in McMahan et al. [2017], Li et al. [2019], and Cai et al. [2020]. FedAvg in McMahan et al. [2017] is the current state-of-the-art algorithm, which averages local updates of all participating devices to improve the global model. As a generalization and reparametrization of FedAvg, FedProx was designed in Li et al. [2019] to tackle the heterogeneity issue in FEEL. With an emphasis on the non-IID data across devices, FedProx added a proximal term for local update. Considering unbalanced data across devices, FedSS was proposed in Cai et al. [2020] to dynamically select the training sample size for local update. However, in each communication round, FedAvg, FedProx, and FedSS schedule the devices uniformly at random, overlooking the straggler issue and device energy constraint in realistic wireless networks.

In communication society, some researches, e.g., Wang et al. [2019], Dinh et al. [2021], and Yang et al. [2021], assumed full participation of all devices to investigate the trade-off between the number of local updates in one round and the total required rounds for global model convergence. With the consideration of heterogeneous devices, some researches concentrated on the importance-aware scheduling policy design, such as Nguyen et al. [2021] and Ren et al. [2020]. In Chen et al. [2021], Shi et al. [2021], and Nishio and Yonetani [2019], straggler issues were addressed by adding a latency constraint in each round when designing the scheduling policy. Particularly, the sequential transmission was adopted in Nishio and Yonetani [2019] for global aggregation, which is an effective way to tackle the asynchrony of local updates. However, in Nguyen et al. [2021], Ren et al. [2020], Chen et al. [2021], Shi et al. [2021], and Nishio and Yonetani [2019], the impacts of time-varying communication conditions and time-correlated energy consumption on the set of scheduled devices were not addressed.

Reinforcement learning and Lyapunov optimization are two main techniques to deal with the optimization problems in time-varying networks. Reinforcement learning always needs many iterative numbers for its own convergence, which may be larger than the required rounds for federated learning convergence and result in suboptimal decision on the device scheduling (Xia et al. [2020]). In this regard, we consider Lyapunov optimization technique as Xu and Wang [2021], to address the energy correlations among communication rounds. In Xu and Wang [2021], more attention are paid on the time patterns for device selection rather than device heterogeneity, resulting in an essential difference with this chapter.

8.3 System Model for FEEL

As shown in Figure 8.1, we consider a general FEEL system with one edge server and K heterogeneous devices, which have unbalanced data and construct the set of devices $\mathcal{K} = \{1, \ldots, K\}$. For device k, its collected data are described by $\{\mathbf{x}_k, \mathbf{y}_k\}$ with size n_k, where $\mathbf{x}_k = [\mathbf{x}_{k1}, \ldots, \mathbf{x}_{kn_k}]$ and $\mathbf{y}_k = [y_{k1}, \ldots, y_{kn_k}]$ are the data points and labels, respectively. In IIoT, these data can be collected through sensors installed at the device and are employed for event prediction. Specifically, a local model with parameters $\mathbf{w}_k$ is trained through local updates at device k using $\mathbf{x}_k$ and $\mathbf{y}_k$.

Through global aggregation of local models in multiple communication rounds, an optimal global model parameters are expected to be found, with the minimum global loss function as follows:

$$\mathbf{w}^{\star} = \arg\min_{\mathbf{w}} F(\mathbf{w}, \mathbf{x}, \mathbf{y}). \tag{8.1}$$

In (8.1), $\mathbf{x} = [\mathbf{x}_1, \ldots, \mathbf{x}_K]$ is all data points, $\mathbf{y} = [\mathbf{y}_1, \ldots, \mathbf{y}_K]$ is all labels, and $F(\mathbf{w}, \mathbf{x}, \mathbf{y})$ is given by

$$F(\mathbf{w}, \mathbf{x}, \mathbf{y}) = \sum_{k=1}^{K} \frac{n_k}{N} F_k(\mathbf{w}, \mathbf{x}_k, \mathbf{y}_k), \tag{8.2}$$

where $N = \sum_{k=1}^{K} n_k$ is the size of all data points to fit a traditional global loss minimization over the entire dataset (McMahan et al. [2017]). Besides, $F_k(\mathbf{w}, \mathbf{x}_k, \mathbf{y}_k)$ can be defined by

$$F_k(\mathbf{w}, \mathbf{x}_k, \mathbf{y}_k) = \frac{1}{n_k} \sum_{n=1}^{n_k} f(\mathbf{w}, \mathbf{x}_{kn}, y_{kn}), \tag{8.3}$$

with a loss function $f(\mathbf{w}, \mathbf{x}_{kn}, y_{kn})$ to capture the gap between the predicted and true label.

Based on these preliminaries, we then elaborate on the flow of FEEL with device scheduling and the involved delay and energy model.

8.3.1 Flow of FEEL with Scheduling

To minimize the global loss function, T communication rounds are executed in FEEL. Particularly, round t comprises the following procedures:

- *Device scheduling*: The edge server collects data, communication, and computation information from each device and determines the set of scheduled devices in the current round, which is denoted by $S(t)$;
- *Global model distribution*: The edge server broadcasts the current global model $\mathbf{w}^t$ to all scheduled devices;
- *Local model update*: All scheduled devices update the local model after receiving the global model $\mathbf{w}^t$. For device k, its local model is updated as

$$\mathbf{w}_k^{t+1} = \mathbf{w}^t - \eta \mathbf{g}_k^t, \tag{8.4}$$

where $\eta > 0$ is the learning rate and $\mathbf{g}_k^t$ is given by

$$\mathbf{g}_k^t = \nabla F_k(\mathbf{w}^t, \mathbf{x}_k, \mathbf{y}_k); \tag{8.5}$$

- *Global aggregation*: All scheduled devices upload their updated local models to the edge server in the uplink multiaccess channel. After receiving local models from all scheduled devices, the edge server computes the global model for the next round, as follows:

$$\mathbf{w}^{t+1} = \frac{\sum_{k \in \mathcal{K}} n_k s_k(t) \mathbf{w}_k^{t+1}}{\sum_{k \in \mathcal{K}} n_k s_k(t)}, \tag{8.6}$$

where $s_k(t) \in \{0,1\}$ indicates whether device k is scheduled or not in the current round. With $s_k(t) = 1$, device k is scheduled, otherwise, $s_k(t) = 0$.

8.3.2 Delay and Energy Model in FEEL

Without loss of generality, we assume the wireless channels and computing capabilities for devices remain unchangeable in one round and varies independently over rounds (Xia et al. [2020], Xu and Wang [2021], Guo et al. [2021]). Then, we give the delay and energy model.

8.3.2.1 Delay Model
There are three key components in the delay model, which are defined as follows:

- *Model distribution delay*: In round t, we assume the edge server employs a spectral bandwidth W^D to broadcast the global model to all scheduled devices. Then, the achievable broadcast rate for global model distribution is denoted by

$$r^D(t) = \min_{k \in S(t)} \left\{ W^D \log_2 \left(1 + \frac{Ph_k^D(t)}{W^D N_0} \right) \right\}, \tag{8.7}$$

where $h_k^D(t)$ is the broadcast channel gain for scheduled device k, N_0 is the power density of noise, and P is the transmit power of the edge server. Further, we calculate the model distribution delay as

$$d^D(t) = \frac{B(\mathbf{w}^t)}{r^D(t)}, \tag{8.8}$$

where $B(\mathbf{w}^t)$ indicates the number of bits that the edge server requires to transmit global model $\mathbf{w}^t$.

- *Local update delay*: We adopt the central processing unit (CPU) frequency to depict the computing capability at one device (Xia et al. [2020], Guo et al. [2021]). In round t, denote the available CPU frequency at device k by $f_k(t)$, and then we can calculate the local update delay at scheduled device k by

$$d_k^L(t) = \frac{\omega_k n_k}{f_k(t)}. \tag{8.9}$$

In (8.9), ω_k is the number of CPU cycles required to process one data point at scheduled device k. Besides, $f_k(t)$ is time varying because background computation load often changes in different rounds (Xia et al. [2020], Guo et al. [2021]).

- *Global aggregation delay*: For global aggregation in round t, scheduled device k is allocated with spectral bandwidth $W_k(t)$ to upload its updated local model. With the uplink channel gain and transmit power for scheduled device k expressed as $h_k^U(t)$ and p_k^U, respectively, we calculate the achievable uplink transmission rate as

$$r_k^U(t) = W_k(t) \log_2 \left(1 + \frac{p_k^U h_k^U(t)}{W_k(t) N_0} \right). \tag{8.10}$$

Note that there is no interference in (8.10) due to our concerned transmission mechanism in this chapter, which is elaborated in Section 8.4. For scheduled device k, we thus give its global aggregation delay, i.e., the uplink transmission delay, by

$$d_k^U(t) = \frac{B(\mathbf{w}_k^{t+1})}{r_k^U(t)}, \tag{8.11}$$

where $B(\mathbf{w}_k^{t+1})$ is the number of bits that requires to transmit updated local model, i.e., $\mathbf{w}_k^{t+1}$, to the edge server. It is observed from (8.4) that the number of elements in global model $\mathbf{w}^t$ is the same with that of updated local model $\mathbf{w}_k^{t+1}$ at scheduled device k. Hence, we have $B(\mathbf{w}^t) = B(\mathbf{w}_k^{t+1})$ for any device k in the tth round.

Based on the defined $d^D(t)$, $d_k^L(t)$, and $d_k^U(t)$, we can further calculate the completion time of global aggregation at the edge server in round t, termed as $\tau^S(t)$, which cannot exceed its maximum allowable delay to tackle the straggler issue. Regardless of the time for global model update at powerful edge server, $\tau^S(t)$ is equal to the arrival time of last scheduled device at the edge server, which highly depends on the adopted transmission mechanism for global aggregation, as elaborated in Section 8.4.

8.3.2.2 Energy Model

In one round, only the scheduled device has energy consumption, which comes from the local update and global aggregation. Specifically, at scheduled device k, we give the energy consumed for the local update by

$$e_k^L(t) = p_k^L(t)d_k^L(t) = \kappa_k \omega_k n_k f_k(t)^2, \tag{8.12}$$

where power consumption for local update is calculated as $p_k^L = \kappa_k f_k(t)^3$ with κ_k as power coefficient (Kaxiras and Martonosi [2008]). Similarly, for scheduled device k, we can give its uplink transmission energy consumption for global aggregation by

$$e_k^U(t) = p_k^U d_k^U(t) = \frac{p_k^U B(\mathbf{w}_k^{t+1})}{r_k^U(t)}. \tag{8.13}$$

Further, we can calculate the total energy consumption at device k as

$$e_k(t) = s_k(t)\left(e_k^L(t) + e_k^U(t)\right), \tag{8.14}$$

which indicates that the total energy consumption is zero for the unscheduled device in the tth round.

8.4 Heterogeneity-Aware Dynamic Scheduling Problem Formulation

In this section, we start with the convergence analysis of FEEL to find a metric, termed as the average scheduled data size, which is in an explicit form with respect to the device scheduling. By maximizing this metric, the minimum global loss function can be obtained when the FEEL converges. Then, to tackle the asynchrony of local updates, we adopt the scheduling policy with sequential transmission for global aggregation. Finally, we formulate a heterogeneity-aware dynamic scheduling problem to maximize the average scheduled data size, under the constraints of maximum allowable delay and average energy level in one communication round.

8.4.1 Convergence of FEEL with Scheduling

Through convergence analysis, we aim to reveal how the set of scheduled devices affects the global loss function. Without loss of generality, we make the following assumptions (Nguyen et al. [2021]):

- *L-Lipschitz continuousness*: We assume that the gradient of global loss function, i.e., $\nabla F(\mathbf{w})$,[1] is uniformly Lipschitz continuous with respect to $\mathbf{w}$. Hence, we have

$$\|\nabla F(\mathbf{w}^{t+1}) - \nabla F(\mathbf{w}^t)\| \leq L\|\mathbf{w}^{t+1} - \mathbf{w}^t\|, \tag{8.15}$$

where L is a positive constant and $\|\cdot\|$ is the l_2-norm of a vector.
- *μ-strongly convexity*: Assuming that $F(\mathbf{w})$ is strongly convex with positive parameter μ, we have

$$F(\mathbf{w}^{t+1}) \geq F(\mathbf{w}^t) + \left(\mathbf{w}^{t+1} - \mathbf{w}^t\right)^T \nabla F(\mathbf{w}^t) + \frac{\mu}{2}\|\mathbf{w}^{t+1} - \mathbf{w}^t\|^2. \tag{8.16}$$

- *ξ-local dissimilarity*: With nonnegative ξ_1 and ξ_2, this assumption makes the following inequality hold:

$$\|\nabla f(\mathbf{w}^t, \mathbf{x}_{kn}, y_{kn})\|^2 \leq \xi_1 + \xi_2\|\nabla F(\mathbf{w}^t)\|^2. \tag{8.17}$$

Then, we incorporate the device scheduling into the convergence result of FEEL and give Theorem 8.1 to reflect the impact of data sizes across devices on the learning performance.

---

1 For simplicity, we use $F(\mathbf{w})$ and $\nabla F(\mathbf{w})$ to replace $F(\mathbf{w}, \mathbf{x}, \mathbf{y})$ and $\nabla F(\mathbf{w}, \mathbf{x}, \mathbf{y})$ in the sequel, respectively.

Theorem 8.1   *Given the learning rate $\eta = 1/L$, we deduce*

$$\mathbb{E}\left\{F(\mathbf{w}^{t+1}) - F(\mathbf{w}^{\star})\right\} \leq C_1^{t+1} \mathbb{E}\left\{F(\mathbf{w}^0) - F(\mathbf{w}^{\star})\right\} + \frac{C_2\left(1 - C_1^{t+1}\right)}{1 - C_1}, \quad (8.18)$$

where C_1 is a descent coefficient and is given by

$$C_1 = 1 - \frac{\mu}{L} + \frac{4\mu\xi_2}{NL}\left(N - \sum_{k \in \mathcal{K}} n_k \phi_k\right), \quad (8.19)$$

and C_2 is expressed as

$$C_2 = \frac{2\xi_1}{LN}\left(N - \sum_k n_k \phi_k\right), \quad (8.20)$$

with ϕ_k denoting the scheduling probability of device k in one communication round, i.e., $\mathbb{E}\{s_k(t)\} = \phi_k$.

Proof: The proof can be completed by adopting (8.6) for the global model update in Appendix A in Chen et al. [2021].   □

From Theorem 8.1, two conclusions can be drawn when t tends to infinity with $0 < \xi_2 < 1/4$ (i.e., $0 < C_1 < 1$): (i) The FEEL training converges and (ii) a gap $C_2/(1 - C_1)$ exists between $\mathbb{E}\left\{F(\mathbf{w}^t)\right\}$ and $F(\mathbf{w}^{\star})$. Particularly, C_1 and $C_2/(1 - C_1)$ impacts on the learning rate and accuracy, respectively. On one hand, the smaller C_1 is, the faster the learning rate is. On the other hand, smaller $C_2/(1 - C_1)$ leads to lower global loss function and higher learning accuracy. These observations motivate us to maximize $\sum_{k \in \mathcal{K}} n_k \phi_k$, i.e., average scheduled data size, for the learning performance improvement.

8.4.2   Scheduling Policy with Sequential Transmission

Due to the device heterogeneity, the completion times of local updates at different devices are asynchronous, which makes the traditional orthogonal frequency division multiplexing access (OFDMA) scheme inefficient (Xu and Wang [2021]). The hidden reason is that the reserved spectrum resources for the late completed local updates cannot be used for the early completed local updates, and the spectrum resources allocated for the early arrival local updates at the edge server cannot be released for the late arrival local updates. Profiting by the flexible numerology and mini-slot scheduling, NR technology supports more flexible scheduling policies to account for the asynchrony of local updates (Yin et al. [2021]). Specifically, the flexible numerology (α) that allows subcarrier spacing to scale as $2^\alpha \times 15$ kHz. With an increase in the numerology index α, the number of slots in a subframe (e.g., 1 ms in long-term evolution [LTE]) increases, thereby facilitating mini-slot scheduling.

That is, the flexible numerology and mini-slot in NR technology can enlarge the subcarrier spacing and shorten the slot duration simultaneously, which makes the sequential transmission applicable to a realistic edge network.

In the sequential transmission, the whole spectral bandwidth is used to upload the updated local models from different devices to the edge server one by one. In this way, within the maximum allowable delay, high-efficient spectrum resource utilization makes more data points participate in each communication round. Alternatively, the sequential transmission mechanism can shorten the global aggregation delay, thereby reducing the energy consumption at scheduled devices. That is, both the learning performance and energy utilization can be improved by the schedule policy with sequential transmission.

Next, we complete the calculation of completion time of global aggregation at the edge server in round t, i.e., $\tau^S(t)$. Denoting the spectral bandwidth used for global aggregation by W^U, we have $W_k(t) = W^U$ in (8.10). Define the set of scheduled devices in round t as $S(t) = \{k|s_k(t) = 1, \forall k \in \mathcal{K}\}$, in which the device indices are arranged according to their scheduled orders. The completion time of local update at scheduled device k is denoted by

$$\tau_k(t) = d^D(t) + d_k^L(t), \tag{8.21}$$

and then, its arrival time at the edge server can be calculated using the following recursion equations:

$$\tau_{\langle k \rangle}^A = \begin{cases} \tau_{\langle k \rangle} + d_{\langle k \rangle}^U, & k = 1 \\ \max\left\{\tau_{\langle k \rangle}, \tau_{\langle k-1 \rangle}^A\right\} + d_{\langle k \rangle}^U, & k = 2, \dots, |S(t)|. \end{cases} \tag{8.22}$$

Note that, in (8.22), $\langle k \rangle$ indicates the kth scheduled device for global aggregation. Hence, in the sequential transmission, both the set of scheduled devices and their schedule orders have significant impact on the arrival times of local updates at the edge server. Regardless of global aggregation time at powerful edge server, $\tau^S(t)$ is calculated as

$$\tau^S(t) = \tau_{\langle |S(t)| \rangle}^A(t), \tag{8.23}$$

which is the arrival time of the last scheduled device at the edge server in the tth round.

8.4.3 Problem Formulation

Subject to the straggler and limited device energy issues, we aim to design a dynamic scheduling policy with sequential transmission to minimize the global loss function after T rounds in heterogeneous FEEL. To this end, we formulate the following dynamic scheduling optimization problem:

$$(\text{P0}) \max_{\mathbf{s}(t), S(t)} \quad \frac{1}{T} \sum_{t=0}^{T-1} \sum_{k \in \mathcal{K}} n_k s_k(t)$$

$$\text{s.t.} \quad C1 : \tau^S(t) \leq \tau^{\max}, \quad \forall t \in \mathcal{T}$$

$$C2 : \frac{1}{T} \sum_{t=0}^{T-1} e_k(t) \leq \bar{e}_k, \quad \forall k \in \mathcal{K}$$

$$C3 : s_k(t) \in \{0,1\}, \quad \forall k \in \mathcal{K}, \quad \forall t \in \mathcal{T}.$$

In problem (P0), we minimize the global loss function by maximizing the average scheduled data size over T rounds, which is motivated by Theorem 8.1. Besides, $\mathbf{s}(t) = [s_1(t), \dots, s_k(t)]$ is constructed by $s_k(t) \in \{0,1\}$ in C3, to determine which devices are scheduled in round t, and $S(t)$ is optimized to generate the optimal orders for scheduled devices. C1 is used to tackle the straggler issue, which stipulates that the completion time of global aggregation in one round cannot exceed its maximum allowable delay $\tau^{\max}$. C2 can prolong the battery life of device k, where $\bar{e}_k$ is the average energy level to impose restriction on the energy consumption at device k. To satisfy C2, higher energy consumption in some rounds must result in lower energy consumption in other rounds. Hence, C2 can account for the energy correlations across rounds.

Problem (P0) falls in the category of dynamic optimization problems. As a result, the Lyapunov optimization framework can be leveraged to deal with the correlations among rounds (Neely [2010]). Nonetheless, the binary variable $s_k(t)$ and couplings among scheduled devices in C1 still make problem (P0) hard to solve. In Section 8.5, we aim to find the special structure of problem (P0) to make an appropriate decision on the set and order of scheduled devices in each round.

8.5 Dynamic Scheduling Algorithm Design and Analysis

In this section, we first introduce an offline benchmark algorithm to achieve the optimal solution of problem (P0). Then, we leverage the Lyapunov technique to devise an online dynamic scheduling algorithm and compare it with the benchmark to present its optimality and feasibility.

8.5.1 Benchmark: *R*-Round Lookahead Algorithm

The offline R-round lookahead algorithm is given as an optimal solution benchmark, where both the communication (i.e., $h_k^D(t)$ and $h_k^U(t)$) and computation (i.e., $f_k(t)$) information in the next R rounds are assumed to be known a prior. Then, the total rounds T are divided into $F \geq 1$ frames, each of which has $R \geq 1$ rounds with $T = FR$ and $F, R \in \mathbb{Z}^+$. In frame $f, f = 0, \ldots, F - 1$, the following problem is addressed:

$$(Q0) \min_{s(t),S(t)} \frac{1}{R} \sum_{t=fR}^{(f+1)R-1} \sum_{k \in \mathcal{K}} n_k s_k(t)$$

$$\text{s.t.} \quad \text{C1 and C3}$$

$$\text{C4:} \frac{1}{R} \sum_{t=fR}^{(f+1)R-1} e_k(t) \leq \bar{e}_k, \quad \forall k \in \mathcal{K},$$

whose optimal value is denoted by $U_f^\star$. Averaging over F frames, the R-round lookahead algorithm achieves the average scheduled data size as $\frac{1}{F} \sum_{f=0}^{F-1} U_f^\star$, which is optimal to problem (P0) with $F = 1$ and $R = T$.

8.5.2 DISCO: Dynamic Scheduling Algorithm

The dynamic scheduling algorithm, i.e., DISCO, is online and proposed by applying the Lyapunov technique. Particularly, to satisfy C2, we construct a virtual energy queue $q_k(t)$ for each device k, as follows:

$$q_k(t + 1) = \max \{q_k(t) + e_k(t) - \bar{e}_k, 0\}. \tag{8.24}$$

Then, in a generic round t, we need to solve the following one-round scheduling problem:

$$(P1) \min_{s(t),S(t)} \quad U(t)$$

$$\text{s.t.} \quad \text{C5:} \tau^S(t) \leq \tau^{\max}$$

$$\text{C6:} s_k(t) \in \{0,1\}, \quad \forall k \in \mathcal{K},$$

where the objective function is given by

$$U(t) = \sum_{k \in \mathcal{K}} q_k(t) e_k(t) - V \sum_{k \in \mathcal{K}} n_k s_k(t)$$

$$= \sum_{k \in \mathcal{K}} s_k(t) \left(q_k(t) \left(e_k^L(t) + e_k^U(t) \right) - V n_k \right). \tag{8.25}$$

In (8.25), $V \geq 0$ can be regarded as a weight factor to strike a balance between the energy consumption at devices and learning performance. Generally, a larger V puts more emphasis on the learning performance improvement, at the expense of energy consumption at devices. Besides, we can observe from (8.25) that the unscheduled devices in the former rounds have lower $q_k(t)$ in the current round, and it is encouraged for these devices to participate in the current round for $U(t)$ minimization. With $U(t)$ minimization as the objective, problem (P1) thus contributes to the fair scheduling, which makes for a good convergent global model (Li et al. [2020]).

In the sequel, we turn our focus on addressing problem (P1), which can be further recast as

$$(\text{P2}) \min_{\mathbf{s}(t)} \quad U(t)$$

$$\text{s.t.} \quad \text{C6}$$

$$\text{C7}: \min_{S(t)} \tau^S(t) \leq \tau^{\max},$$

by applying the primal decomposition (Palomar and Chiang [2006]). We observe that the following two special properties are hidden in problem (P2):

- *Optimality of FCFS policy*: Given the set of scheduled devices $\mathbf{s}(t)$, Theorem 8.2 unveils that the optimal schedule orders in $S(t)$ follow the FCFS policy. That is, the early completed local update shall be uploaded first.
- *Monotonicity of* $\min_{S(t)} \tau^S(t)$: It is effortless to deduce that $\min_{S(t)} \tau^S(t)$ is nondescending with one more device added into $S(t)$. On this basis, we obtain the other conclusion that for any $|S(t)|$ scheduled devices, $\min_{S(t)} \tau^S(t) > \tau^{\max}$ makes C7 not satisfied for any $|S(t)| + 1$ scheduled devices.

Theorem 8.2 *Given the set of scheduled devices, their optimal schedule orders follow the FCFS policy.*

Proof: See Section 8.A.1. □

The special properties in problem (P2) make the decision on the set and order of scheduled devices efficient and novel. Specifically, these properties guide a one-round scheduling algorithm in Algorithm 8.1, by which the optimal set and order of scheduled devices can be achieved for each communication round in an exhaustive search way. On one hand, the monotonicity of $\min_{S(t)} \tau^S(t)$ is leveraged to reduce the search space on the possible set of scheduled devices and to set the algorithm termination conditions. On the other hand, the optimality of FCFS policy helps to fast calculate $\min \tau^S(t)$ for each possible set of scheduled devices and judge whether the possible set is feasible or not. Then, we elaborate on Algorithm 8.1. Specifically, we gradually increase the number of scheduled devices until it exceeds the number of devices in the system or all the *possible* set of scheduled devices cannot satisfy C7, as given in line 12. For each number of scheduled devices, we define the *feasible* set of scheduled devices in line 7, in which the one with the maximum $U(t)$ becomes an element in *candidate* set of

---

Algorithm 8.1 One-round scheduling algorithm for problem (P1)

1: Obtain V, $q_k(t)$, $e_k^L(t)$, $e_k^U(t)$, and n_k;
2: Initialize the number of scheduled devices as $i = 1$;
3: **repeat**
4:    **for** Each *possible* $S_i(t)$ **do**
5:       Obtain $\min_{S_i(t)} \tau^S(t)$ with the FCFS policy.
6:    **end for**
7:    Define *feasible* set with i scheduled devices as $S_i^F(t) = \{ S_i(t) | \min_{S_i(t)} \tau^S(t) \leq \tau^{\max} \}$;
8:    **if** $S_i^F(t) \neq \emptyset$ **then**
9:       Find the optimal set with i scheduled devices as $S_i^\star(t) = \arg\min_{S_i(t) \in S_i^F(t)} U(t)$;
10:      Update $i = i + 1$;
11:    **end if**
12: **until** $S_i^F(t) = \emptyset$ or $i = K + 1$
13: Generate the *candidate* set of scheduled devices as $S^C(t) = \{ S_j^\star(t) | 1 \leq j \leq i - 1 \}$;
14: Output the optimal solution for problem (P1) as $S^\star(t) = \arg\min_{S(t) \in S^C(t)} U(t)$ and $\mathbf{s}^\star(t)$, with $s_k(t) = 1, \forall k \in S^\star(t)$ and $s_k(t) = 0, \forall k \notin S^\star(t)$;

---

scheduled device. Finally, we search the candidate set to find out the optimal scheduling decision in line 14.

Based on Algorithm 8.1, we devise online dynamic scheduling algorithm, termed as DISCO, to solve the original problem (P0), as summarized in Algorithm 8.2. The performance of the DISCO is analyzed, its computational complexity is further reduced, and its implementation is discussed in the sequel.

Algorithm 8.2 DISCO: Dynamic scheduling algorithm for problem (P0)

1: Initialize R, F, and V_f, $f = 0, \ldots, F - 1$;
2: **for** $t = 0, 1, \ldots, T - 1$ **do**
3:    **if** $t \bmod R = 0$ **then**
4:       Set $q_k(t) = 0, \forall k \in \mathcal{K}$ and $V = V_f$ with $f = t/R$;
5:    **end if**
6:    Observe the current channel state $h_k^U(t)$ and $h_k^D(t)$, as well as, CPU frequency state $f_k(t)$ for any device k;
7:    Involve Algorithm 8.1 to solve problem (P1) for scheduling decision in round t;
8:    Update $q_k(t + 1)$ based on (8.24).
9: **end for**

8.5.3 Algorithm Analysis, Complexity Reduction, and Implementation Discussion

In this section, we analyze the performance of DISCO, reduce its computational complexity, and discuss its implementation, subsequently.

8.5.3.1 Algorithm Analysis

Following Xu and Wang [2021] and Theorem 8.1, we prove the feasibility, optimality, and convergence of our proposed DISCO in the following Theorem.

Theorem 8.3 *For any $R, F \in \mathbb{Z}^+$ satisfying $T = FR$, the following statements hold for our proposed DISCO:*

- *The average energy consumption of device $k, \forall k \in \mathcal{K}$ is upper bounded by*

$$\frac{1}{T}\sum_{t=0}^{T-1}e_k(t) \leq \bar{e}_k + \frac{1}{F}\sum_{f=0}^{F-1}\sqrt{\frac{2(C_1 + V_f N)}{R}}, \tag{8.26}$$

with $C_1 = \frac{1}{2}KE_{\max}^2$ and $E_{\max} = \max_{k,t}(e_k(t) - \bar{e}_k)$.
- *The average scheduled data size satisfies*

$$\frac{1}{T}\sum_{t=0}^{T-1}\sum_{k\in\mathcal{K}}n_k s_k(t) \geq \frac{1}{F}\sum_{f=1}^{F-1}U_f^{\star} - \frac{1}{F}\sum_{f=1}^{F-1}\frac{C_2}{V_f}, \tag{8.27}$$

with $C_2 = RC_1$. Recall that $U_f^{\star}$ is the optimal value of problem (Q0) in the fth frame and is achieved by the R-round lookahead algorithm.
- *When T tends to infinity with $0 < \xi_2 < 1/4$, the DISCO makes the FEEL training convergent under the three Assumptions in Section 8.4.1. This is because $0 \leq N - \sum_k n_k \mathbb{E}\{s_k(t)\} \leq N$ holds with the DISCO, resulting in $0 < C_1 < 1$ under this condition.*

Proof: See Section 8.A.2 for the proofs of the former two conclusions. □

Based on the relationship between problems (P0) and (Q0), the optimal value of problem (P0) is reached as $\frac{1}{F}\sum_{f=1}^{F-1}U_f^{\star}$ with $F = 1$ and $R = T$. Then, we conclude from Theorem 8.3 the following insights: (i) The energy constraint at each device is approximately satisfied with $\mathcal{O}(\sqrt{V})$-bounded factor. (ii) The DISCO is $\mathcal{O}(1/V)$-optimal with respect to the performance of R-round lookahead algorithm. (iii) Under certain conditions, the DISCO can guarantee the FEEL convergence. Hence, by adjusting the weight factor V, we can strike a balance between the learning performance and energy consumption at devices. Specifically, with larger V, more emphasis is put on the average scheduled data size to improve the learning performance while more energy is consumed at devices, and vice versa.

8.5.3.2 Complexity Reduction

In the DISCO, the one-round scheduling problem (P1) is solved using Algorithm 8.1 with the concept of exhaustive search on all possible set of scheduled devices. Although we have exploited some special properties to fast the search, there is still space to further fast the running of Algorithm 8.1. To this end, we introduce a random device selection procedure before line 6 in the DISCO. Particularly, we introduce a parameter $\theta \in (0,1]$ and select $\lceil \theta K \rceil$ devices to participate in the search in Algorithm 8.1. In this way, the search space on the possible set of scheduled devices is shrunken, along with reduced computation complexity in the DISCO. Besides, some randomness is introduced to improve the diversity of participating datasets over communication rounds, which is beneficial to enhance the learning performance.

8.5.3.3 Implementation Discussion

Our proposed DISCO can be implemented in an online manner. In any round t, the edge server randomly selects $\lceil \theta K \rceil$ devices and collects their information of current energy, communication, and computation, including $q_k(t)$, $h_k^U(t)$, $h_k^D(t)$, and $f_k(t)$. Then, the edge server executes Algorithm 8.1 to make a decision on the set and order of the scheduled devices and broadcasts the global model to these scheduled devices. After receiving the global model, the scheduled device starts to update its local model using its own dataset. At the same time, the edge server sends a signal to the first scheduled device to upload its local model, once it finishes the local model update. After receiving the updated local model from the first scheduled device, the edge server sends the same signal to the second scheduled device. This procedure is executed in accordance with the assigned orders of scheduled devices until the edge server receives all updated local models from the scheduled devices or the maximum allowable delay is triggered. Note that the DISCO can be terminated before the predefined round T, as long as the global model is convergent.

8.6 Evaluation Results

In this section, we demonstrate the efficacy of our proposed DISCO. First, we give numerical results to show the average scheduled data size of the DISCO, under the constraints of maximum allowable delay and average energy level in one round. Then, we conduct experiments to present the learning performance of the DISCO, in terms of the test accuracy and training loss, respectively.

To this end, we compare the DISCO with different θ and adopt the *O*nline *C*lient *s*E*lection and *BA*ndwidth allocatio*N* algorithm in Xu and Wang [2021], termed as OCEAN, as the baseline. In the OCEAN, the spectral bandwidth is orthogonally allocated to the scheduled devices for global aggregation in each communication round, regardless of the asynchronous completion times of local updates. Besides, we consider the other baseline, which combines the round robin (RR) policy and the FCFS policy, referred to as RR + FCFS.

In the RR + FCFS, a set of participating devices (with the size as 12) is first determined by the RR policy, which contributes to the fairness among devices. Then, the device without sufficient energy to support its current local update and uplink transmission is moved from the predetermined set of participating devices. Further, the FCFS policy is used to pick out the possible devices, which cannot violate the maximum allowable delay in one round. Before presenting numerical and experimental results, we elaborate on the parameter settings.

8.6.1 Parameter Settings

We consider FEEL with $K = 20$ heterogeneous devices, which have unbalanced and low non-IID data, as well as diverse communication and computation capabilities. The related parameter settings are detailed in the following aspects:

- *Dataset parameters*: We consider two kinds of datasets: MNIST and CIFAR-10. The MNIST dataset of handwritten digits (LeCun et al. [1998]), which are labeled from 0 to 9 and comprises 60,000 and 10,000 data points for model training and test, respectively. Each data point is a handwritten digit and is displayed as a gray-valued image of 28×28 pixels. The CIFAR-10 dataset consists of 60,000 32×32 color images in 10 classes, which includes 50,000 training images and 10,000 test images (Krizhevsky [2009]). Then, we distribute all data points in the training set across devices to train the classifier on MNIST and CIFAR-10 dataset. Particularly, we sort all data points in the training set in an ascend order of their labels and divide them as 1200 and 1000 shards for MNIST and CIFAR-10 dataset, respectively, each of which has 50 images. For each device, its number of shards is randomly generated while satisfying the condition that the total number of shards for all devices are equal to 1200 for MNIST dataset and to 1000 for CIFAR-10 dataset. In this way, the randomness of image assignment among devices is large and the resultant datasets among devices are low non-IID, along with small images in each shard and unequal data sizes at devices. In Table 8.1, we give the data sizes for all devices, which are adopted to obtain the numerical results and experimental results in Sections 8.6.2 and 8.6.3.
- *Convolutional neural network (CNN) model parameters*: To train the classifier on MNIST dataset, we consider two convolution layers. There are 10 $5 \times 5 \times 1$ filters and 20 $5 \times 5 \times 10$ filters in the first and second convolution layers. Besides, each convolution layer is followed by one 2×2 max pooling layer and

Table 8.1 Unbalanced data sizes for $K = 20$ devices, with MNIST and CIFAR-10 datasets.

Device index	1	2	3	4	5	6	7	8	9	10
MNIST data size	1300	850	5200	5850	3700	4350	1500	450	6050	4550
CIFAR-10 data size	1350	3550	3700	600	4450	2500	300	3850	450	1900

Device index	11	12	13	14	15	16	17	18	19	20
MNIST data size	3700	6300	200	1500	1950	4750	1300	2400	450	3650
CIFAR-10 data size	2800	4150	3550	3250	750	1500	2650	4300	3400	1000

activated by ReLU. After the second max pooling layer, two fully connected layers are followed. One is composed of 320 and 50 units with ReLU activation, and the other composed of 50 units for the input and 10 units for the final softmax output. In the concerned CNN, 21,840 model parameters are involved in MNIST dataset, which generates $B(\mathbf{w}^t) = B(\mathbf{w}_k^{t+1}) = 698,880$ bits (in 32-bit float) for global aggregation and model distribution. The same structures of convolution layers and max pooling layers are adopted for the classifier training in CIFAR-10 dataset. The differences are that there are 6 $5 \times 5 \times 3$ filters and 16 $5 \times 5 \times 6$ filters in the first and second convolution layers for CIFAR-10 dataset. Moreover, three fully connected layers are followed after the second max pooling layer. The former two are, respectively, composed of 400×120 units and 120×84 units with ReLU activation. The last one includes 84 units for the input and 10 units for the final softmax output. In the CNN adopted for CIFAR-10 dataset, 62,006 model parameters are introduced with $B(\mathbf{w}^t) = B(\mathbf{w}_k^{t+1}) = 1,984,192$ bits (in 32-bit float).

- *Other parameters*: If not specified, these parameters related to communication, computation, and algorithm are set as shown in Table 8.2. What is more, we construct a circle area with the edge server located at the center and the device randomly distributed around the edge server. With the distance between the device and edge server denoted by d in meter, we generate d in each communication round following the uniform distribution in [30,100] and the pathloss following $L(\text{dB}) = 128.1 + 37.6\log_{10}(10^{-3}d)$. The normalized Rayleigh fading is considered as the small-scale channel fading across different rounds. Considering time-varying computation capabilities at device $k, \forall k \in \mathcal{K}$, we randomly and uniformly generate its computation capacity in one round from $[100,800] \times 10^6$ cycles/s with interval 100×10^6 cycles/s. Besides, we consider the same average energy level for all devices, i.e., $\bar{e}_k = \bar{e}, \forall k \in \mathcal{K}$.

Based on the above parameters, we then present the effectiveness of our proposed DISCO through numerical and experimental results, respectively.

Table 8.2 Parameter settings related to the communication, computation, and algorithm.

Parameters	Values
P and p_k^U	1 and 0.1 W
W^U/W^D	1.4 MHz
N_0	-174 dBm
κ_k and ω_k	10^{-26} and 2×10^4 cycles
T and η for MNIST	150 and 0.001
T and η for CIFAR-10	200 and 0.01
F and R	1 and T
V_0	10^{-5}

8.6.2 Numerical Results

In this section, we aim to verify the superiority of our proposed DISCO from the perspective of numerical results. That is, the concern metric is the average scheduled data size per round, as proved in Section 8.4. Besides, the proposed DISCO should outperform the baseline algorithm in terms of this metric, under the constraints of maximum allowable delay and average energy level per round. To avoid redundancy, we give the numerical comparisons only using the MNIST dataset.

In Figures 8.2 and 8.3, we give the impact of maximum allowable delay $\tau^{\max}$ and average energy level $\bar{e}$ on the average scheduled data size, respectively. We observe that with the relaxed conditions, such as higher $\tau^{\max}$ and $\bar{e}$, more data points can participate in one round for model training. Besides, the DISCO with $\theta = 1$ is the best one, in terms of the average scheduled data size. That the less data points participate in the model training for the DISCO with $\theta = 0.6$ is derived from the shrunken search space when finding out the optimal solution of problem (P1). The DISCO with $\theta = 0.6$ is sometimes better and sometimes worse compared with the RR + FCFS. This is reasonable due to that the initial sizes of search space are 12 for both algorithms, but with changeable initial devices. The reason why the DISCO outperforms the baseline OCEAN is that the DISCO can sufficiently utilize spectrum resources to schedule more devices in one round. On the other hand, the DISCO can schedule the devices frequently with high-efficient energy usage, which is explained by combining Figures 8.4 and 8.5.

Note that Figures 8.4 and 8.5 are instantiated with $\tau^{\max} = 0.3$ s and $\bar{e} = 0.08$ J to show the feasibility of our proposed DISCO. Specifically, the observation in Figure 8.5 corresponds to the conclusion in Theorem 8.3 that an appropriate V_f can make the average energy constraints satisfied. What is more, the results in

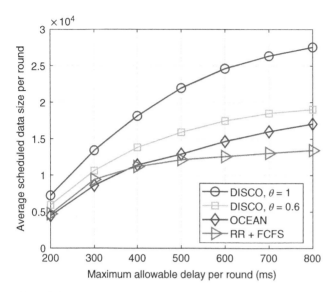

Figure 8.2 Average scheduled data size per round versus τ^{max}.

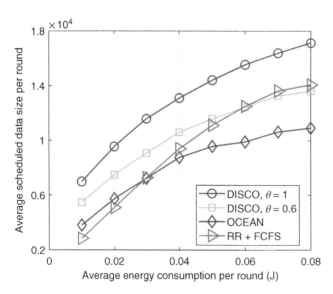

Figure 8.3 Average scheduled data size per round versus $\bar{e}$.

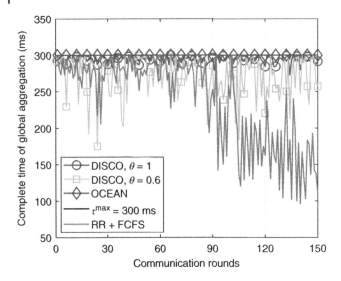

Figure 8.4 Completion time of global aggregation versus rounds.

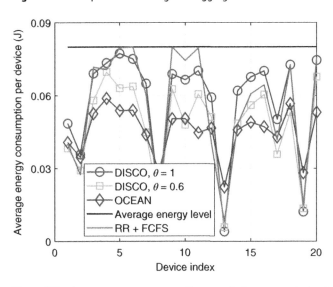

Figure 8.5 Average energy consumption per device versus device index.

Figures 8.4 and 8.5 could reveal the reason why the DISCO schedules the devices more frequently than the OCEAN. For the OCEAN, the triggered maximum allowable delay in Figure 8.4 and less data points participating in one round means that the global aggregation delay of scheduled devices is prolonged. Hence, the energy consumption of scheduled devices in one round becomes larger, resulting in lower schedule frequency and lower energy usage, as shown in Figure 8.5. On the contrary, the DISCO has higher energy usage, thereby scheduling the devices more frequently. As for the RR + FCFS, an interesting phenomenon is in Figure 8.4 that less devices are scheduled in the later rounds with the global aggregation time far away from the maximum allowable delay. This is because some devices almost exhaust their energy in the former rounds, as shown in Figure 8.5, and they are not scheduled in the later rounds.

8.6.3 Experimental Results

In this section, we conduct experiments for performing the image classification on MNIST and CIFAR-10 dataset to present the efficacy of our proposed DISCO, in terms of the test accuracy and training loss, respectively.

Considering MNIST dataset, we give the mean and standard deviation values of test accuracy in Table 8.3. To this end, we run each experiment 3 times and give the mean and standard deviation value using the test accuracy over the last 10 communication rounds of 3-time experiments. From Figure 8.2, we observe that the average scheduled data size with $\tau^{max} = 0.8\,s$ and $\bar{e} = 0.04\,J$ is larger than that under the setting of $\tau^{max} = 0.3\,s$ and $\bar{e} = 0.02\,J$. Correspondingly, the convergence results of all algorithms are improved with more participating data points. To be more specific, the test accuracy is higher and more stable with more data participating, due to the higher mean and smaller standard deviation.

In Figures 8.6 and 8.7, we give test accuracy and training loss over communication rounds, with the consideration of MNIST dataset. It is observed that our proposed DISCO outperforms the baseline OCEAN, no matter $\theta = 1$ or $\theta = 0.6$. This is because the DISCO can make more data points participate for the model training in each round. Compared with the RR + FCFS, the DISCO has a little worse test accuracy and training loss in the former rounds. However, in the later rounds,

Table 8.3 Test accuracy comparisons among different algorithms.

	MNIST dataset		CIFAR-10 dataset	
	$\tau^{max} = 0.3\,s$	$\tau^{max} = 0.8\,s$	$\tau^{max} = 0.3\,s$	$\tau^{max} = 0.8\,s$
Algorithm	$\bar{e} = 0.02\,J$	$\bar{e} = 0.04\,J$	$\bar{e} = 0.02\,J$	$\bar{e} = 0.04\,J$
DISCO, $\theta = 1$	$93.46\% \pm 0.24\%$	$94.64\% \pm 0.24\%$	$55.08\% \pm 3.17\%$	$62.00\% \pm 1.96\%$
DISCO, $\theta = 0.6$	$93.37\% \pm 0.20\%$	$94.65\% \pm 0.24\%$	$54.72\% \pm 3.04\%$	$61.31\% \pm 2.10\%$
OCEAN	$92.74\% \pm 0.58\%$	$94.06\% \pm 0.36\%$	$51.41\% \pm 4.13\%$	$59.66\% \pm 1.98\%$
RR + FCFS	$79.39\% \pm 0.57\%$	$92.80\% \pm 0.25\%$	$35.38\% \pm 2.54\%$	$52.21\% \pm 3.41\%$

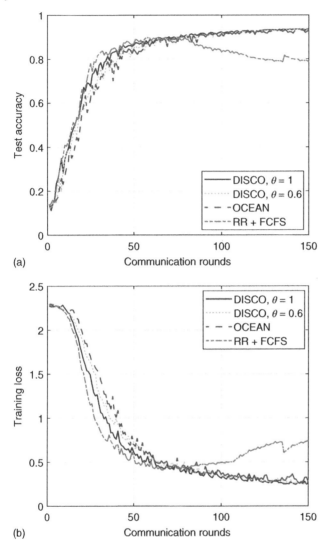

Figure 8.6 Learning performance with MNIST dataset, $\tau^{max} = 0.3\,s$, $\bar{e} = 0.02\,J$. (a) Test accuracy versus communication rounds and (b) training loss versus communication rounds.

the learning performance of RR + FCFS becomes deteriorated. This is because more devices exhaust their energy in the former rounds and only a few devices with residual energy can be scheduled in the later rounds, as shown in Figure 8.4. Finally, the learnt global model in the RR + FCFS is biased to the local models of a few devices with residual energy.

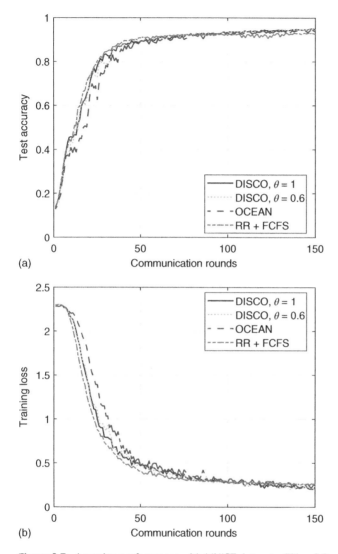

Figure 8.7 Learning performance with MNIST dataset, $\tau^{max} = 0.8\,s, \bar{e} = 0.04\,J.$ (a) Test accuracy versus communication rounds and (b) training loss versus communication rounds.

Further, we give experimental results on CIFAR-10 dataset in Table 8.3 and Figures 8.8 and 8.9. From Table 8.3, the same conclusion is drawn in CIFAR-10 dataset as that in MNIST dataset. That is, more data participating can make the learning performance better, in terms of the test accuracy and stability. In Figure 8.8, our proposed DISCO has the best test accuracy and training loss

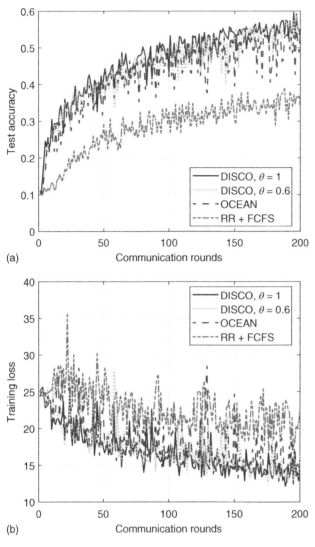

Figure 8.8 Learning performance with CIFAR-10 dataset, $\tau^{max} = 0.3\,\text{s}$, $\bar{e} = 0.02\,\text{J}$. (a) Test accuracy versus communication rounds and (b) training loss versus communication rounds.

as expected. Whereas, the least data participating in the RR + FCFS results in the slowest and worst convergence. In Figure 8.9, the leaning performance of RR + FCFS first becomes improved and then deteriorated, which is the same as that in Figures 8.6 and 8.7, and is caused by the unreasonable energy usage. By

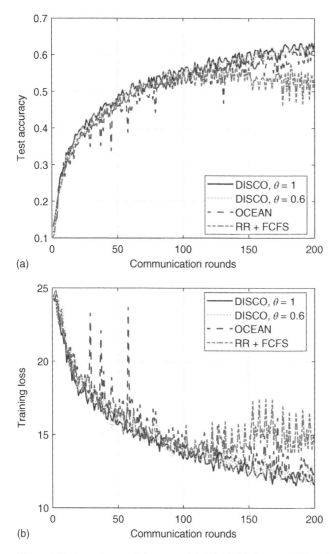

Figure 8.9 Learning performance with CIFAR-10 dataset, $\tau^{max} = 0.8\,s$, $\bar{e} = 0.04\,J$. (a) Test accuracy versus communication rounds and (b) training loss versus communication rounds.

taking advantage of spectrum resources, the proposed DISCO is much better than the OCEAN for CIFAR-10 dataset as well.

Surprisingly, we observe from the above experimental results that, with smaller scheduled data size in the DISCO with $\theta = 0.6$, yet the resultant learning performance approaches that in the DISCO with $\theta = 1$. It means that the randomness brought by $0 < \theta < 1$ sometimes is beneficial to the global model convergence. The reason is that the randomness is always powerful to increase the diversity of participating data points for learning performance improvement. In this regard, with unbalanced and non-IID data distributed across devices, our proposed DISCO could fast learn a good model by adjusting the parameter $\theta \in (0,1]$.

8.7 Conclusions

In this chapter, we have proposed a dynamic scheduling algorithm, termed as DISCO, for heterogeneous federated edge learning, under the constraints of maximum allowable delay and average energy level in one communication round. The design of the DISCO has been motivated by the theoretical result that the more the data points participating in the model training in each communication round are, the better the learning performance is. Besides, we have introduced a parameter $\theta \in (0,1]$ in the DISCO to reduce the computational complexity while improving the diversity of data points participating in the model training. In consequence, the DISCO is able to account for unbalanced data across devices. On the other hand, we have considered the diverse communication and computation capabilities of devices when designing the DISCO. Specifically, we have explored the asynchronous completion times of local updates and adopted the sequential transmission for more data points participating in each communication round. Through theoretical analysis, we have concluded that the learning performance and feasibility of the DISCO can be guaranteed under certain conditions. Finally, numerical and experimental results have been presented to corroborate the efficacy of our proposed DISCO.

8.A Appendices

8.A.1 Proof of Theorem 8.2

Given the set of scheduled devices, we express the number of scheduled devices as S for simplicity. Then, we adopt the proof by contradiction to prove the optimal scheduling orders follow the FCFS policy. To this end, we express the FCFS scheduling order as $S(t) = \{\langle 1 \rangle, \ldots, \langle k-2 \rangle, \langle k-1 \rangle, \langle k \rangle, \ldots, \langle S \rangle\}$ and the other scheduling order as $S^\star(t) = \{\langle 1 \rangle, \ldots, \langle k-2 \rangle, \langle k \rangle, \langle k-1 \rangle, \ldots, \langle S \rangle\}$, which is assumed to be optimal with $\tau_{\langle k \rangle}(t) > \tau_{\langle k-1 \rangle}(t)$. In the sequel, we prove that $\tau^S(t)$ of $S(t)$ is not larger than that of $S^\star(t)$.

Considering $S(t)$, we calculate the arrival time at the edge server of device $\langle k \rangle$ as follows:

$$\tau^A_{\langle k \rangle} = \max\left\{\tau_{\langle k \rangle}, \tau^A_{\langle k-1 \rangle}\right\} + d^U_{\langle k \rangle}$$

$$= \max\left\{\tau_{\langle k \rangle} + d^U_{\langle k \rangle}, \tau_{\langle k-1 \rangle} + d^U_{\langle k-1 \rangle} + d^U_{\langle k \rangle}, \tau^A_{\langle k-2 \rangle} + d^U_{\langle k-1 \rangle} + d^U_{\langle k \rangle}\right\}, \quad (8.A.1)$$

which is based on (8.22). With $\tau_{\langle k \rangle}(t) > \tau_{\langle k-1 \rangle}(t)$, $S^\star(t)$ makes the arrival time at the edge server of device $\langle k - 1 \rangle$ equal to

$$\tau^A_{\langle k-1 \rangle} = \tau^A_{\langle k \rangle} + d^U_{\langle k-1 \rangle}$$

$$= \left\{\tau^A_{\langle k-2 \rangle} + d^U_{\langle k-1 \rangle} + d^U_{\langle k \rangle}, \tau_{\langle k \rangle} + d^U_{\langle k-1 \rangle} + d^U_{\langle k \rangle}\right\}. \quad (8.A.2)$$

Comparing (8.A.1) and (8.A.2), we can readily conclude that $\tau^A_{\langle k \rangle}$ with $S(t)$ is not larger than $\tau^A_{\langle k-1 \rangle}$ with $S^\star(t)$. It means that arrival times at the edge server of the subsequently scheduled devices with $S(t)$ are not larger than that with $S^\star(t)$. Hence, $\tau^S(t)$ of $S(t)$ is not larger than that of $S^\star(t)$. In the same way, we can prove that for any scheduling order, gradually rearranging the orders of scheduled devices following the FCFS policy can always find a better scheduling order. Finally, we can conclude that given the set of scheduled devices, their optimal scheduling orders are given by the FCFS policy.

8.A.2 Proof of Theorem 8.3

Without loss of generality, we define the Lyapunov drift as

$$\Delta_1(t) = L(\mathbf{q}(t+1)) - L(\mathbf{q}(t)), \quad (8.A.3)$$

with $\mathbf{q}(t) = [q_1(t), \ldots, q_K(t)]$ and

$$L(\mathbf{q}(t)) = \frac{1}{2}\sum_{k \in \mathcal{K}} q_k^2(t). \quad (8.A.4)$$

Since $\Delta_1(t)$ plays a key role in the proof of Theorem 8.3, we then turn our focus on bounding $\Delta_1(t)$.

Particularly, we derive from (8.24) that

$$\frac{1}{2}\sum_{k \in \mathcal{K}} q_k^2(t+1) \leq \frac{1}{2}\sum_{k \in \mathcal{K}} (q_k(t) + e_k(t) - \bar{e}_k)^2$$

$$\leq C_1 + \frac{1}{2}\sum_{k \in \mathcal{K}} q_k^2(t) + \sum_{k \in \mathcal{K}} q_k(t)(e_k(t) - \bar{e}_k), \quad (8.A.5)$$

where we give $C_1 = \frac{1}{2}KE_{\max}^2$ with $E_{\max} = \max_{k,t}(e_k(t) - \bar{e}_k)$. Rearranging (8.A.5), we can bound $\Delta_1(t)$ as

$$\Delta_1(t) \leq C_1 + \sum_{k \in \mathcal{K}} q_k(t)(e_k(t) - \bar{e}_k), \quad (8.A.6)$$

which makes the following inequality hold:

$$\Delta_1(t) - V_f \sum_{k \in \mathcal{K}} n_k s_k(t) \leq C_1 + \sum_{k \in \mathcal{K}} q_k(t)(e_k(t) - \bar{e}_k) - V_f \sum_{k \in \mathcal{K}} n_s s_k(t). \tag{8.A.7}$$

On this basis, we then prove the feasibility and optimality of the DISCO.

8.A.2.1 Feasibility Proof

Since the DISCO minimizes the right-hand term in (8.A.7) in frame $f, f = 0, \ldots, F - 1$, its solution makes the right-hand term not larger than that achieved by any feasible solution. Denoting the solution of DISCO by $\hat{s}_k(t), \forall k \in \mathcal{K}$ and considering a feasible solution with $s_k(t) = 0$ and $e_k(t) = 0, \forall k \in \mathcal{K}$, we get from (8.A.7) that

$$\Delta_1(t) - V_f \sum_{k \in \mathcal{K}} n_k \hat{s}_k(t) \leq C_1. \tag{8.A.8}$$

Further, with $N = \sum_{k \in \mathcal{K}} n_k$, we have

$$\Delta_1(t) \leq C_1 + V_f N. \tag{8.A.9}$$

Summing over $t = fR, \ldots, (f + 1)R - 2$, we deduce the following relationship:

$$\sum_{t=fR}^{(f+1)R-2} \Delta_1(t) = \frac{1}{2} \sum_{k \in \mathcal{K}} q_k^2((f + 1)R - 1) \leq R(C_1 + V_f N), \tag{8.A.10}$$

where the equation holds due to $q_k(fR) = 0$ and the inequality is derived from (8.A.9). From (8.A.10), we thus obtain

$$q_k((f + 1)R - 1) \leq \sqrt{2R(C_1 + V_f N)}. \tag{8.A.11}$$

Next, we aim to connect (8.A.11) and the energy constraints. To this end, we derive from (8.24) that

$$e_k(t) - \bar{e}_k \leq q_k(t + 1) - q_k(t). \tag{8.A.12}$$

Summing over R rounds in the fth frame, we deduce

$$\frac{1}{R} \sum_{t=fR}^{(f+1)R-1} \left(e_k(t) - \bar{e}_k\right) \leq \frac{1}{R} \sum_{t=fR}^{(f+1)R-1} \left(q_k(t + 1) - q_k(t)\right)$$

$$= \frac{q_k((f + 1)R - 1) - q_k(fR)}{R}$$

$$= \frac{q_k((f + 1)R - 1)}{R}, \tag{8.A.13}$$

where the first equation comes from the telescope summing and the second equation holds due to $q_k(fR) = 0$. Further, summing over F frames, we get from (8.A.13) that

$$\frac{1}{T} \sum_{t=0}^{T-1} \left(e_k(t) - \bar{e}_k \right) \leq \frac{1}{F} \sum_{f=0}^{F-1} \frac{q_k((f+1)R - 1)}{R}. \tag{8.A.14}$$

Plugging (8.A.11) into (8.A.14) finally yields the following result:

$$\frac{1}{T} \sum_{t=0}^{T-1} e_k(t) \leq \bar{e}_k + \frac{1}{F} \sum_{f=0}^{F-1} \sqrt{\frac{2(C_1 + V_f N)}{R}}. \tag{8.A.15}$$

8.A.2.2 Optimality Proof

Based on (8.A.7), we sum over R rounds in the fth frame and obtain the inequality as follows:

$$\sum_{t=fR}^{(f+1)R-1} \Delta_1(t) - V_f \sum_{t=fR}^{(f+1)R-1} \sum_{k \in \mathcal{K}} n_k s_k(t)$$

$$\leq RC_1 + \sum_{t=fR}^{(f+1)R-1} \sum_{k \in \mathcal{K}} q_k(t)(e_k(t) - \bar{e}_k) - V_f \sum_{t=fR}^{(f+1)R-1} \sum_{k \in \mathcal{K}} n_k s_k(t). \tag{8.A.16}$$

Since $\sum_{t=fR}^{(f+1)R-1} \Delta_1(t)$ is positive and the DISCO minimizes the right-hand term in (8.A.7) and (8.A.16), we have

$$- V_f \sum_{t=fR}^{(f+1)R-1} \sum_{k \in \mathcal{K}} n_k \hat{s}_k(t)$$

$$\leq RC_1 + \sum_{t=fR}^{(f+1)R-1} \sum_{k \in \mathcal{K}} q_k(t)(e_k^\star(t) - \bar{e}_k) - V_f \sum_{t=fR}^{(f+1)R-1} \sum_{k \in \mathcal{K}} n_k s_k^\star(t). \tag{8.A.17}$$

In (8.A.17), $e_k^\star(t)$ and $s_k^\star(t)$ are solutions from the R-round lookahead algorithm. Next, we bound the second term in the right-hand term of (8.A.17) as follows:

$$\sum_{t=fR}^{(f+1)R-1} \sum_{k \in \mathcal{K}} q_k(t)(e_k^\star(t) - \bar{e}_k) = \sum_{t=fR}^{(f+1)R-1} \sum_{k \in \mathcal{K}} (q_k(t) - q_k(fR))(e_k^\star(t) - \bar{e}_k)$$

$$\leq \sum_{t=fR}^{(f+1)R-1} \sum_{k \in \mathcal{K}} (t - fR)E_{\max}(e_k^\star(t) - \bar{e}_k)$$

$$\leq \frac{R(R-1)KE_{\max}^2}{2}, \tag{8.A.18}$$

where the equation holds due to $q_k(fR) = 0$ and the following two inequalities are derived from the fact that $q_k(t + 1) - q_k(t) \leq E_{max}$ holds with $E_{max} = \max_{k,t}(e_k(t) - \bar{e}_k)$. Plugging (8.A.18) into (8.A.17), we further have

$$\sum_{t=fR}^{(f+1)R-1} \sum_{k \in \mathcal{K}} n_k \hat{s}_k(t) \geq -\frac{RC_1}{V_f} - \frac{R(R-1)KE_{max}^2}{2V_f} + \sum_{t=fR}^{(f+1)R-1} \sum_{k \in \mathcal{K}} n_k s_k^\star(t). \qquad (8.A.19)$$

Summing over F frames and dividing by T on the both sides of (8.A.19), we finally attain

$$\frac{1}{T} \sum_{k \in \mathcal{K}} n_k \hat{s}_k(t) \geq \frac{1}{F} \sum_{f=1}^{F-1} U_f^\star - \frac{1}{F} \sum_{f=1}^{F-1} \frac{C_2}{V_f}, \qquad (8.A.20)$$

where $C_2 = C_1 + (R-1)KE_{max}^2/2 = RC_1$ is a constant.

References

3GPP. 5G; NR; physical layer procedures for data. Technical Specification (TS) 38.214. *3rd Generation Partnership Project (3GPP)*, Oct. 2018. Version 15.3.0.

L. Cai, D. Lin, J. Zhang, and S. Yu. Dynamic sample selection for federated learning with heterogeneous data in fog computing. In *Proceedings of the ICC*, pages 1–6, Dublin, Ireland, Jul. 2020. doi: 10.1109/ICC40277.2020.9148586.

M. Chen, Z. Yang, W. Saad, C. Yin, H. V. Poor, and S. Cui. A joint learning and communications framework for federated learning over wireless networks. *IEEE Transactions on Wireless Communications*, 20(1):269–283, Jan. 2021. doi: 10.1109/TWC.2020.3024629.

C. T. Dinh, N. H. Tran, M. N. H. Nguyen, C. S. Hong, W. Bao, A. Y. Zomaya, and V. Gramoli. Federated learning over wireless networks: Convergence analysis and resource allocation. *IEEE/ACM Transactions on Networking*, 29(1):398–409, Feb. 2021. doi: 10.1109/TNET.2020.3035770.

K. Guo and T. Q. S. Quek. On the asynchrony of computation offloading in multi-user MEC systems. *IEEE Transactions on Communications*, 68(12):7746–7761, Dec. 2020. doi: 10.1109/TCOMM.2020.3024577.

K. Guo, R. Gao, W. Xia, and T. Q. S. Quek. Online learning based computation offloading in MEC systems with communication and computation dynamics. *IEEE Transactions on Communications*, 69(2):1147–1162, Feb. 2021. doi: 10.1109/TCOMM.2020.3038875.

Yun Chao Hu, Milan Patel, Dario Sabella, Nurit Sprecher, and Valerie Young. Mobile Edge Computing: A key technology towards 5G. *White Paper*, European Telecommunications Standards Institute (ETSI), Sep. 2015.

S. Kaxiras and M. Martonosi. *Computer Architecture Techniques for Power-Efficiency*. Morgan & Claypool, 2008.

A. Krizhevsky. Learning multiple layers of features from tiny images. Master's thesis, Department of Computer Science, University of Toronto, 2009.

Y. LeCun, L. Bottou, Y. Bengio, and P. Haffner. Gradient-based learning applied to document recognition. *Proceedings of the IEEE*, 86(11):2278–2324, Nov. 1998.

Tian Li, Anit Kumar Sahu, Manzil Zaheer, Maziar Sanjabi, Ameet Talwalkar, and Virginia Smith. Federated optimization in heterogeneous networks. In *Proceedings of the AMTL Workshop*, pages 1–16, Long Beach, CA, USA, Jun. 2019.

Tian Li, Maziar Sanjabi, Ahmad Beirami, and Virginia Smith. Fair resource allocation in federated learning. In *Proceedings of the ICLR*, pages 1–27, Addis Ababa, Ethiopia, Apr. 2020.

D. Liu, G. Zhu, J. Zhang, and K. Huang. Data-importance aware user scheduling for communication-efficient edge machine learning. *IEEE Transactions on Cognitive Communications and Networking*, 7(1):265–278, 2021. doi: 10.1109/TCCN.2020.2999606.

Brendan McMahan, Eider Moore, Daniel Ramage, Seth Hampson, and Blaise Aguera y Arcas. Communication-efficient learning of deep networks from decentralized data. In *Proceedings of the AISTATS*, pages 1273–1282, Fort Lauderdale, FL, USA, Apr. 2017.

M. Neely. *Stochastic Network Optimization with Application to Communication and Queueing Systems*. Morgan & Claypool, 2010.

H. T. Nguyen, V. Sehwag, S. Hosseinalipour, C. G. Brinton, M. Chiang, and H. Vincent Poor. Fast-convergent federated learning. *IEEE Journal on Selected Areas in Communications*, 39(1):201–218, Jan. 2021. doi: 10.1109/JSAC.2020.3036952.

T. Nishio and R. Yonetani. Client selection for federated learning with heterogeneous resources in mobile edge. In *Proceedings of the IEEE ICC*, pages 1–7, Shanghai, China, Jul. 2019. doi: 10.1109/ICC.2019.8761315.

D. P. Palomar and Mung Chiang. A tutorial on decomposition methods for network utility maximization. *IEEE Journal on Selected Areas in Communications*, 24(8):1439–1451, Aug. 2006. ISSN 0733-8716. doi: 10.1109/JSAC.2006.879350.

J. Ren, Y. He, D. Wen, G. Yu, K. Huang, and D. Guo. Scheduling for cellular federated edge learning with importance and channel awareness. *IEEE Transactions on Wireless Communications*, 19(11):7690–7703, Nov. 2020. doi: 10.1109/TWC.2020.3015671.

W. Shi, S. Zhou, Z. Niu, M. Jiang, and L. Geng. Joint device scheduling and resource allocation for latency constrained wireless federated learning. *IEEE Transactions on Wireless Communications*, 20(1):453–467, Jan. 2021. doi: 10.1109/TWC.2020.3025446.

S. Wang, T. Tuor, T. Salonidis, K. K. Leung, C. Makaya, T. He, and K. Chan. Adaptive federated learning in resource constrained edge computing systems. *IEEE Journal on Selected Areas in Communications*, 37(6):1205–1221, Jun. 2019. doi: 10.1109/JSAC.2019.2904348.

X. Wang, Y. Han, V. C. M. Leung, D. Niyato, X. Yan, and X. Chen. Convergence of edge computing and deep learning: A comprehensive survey. *IEEE Communication Surveys & Tutorials*, 22(2):869–904, 2020. doi: 10.1109/COMST.2020.2970550.

W. Xia, T. Q. S. Quek, K. Guo, W. Wen, H. H. Yang, and H. Zhu. Multi-armed bandit-based client scheduling for federated learning. *IEEE Transactions on Wireless Communications*, 19(11):7108–7123, Nov. 2020. doi: 10.1109/TWC.2020.3008091.

J. Xu and H. Wang. Client selection and bandwidth allocation in wireless federated learning networks: A long-term perspective. *IEEE Transactions on Wireless Communications*, 20(2):1188–1200, Feb. 2021. doi: 10.1109/TWC.2020.3031503.

H. H. Yang, Z. Liu, T. Q. S. Quek, and H. V. Poor. Scheduling policies for federated learning in wireless networks. *IEEE Transactions on Communications*, 68(1):317–333, Jan. 2020. doi: 10.1109/TCOMM.2019.2944169.

Z. Yang, M. Chen, W. Saad, C. S. Hong, and M. Shikh-Bahaei. Energy efficient federated learning over wireless communication networks. *IEEE Transactions on Wireless Communications*, 20(3):1935–1949, Mar. 2021. doi: 10.1109/TWC.2020.3037554.

H. Yin, L. Zhang, and S. Roy. Multiplexing URLLC traffic within eMBB services in 5G NR: Fair scheduling. *IEEE Transactions on Communications*, 69(2):1080–1093, Feb. 2021. doi: 10.1109/TCOMM.2020.3035582.

9

Robust Federated Learning with Real-World Noisy Data

Jingyi Xu, Zihan Chen, Tony Q. S. Quek, and Kai Fong Ernest Chong

Information Systems Technology and Design (ISTD) Pillar, Singapore University of Technology and Design, Singapore

9.1 Introduction

In real-world federated learning (FL) implementations over heterogeneous networks, there may be differences in the characteristics of different clients due to diverse annotators' skill, bias, and hardware reliability (Chen et al. [2020], Yang et al. [2022]). Client data is rarely IID and frequently imbalanced. Also, some clients would have clean data, while other clients may have data with label noise at different noise levels. Hence, the deployment of practical FL systems would face challenges brought by discrepancies in two aspects (i): local data statistics (McMahan et al. [2017], Li et al. [2020b], Hsu et al. [2019], Chen et al. [2021]) and (ii): local label quality (Chen et al. [2020], Yang et al. [2022]). In comparison to recent works that separately explored the discrepancy in local data statistics in FL, and learning with label noise in centralized learning (CL), `FedCorr` is the first unified approach for tackling both challenges simultaneously in FL.

9.1.1 Work Prior to `FedCorr`

The first challenge has been explored in recent FL works, with a focus on performance with convergence guarantees (Li et al. [2019b], Reddi et al. [2021]). However, these works have the common implicit assumption that the given labels of local data are completely correct, which is rarely the case in real-world datasets.

The second challenge can be addressed by reweighting (Fu et al. [2021], Chen et al. [2020], Wan and Chen [2021]) or discarding (Xu and Lyu [2021]) those

Federated Learning for Future Intelligent Wireless Networks, First Edition.
Edited by Yao Sun, Chaoqun You, Gang Feng, and Lei Zhang.

client updates that are most dissimilar. In these methods, the corresponding clients are primarily treated as malicious agents. However, dissimilar clients are not necessarily malicious and could have label noise in local data that would otherwise still be useful after label correction. For FL systems, the requirement of data privacy poses an inherent challenge for any label correction scheme. *How can clients identify their noisy labels to be corrected without needing other clients to reveal sensitive information?* For example, Yang et al. [2022] propose label correction for identified noisy clients with the guidance of extra data feature information exchanged between clients and server, which may lead to privacy concerns.

Label correction and, more generally, methods to deal with label noise are well studied in CL. Yet, even state-of-the-art CL methods for tackling label noise (Li et al. [2020a], Arazo et al. [2019], Tanaka et al. [2018], Han et al. [2018, 2020], Yu et al. [2019], Xia et al. [2020], Xu et al. [2021]), when applied to local clients, are inadequate in mitigating the performance degradation in the FL setting, due to the limited sizes of local datasets. These CL methods cannot be applied on the global sever or across multiple clients due to FL privacy requirements. Consequently, it is necessary and natural to adopt a more general framework that jointly considers the two discrepancies, for a better emulation of real-world data heterogeneity. Most importantly, as an main motivation for `FedCorr`, privacy-preserving label correction should be incorporated in training to improve robustness to data heterogeneity in FL.

`FedCorr` is a multistage FL framework that simultaneously deals with both discrepancy challenges; see Figure 9.1 for an overview. To ensure privacy, a dimensionality-based filter is introduced to identify noisy clients, by measuring the local intrinsic dimensionality (LID) (Houle [2013]) of local model prediction subspaces. Extensive experiments have shown that clean datasets can be distinguished from noisy datasets by the behavior of LID scores during training (Ma et al. [2018a, 2018b]). Hence, in addition to the usual local weight updates, each client also sends an LID score to the server, which is a single scalar representing the discriminability of the predictions of the local model. The noisy samples are then filtered based on per-sample training losses independently for each identified noisy client, and the large-loss samples are relabeled with the predicted labels of the global model. To improve training stability and alleviate the negative impact caused by noisy clients, a weighted proximal regularization term is introduced, where the weights are based on the estimated local noise levels. Furthermore, the global model is fine-tuned on the identified clean clients, and the local data for the remaining noisy clients is relabeled.

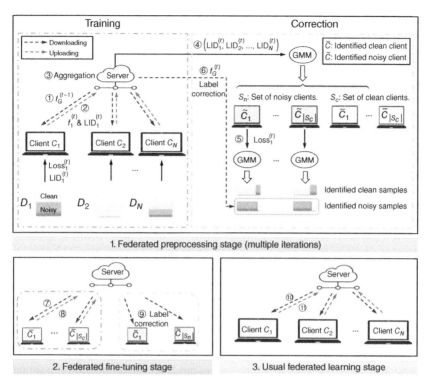

Figure 9.1 An overview of `FedCorr`, organized into three stages. Algorithm steps are numbered accordingly.

9.2 Related Work

9.2.1 Federated Methods

In this chapter, we focus on three closely related aspects of FL: generation of non-IID federated datasets, methods to deal with non-IID local data, and methods for robust FL.

The generation of non-IID local data partitions for FL was first explored in McMahan et al. [2017], based on dividing a given dataset into shards. More recent non-IID data partitions are generated via Dirichlet distributions (Hsu et al. [2019], Acar et al. [2020], Wan and Chen [2021]).

Recent federated optimization work mostly focus on dealing with the discrepancy in data statistics of local clients and related inconsistency issues

(Li et al. [2020b], Wang et al. [2020], Acar et al. [2020]). For instance, `FedProx` deals with non-IID local data, by including a proximal term in the local loss functions (Li et al. [2020b]), while `FedDyn` uses a dynamic proximal term based on selected clients (Acar et al. [2020]). `SCAFFOLD` (Karimireddy et al. [2020]) is another method suitable for non-IID local data that uses control variates to reduce client drift. In Hsu et al. [2019] and Reddi et al. [2021], adaptive FL optimization methods for the global server are introduced, which are compatible with non-IID data distributions. Moreover, the power-of-choice (`PoC`) strategy (Cho et al. [2020]), a biased client selection scheme that selects clients with higher local losses, can be used to increase the rate of convergence.

There are numerous works on improving the robustness of FL; these include robust aggregation methods (Wan and Chen [2021], Li et al. [2021], Fu et al. [2021]), reputation mechanism-based contribution examining (Xu and Lyu [2021]), credibility-based reweighting (Chen et al. [2020]), distillation-based semi-supervised learning (Itahara et al. [2021]), and personalized multitask learning (Li et al. [2021]). However, these methods are not designed for identifying noisy labels. Even when these methods are used to detect noisy clients, either there is no mechanism for further label correction at the noisy clients (Wan and Chen [2021], Li et al. [2021], Xu and Lyu [2021], Fu et al. [2021]) or the effect of noisy labels is mitigated with the aid of an auxiliary dataset, without any direct label correction (Chen et al. [2020], Itahara et al. [2021]). One notable exception is Yang et al. [2022], which carries out label correction during training by exchanging feature centroids between clients and server. This exchange of centroids may lead to privacy concerns, since centroids could potentially be used as part of reverse engineering to reveal nontrivial information about raw local data.

In contrast to these methods, `FedCorr` incorporates the generation of diverse local data distributions with synthetic label noise, together with noisy label identification and correction, without privacy leakage.

9.2.2 Local Intrinsic Dimension (LID)

Informally, LID (Houle [2013]) is a measure of the intrinsic dimensionality of the data manifold. In comparison to other measures, LID has the potential for wider applications as it makes no further assumptions on the data distribution beyond continuity. The key underlying idea is that at each datapoint, the number of neighboring datapoints would grow with the radius of neighborhood, and the corresponding growth rate would then be a proxy for "local" dimension.

LID builds upon this idea (Houle [2017]) via the geometric intuition that the volume of an m-dimensional Euclidean ball grows proportionally to r^m when its radius is scaled by a factor of r. Specifically, when we have two m-dimensional

Euclidean balls with volumes V_1, V_2, and with radii r_1, r_2, we can compute m as follows:

$$\frac{V_2}{V_1} = \left(\frac{r_2}{r_1}\right)^m \Rightarrow m = \frac{\log\left(V_2/V_1\right)}{\log\left(r_2/r_1\right)}. \tag{9.1}$$

We shall now formally define LID. Suppose we have a dataset consisting of vectors in $\mathbb{R}^n$. We shall treat this dataset as samples drawn from an n-variate distribution $\overline{D}$. For any $x \in \mathbb{R}^n$, let Y_x be the random variable representing the (non-negative) distance from x to a randomly selected point y drawn from $\overline{D}$, and let $F_{Y_x}(t)$ be the cumulative distribution function of Y_x. Given $r > 0$ and a sample point x drawn from $\overline{D}$, define the *LID of x at distance r* to be

$$LID_x(r) := \lim_{\varepsilon \to 0} \frac{\log F_{Y_x}((1 + \varepsilon)r) - \log F_{Y_x}(r)}{\log(1 + \varepsilon)},$$

provided that it exists, i.e., provided that $F_{Y_x}(t)$ is positive and continuously differentiable at $t = r$. The *LID at x* is defined to be the limit $LID_x = \lim_{r \to 0} LID_x(r)$. Intuitively, the LID at x is an approximation of the dimension of a smooth manifold containing x that would "best" fit the distribution $\overline{D}$ in the vincinity of x.

9.2.2.1 Estimation of LID

By treating the smallest neighbor distances as "extreme events" associated to the lower tail of the underlying distance distribution, Amsaleg et al. [2015] propose several estimators of LID based on extreme value theory. In particular, given a set of points $\mathcal{X}$, a reference point $x \in \mathcal{X}$, and its k nearest neighbors in $\mathcal{X}$, the maximum-likelihood estimate (MLE) of x is

$$\widehat{LID}(x) = -\left(\frac{1}{k}\sum_{i=1}^{k}\log\frac{r_i(x)}{r_{\max}(x)}\right)^{-1}, \tag{9.2}$$

where $r_i(x)$ denotes the distance between x and its ith nearest neighbor, and $r_{\max}(x)$ is the maximum distance from x among the k nearest neighbors.

9.3 FedCorr

In this section, we introduce `FedCorr`, a multistage training method to tackle heterogeneous label noise in FL systems (see Algorithm 9.1). `FedCorr` comprises three stages: preprocessing, fine-tuning, and usual training. In the first stage, we sample the clients without replacement using a small fraction to identify noisy clients via LID scores and noisy samples via per-sample losses, after which we relabel the identified noisy samples with the predicted labels of the global model. The noise level of each client is also estimated in this stage. In the second stage, we fine-tune the model with a typical fraction on relatively clean clients, and use the fine-tuned model to further correct the samples for the remaining clients. Finally, in the last stage, we train the model via the usual FL method (`FedAvg` (McMahan et al. [2017])) using the corrected labels at the end of the second stage.

Algorithm 9.1 `FedCorr`.

Inputs: N (number of clients), T_1, T_2, T_3, $\mathcal{D} = \{\mathcal{D}_i\}_{i=1}^{N}$ (dataset), $w^{(0)}$ (initialized global model weights).
Output: Global model f_G^{final}

    // Federated Preprocessing Stage
1: $(\hat{\mu}_1^{(0)}, \dots, \hat{\mu}_N^{(0)}) \leftarrow (0, \dots, 0)$ *// estimated noise levels*
2: **for** $t = 1$ **to** T_1 **do**
3:     $S =$ Shuffle($\{1, \dots, N\}$)
4:     $w_{inter} \leftarrow w^{(t-1)}$ *// intermediary weights*
5:     **for** $k \in S$ **do**
6:         $w_k^{(t)} \leftarrow$ weights that minimize loss function (9.5)
7:         Upload weights $w_k^{(t)}$ and LID score to server
8:     Update global model $w^{(t)} \leftarrow w_{inter}$
9:     Divide all clients into clean set S_c and noisy set S_n based on cumulative LID scores via GMM
10:     **for noisy client** $k \in S_n$ **do**
11:         Divide $\mathcal{D}_k$ into clean subset $\mathcal{D}_k^c$ and noisy subset $\mathcal{D}_k^n$ based on per-sample losses via GMM
12:         $\hat{\mu}_k^{(t)} \leftarrow \frac{|\mathcal{D}_k^n|}{|\mathcal{D}_k|}$ *// update estimated noise level*
13:         $y_k^{(i)} \leftarrow \arg\max f(x_k^{(i)}; w^{(0)}), \forall(x_k^{(i)}, y_k^{(i)}) \in \mathcal{D}_k^n$

    // Federated Finetuning Stage
14: $S_c \leftarrow \{k | k \in S, \mu_k < 0.1\}, S_n \leftarrow S \setminus S_c$.
15: **for** $t = T_1 + 1$ **to** $T_1 + T_2$ **do**
16:     Update $w_k^{(t)}$ by usual `FedAvg` among clients in S_c
17: **for Noisy client** $k \in S_n$ **do**
18:     $y_k^{(i)} \leftarrow \arg\max f(x_k^{(i)}; w^{(i)}), \forall(x_k^{(i)}, y_k^{(i)}) \in \mathcal{D}_k$

    // Usual Federated Learning Stage
19: **for** $t = T_1 + T_2 + 1$ **to** $T_1 + T_2 + T_3$ **do**
20:     Update $w_k^{(t)}$ by usual `FedAvg` among all clients
21: **return** $f_G^{\text{final}} := f(\cdot; w^{(T_1 + T_2 + T_3)})$

9.3.1 Preliminaries

Consider an FL system with N clients and an M-class dataset $\mathcal{D} = \{\mathcal{D}_k\}_{k=1}^N$, where each $\mathcal{D}_k = \{(x_k^i, y_k^i)\}_{i=1}^{n_k}$ denotes the local dataset for client k. Let S denotes the set of all N clients, and let $w_k^{(t)}$ (resp. $w^{(t)}$) denotes the local model weights of client k (resp. global model weights obtained by aggregation) at the end of communication round t. At the end of round t, the global model $f_G^{(t)}$ would have its weights $w^{(t)}$ updated as follows:

$$w^{(t)} \leftarrow \sum_{k \in S_t} \frac{|\mathcal{D}_k|}{\sum_{i \in S_t} |\mathcal{D}_i|} w_k^{(t)}, \tag{9.3}$$

where $S_t \subseteq S$ is the subset of selected clients in round t.

For the rest of this subsection, we shall give details on client data partition, noise model simulation, and LID score computation. These are three major aspects of `FedCorr` to emulate data heterogeneity and to deal with the discrepancies in both local data statistics and label quality.

9.3.1.1 Data Partition

In Xu et al. [2022], a new non-IID heterogeneous data partition was introduced. To describe this partition, an $N \times M$ indicator matrix Φ is first generated, where each entry Φ_{ij} indicates whether the local dataset of client i contains class j. Each Φ_{ij} shall be sampled from the Bernoulli distribution with a fixed probability p. For each $1 \leq j \leq M$, let v_j be the sum of entries in the jth column of Φ; this equals the number of clients whose local datasets contain class j. Let $\boldsymbol{q}_j$ be a vector of length v_j, sampled from the symmetric Dirichlet distribution with the common parameter $\alpha_{Dir} > 0$. Using $\boldsymbol{q}_j$ as a probability vector, the samples within class j are then randomly allocated to these v_j clients. Note that this non-IID data partition method is significant, since it provides a general framework to control the variability in both class distribution and the sizes of local datasets (see Figure 9.2).

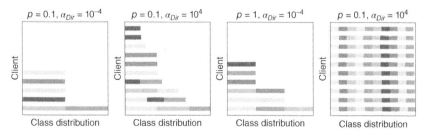

Figure 9.2 Depiction of non-IID partitions for different parameters.

9.3.1.2 Noise Model

To emulate label noise in real-world data, Xu et al. [2022] introduced a general federated noise model framework, under the assumption of instance-independent label noise. This framework has two parameters ρ and τ, where ρ denotes the system noise level (ratio of noisy clients) and τ denotes the lower bound for the noise level of a noisy client. Every client has a probability ρ of being a noisy client, in which case the local noise level for this noisy client is determined randomly, by sampling from the uniform distribution $U(\tau, 1)$. Succinctly, the noise level of client k (for $k = 1, \ldots, N$) is

$$
\mu_k = \begin{cases} u \sim U(\tau, 1), & \text{with probability } \rho; \\ 0, & \text{with probability } 1 - \rho. \end{cases}
\tag{9.4}
$$

When $\mu_k \neq 0$, the $100 \cdot \mu_k\%$ noisy samples are chosen uniformly at random and are assigned random labels, selected uniformly from the M classes.

9.3.1.3 LID Scores for Local Models

In this framework, LID scores are associated to local models. Consider an arbitrary client with local dataset D and current local model $f(\cdot)$. Let $\mathcal{X} := \{f(x)\}_{x \in D}$ be the set of prediction vectors, and for each $x \in D$, compute $\widehat{LID}(f(x))$ w.r.t. the k nearest neighbors in $\mathcal{X}$, as given in (9.2). We define the *LID score* of (D, f) to be the average value of $\widehat{LID}(f(x))$ over all $x \in D$. Note that as the local model $f(\cdot)$ gets updated with each round, the corresponding LID score will change accordingly.

Experiments have shown that given the same training process, models trained on a dataset with label noise tend to have larger LID scores as compared to models trained on the same dataset with clean labels (Ma et al. [2018a, 2018b]). Intuitively, the prediction vectors of a well-trained model, trained on a clean dataset, would cluster around M possible one-hot vectors, corresponding to the M classes. However, as more label noise is added to the clean dataset, the prediction vector of a noisy sample would tend to be shifted toward the other clusters, with different noisy samples shifted in different directions. Hence, the prediction vectors near each one-hot vector would become "more diffuse" and would on average span a higher dimensional space.

9.3.2 Federated Preprocessing Stage

FedCorr begins with the preprocessing stage, which iteratively evaluates the quality of the dataset of each client, and relabels identified noisy samples. This preprocessing stage differs from traditional FL in the following aspects:

- All clients will participate in each iteration. Clients are selected without replacement, using a small fraction.

- An adaptive local proximal term is added to the loss function, and mixup data augmentation is used.
- Each client computes its LID score and per-sample cross-entropy loss after local training and sends its LID score together with local model updates to the server.

9.3.2.1 Client Iteration and Fraction Scheduling

The preprocessing stage is divided into T_1 iterations. In each iteration, every client participates exactly once. Every iteration is organized by communication rounds, similar to the usual FL, but with two key differences: a small fraction is used, and clients are selected without replacement. Each iteration ends when all clients have participated.

It is known that large fractions could help improve the convergence rate (McMahan et al. [2017]), and a linear speedup could even be achieved in the case of convex loss functions (Stich [2018]). However, large fractions have a weak effect in non-IID settings, while intuitively, small fractions would yield aggregated models that deviate less from local models; cf. Li et al. [2019a]. Motivated by these observations, a fraction scheduling scheme is proposed in `FedCorr`, which combines the advantages of both small and large fractions. Specifically, client selection using a small fraction without replacement is adopted in the preprocessing stage, while in the latter two stages, a typical larger fraction with replacement is used. By sampling without replacement during preprocessing, it is ensured all clients participate equally for the evaluation of the overall quality of labels in local datasets.

9.3.2.2 Mixup and Local Proximal Regularization

Throughout the preprocessing stage, for client k with batch $\left(X_b, Y_b\right) = \{\left(x_i, x_j\right)\}_{i=1}^{n_b}$ (where n_b denotes batch size), the following loss function is used:

$$L\left(X_b\right) = L_{CE}\left(f_k^{(t)}\left(\tilde{X}_b\right), \tilde{Y}_b\right) + \beta \hat{\mu}_k^{(t-1)} \left\|w_k^{(t)} - w^{(t-1)}\right\|^2. \tag{9.5}$$

Here, $f_k^{(t)} = f\left(\cdot; w_k^{(t)}\right)$ denotes the local model of client k in round t, and $w^{(t-1)}$ denotes the weights of the global model obtained in the previous round $t - 1$. The first term in (9.5) represents the cross-entropy loss on the mixup augmentation of $\left(X_b, Y_b\right)$, while the second term in (9.5) is an adaptive local proximal regularization term, where $\hat{\mu}_k^{(t-1)}$ is the estimated noise level of client k to be defined later. It should be noted that this local proximal regularization term is only applied in the preprocessing stage.

Recall that mixup (Zhang et al. [2018]) is a data augmentation technique that favors linear relations between samples and that has been shown to exhibit strong robustness to label noise (Arazo et al. [2019], Li et al. [2020a]). Mixup generates new samples $(\tilde{x}, \tilde{y})$ as convex combinations of randomly selected pairs of samples $\left(x_i, y_i\right)$ and $\left(x_j, y_j\right)$, given by $\tilde{x} = \lambda x_i + (1 - \lambda) x_j$, $\tilde{y} = \lambda y_i + (1 - \lambda) y_j$, where $\lambda \sim \text{Beta}(\alpha, \alpha)$, and $\alpha \in (0, \infty)$. Intuitively, mixup achieves robustness to label noise due to random interpolation. For example, if $(x_i, \hat{y}_i)$ is a noisy sample and if y_i is the true label, then the negative impact caused by an incorrect label $\hat{y}_i$ is alleviated when paired with a sample whose label is y_j.

In FedCorr, an adaptive local proximal regularization term is scaled by $\hat{\mu}_k^{(t-1)}$, which is the estimated noise level of client k computed at the end of round $t - 1$. (In particular, this term would vanish for clean clients.) The hyperparameter β is also incorporated to control the overall effect of this term. Intuitively, if a client's dataset has a larger discrepancy from other local datasets, then the corresponding local model would deviate more from the global model, thereby contributing a larger loss value for the local proximal term.

9.3.2.3 Identification of Noisy Clients and Noisy Samples

To address the challenge of heterogeneous label noise, FedCorr iteratively identify and relabel the noisy samples. In each iteration of the preprocessing stage, where all clients will participate, every client will compute the LID score and per-sample loss for its current local model (see Algorithm 9.1, lines 3–9). Specifically, when client k is selected in round t, the model $f_k^{(t)}$ is trained on the local dataset D_k, after which the LID score of $\left(D_k, f_k^{(t)}\right)$ is computed via (9.2). Note that FedCorr preserves the privacy of client data, since in comparison to the usual FL, there is only an additional LID score sent to the server, which is a single scalar that reflects only the predictive discriminability of the local model. Since the LID score is computed from the predictions of the output layer (of the local model), knowing this LID score does not reveal information about the raw input data. This additional LID score is a single scalar, hence it has a negligible effect on communication cost.

At the end of iteration t, the following three steps are performed:

1. The server first computes a Gaussian mixture model (GMM) on the cumulative LID scores of all N clients. Using this GMM, the set of clients S is partitioned into two subsets: S_n (noisy clients) and S_c (clean clients).
2. Each noisy client $k \in S_n$ locally computes a new GMM on the per-sample loss values for all samples in the local dataset D_k. Using this GMM, D_k is partitioned into two subsets: a clean subset D_k^c, and a noisy subset D_k^n. We observe that the large-loss samples are more likely to have noisy labels. The local noise level of client k can then be estimated by $\hat{\mu}_k^{(t)} = |D_k^n|/|D_k|$ if $k \in S_n$ and $\hat{\mu}_k^{(t)} = 0$ otherwise.
3. Each noisy client $k \in S_n$ performs relabeling of the noisy samples by using the predicted labels of the global model as the new labels. In order to avoid overcorrection, only those samples that are identified to be noisy with high confidence would be relabeled. This partial relabeling is controlled by a relabel ratio π

and a confidence threshold θ. Take noisy client k for example: Samples are first chosen from $\mathcal{D}_k^n$ that correspond to the top-$\pi \cdot |\mathcal{D}_k^n|$ largest per-sample cross-entropy losses. Next, we obtain the prediction vectors of the global model and relabel a sample only when the maximum entry of its prediction vector exceeds θ. Thus, the subset $\widetilde{\mathcal{D}}_k^{n'}$ of samples to be relabeled is given by

$$\widetilde{\mathcal{D}}_k^n = \underset{\substack{\hat{D} \subseteq \mathcal{D}_k^n \\ |\hat{D}| = \pi \cdot |\mathcal{D}_k^n|}}{\arg\max} \, L_{CE}\left(\tilde{D}; f_G^{(t)}\right); \tag{9.6}$$

$$\widetilde{\mathcal{D}}_k^{n'} = \left\{ (x,y) \in \widetilde{\mathcal{D}}_k^n \, \Big| \, \max\left(f_G^{(t)}(x)\right) \geq \theta \right\}; \tag{9.7}$$

where $f_G^{(t)}$ is the global model at the end of iteration t.

Intuition for Using Cumulative LID Scores in Step 1 In deep learning, it has been empirically shown that when training on a dataset with label noise, the evolution of the representation space of the model exhibits two distinct phases: (i) an early phase of dimensionality compression, where the model tends to learn the underlying true data distribution and (ii) a later phase of dimensionality expansion, where the model overfits to noisy labels (Ma et al. [2018b]).

As observed in Xu et al. [2022], clients with larger noise levels tend to have larger LID scores. Also, the overlap of LID scores between clean and noisy clients would increase during training. This increase could be due to two reasons: (i) the model may gradually overfit to noisy labels and (ii) the identified noisy samples are corrected after each iteration, thereby making the clients with low noise levels less distinguishable from clean clients. Hence, it is concluded in Xu et al. [2022] that the cumulative LID score (i.e., the sum of LID scores in all past iterations) is a better metric for distinguishing noisy clients from clean clients; see the top two plots in Figure 9.3 for a comparison of using LID score versus cumulative LID score. Furthermore, the bottom two plots in Figure 9.3 show that cumulative LID score has a stronger linear relation with local noise level.

9.3.3 Federated Fine-Tuning Stage

FedCorr aims to fine-tune the global model f_G on relatively clean clients over T_2 rounds and further relabel the remaining noisy clients. The aggregation at the end of round t is given by the same Eq. (9.3), with one key difference: S_t is now a subset of $S_c = \{k | 1 \leq k \leq N, \hat{\mu}_k^{(T_1)} \leq \kappa\}$, where κ is the threshold used to select relatively clean clients based on the estimated local noise levels $\hat{\mu}_1^{(T_1)}, \ldots, \hat{\mu}_N^{(T_1)}$.

At the end of the fine-tuning stage, the remaining noisy clients $S_n = S \backslash S_c$ are relabeled with the predicted labels of f_G. Similar to the correction process in the preprocessing stage, the same confidence threshold θ is used to control the subset of samples to be relabeled; see (9.7).

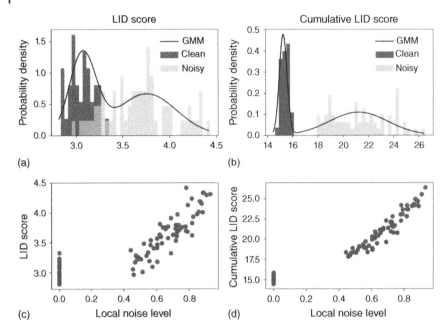

Figure 9.3 Empirical evaluation of LID score (a) and cumulative LID score (b) after 5 iterations on CIFAR-10 with noise model $(\rho, \tau) = (0.6, 0.5)$, and with IID data partition, over 100 clients. (c) Probability density function and estimated GMM; (d) LID/cumulative LID score vs. local noise level for each client.

9.3.4 Federated Usual Training Stage

In the final stage of `FedCorr`, the global model is trained over T_3 rounds via the usual FL (`FedAvg`) on all the clients, using the labels corrected in the previous two training stages. It should be highlighted that this `FedAvg` process could in principle be replaced by any state-of-the-art federated optimization methods. Later in Section 9.4.2, we will discuss how `FedCorr` can be well incorporated with three FL methods.

9.4 Experiments

In this section, we shall review experiments conducted in Xu et al. [2022] with respect to both IID (CIFAR-10/100 (Krizhevsky and Hinton [2009])) and non-IID (CIFAR-10, Clothing1M (Xiao et al. [2015])) data settings, at multiple noise levels. These experiments show that `FedCorr` is simultaneously robust to both local label quality discrepancy and data statistics discrepancy. The versatility of

FedCorr is also demonstrated: various FL methods can have their performances further improved by incorporating the first two stages of FedCorr. This section ends with an ablation study to show the effects of different components of FedCorr. Details on data partition and the noise model used have already been given in Section 9.3.1.

9.4.1 Experimental Setup

9.4.1.1 Baselines
There are two groups of experiments.

In the first group, FedCorr is demonstrated to be robust to discrepancies in both data statistics and label quality. For comparison, FedCorr is compared with the following state-of-the-art methods from three categories: (i) methods to tackle label noise in centralized learning (CL) (JointOpt (Tanaka et al. [2018]) and DivideMix (Li et al. [2020a])) applied to local clients; (ii) classic FL methods (FedAvg (McMahan et al. [2017]) and FedProx (Li et al. [2020b])); and (iii) FL methods designed to be robust to label noise (RoFL (Yang et al. [2022]) and ARFL (Fu et al. [2021])). For reference, experimental results on JointOpt and DivideMix in CL are also provided, so as to show the performance reduction of these two methods when used in FL.

In the second group, the versatility of FedCorr is demonstrated. The performance improvements of three state-of-the-art methods are examined in the context that the first two stages of FedCorr are incorporated. These methods are chosen from three different aspects to improve FL: local optimization (FedDyn (Acar et al. [2020])), aggregation (Median (Yin et al. [2018])), and client selection (PoC (Cho et al. [2020])).

9.4.1.2 Implementation Details
Different models and number of clients N are chosen for each dataset; see Table 9.1. For data preprocessing, normalization, and image augmentation using

Table 9.1 List of datasets used in our experiments.

Dataset	CIFAR-10	CIFAR-100	Clothing1M
Size of D_{train}	50,000	50,000	1,000,000
# of classes	10	100	14
# of clients	100	50	500
Fraction γ	0.1	0.1	0.02
Architecture	ResNet-18	ResNet-34	Pretrained ResNet-50

random horizontal flipping and random cropping with padding = 4 are performed. An stochastic gradient descent (SGD) local optimizer with a momentum of 0.5 is applied, with a batch size of 10 for CIFAR-10/100 and 16 for Clothing1M. With the exception of `JointOpt` and `DivideMix` used in FL settings, five local epochs are used across all experiments. For `FedCorr`, we always use the same hyperparameters on the same dataset. In particular, we use $T_1 = 5, 10$, and 2 for CIFAR-10, CIFAR-100, and Clothing1M, respectively. For fraction scheduling, the fraction $\gamma = \frac{1}{N}$ is used in the preprocessing stage, and then the fractions specified in Table 9.1 are used for the latter two stages.

9.4.2 Comparison with State-of-the-Art Methods

9.4.2.1 IID Settings

`FedCorr` is compared with multiple baselines at different noise levels, using the same configuration. Tables 9.2–9.4 show the results on CIFAR-10, CIFAR-100, and Clothing1M, respectively. Note that no synthetic label noise is added to Clothing1M, since it already contains real-world label noise. In summary, `FedCorr` achieves best test accuracies across all noise settings tested on both datasets, with particularly significant outperformance in the case of high noise levels. Note that `JointOpt` and `DivideMix` are implemented in both centralized and federated settings to show the performance reduction (10% ∼ 30% lower for best accuracy) when these CL methods are applied to local clients in FL. Furthermore, the accuracies in CL can also be regarded as upper bounds for the accuracies in FL. Remarkably, the accuracy gap between `DivideMix` in CL and `FedCorr` in FL is < 4% even in the extreme noise setting $(\rho, \tau) = (0.8, 0.5)$. In the centralized setting, the adopted dataset is corrupted with exactly the same scheme as in the federated setting. For the federated setting, the global model is warmed up for 20 rounds with `FedAvg` to avoid introducing additional label noise during the correction process in the early training stage, and we then apply `JointOpt` or `DivideMix` locally on each selected client, using 20 local training epochs. For Clothing1M, `FedCorr` also achieves the highest accuracy in FL, and this accuracy is even higher than the reported accuracy of `JointOpt` in CL.

9.4.2.2 Non-IID Settings

To evaluate `FedCorr` in more realistic heterogeneous data settings, experiments are conducted via using the non-IID settings as described in Section 9.3.1, over different values for (p, α_{Dir}). Table 9.5 shows the results on CIFAR-10, where `FedCorr` consistently outperforms all baselines by at least 7%.

Table 9.2 Average (five trials) and standard deviation of the best test accuracies of various methods on CIFAR-10 with IID setting at different noise levels (ρ: ratio of noisy clients, τ: lower bound of client noise level).

Method		Best test accuracy (%) ± Standard Deviation (%)						
		$\rho = 0.0$	$\rho = 0.4$		$\rho = 0.6$		$\rho = 0.8$	
		$\tau = 0.0$	$\tau = 0.0$	$\tau = 0.5$	$\tau = 0.0$	$\tau = 0.5$	$\tau = 0.0$	$\tau = 0.5$
CL	JointOpt	93.73±0.21	92.29±0.37	92.11±0.21	91.26±0.46	88.42±0.33	89.18±0.29	85.62±1.17
	DivideMix	95.64±0.05	96.39±0.09	96.17±0.05	96.07±0.06	94.59±0.09	94.21±0.27	94.36±0.16
FL	FedAvg	93.11±0.12	89.46±0.39	88.31±0.80	86.09±0.50	81.22±1.72	82.91±1.35	72.00±2.76
	FedProx	92.28±0.14	88.54±0.33	88.20±0.63	85.80±0.41	85.25±1.02	84.17±0.77	80.59±1.49
	RoFL	88.33±0.07	88.25±0.33	87.20±0.26	87.77±0.83	83.40±1.20	87.08±0.65	74.13±3.90
	ARFL	92.76±0.08	85.87±1.85	83.14±3.45	76.77±1.90	64.31±3.73	73.22±1.48	53.23±1.67
	JointOpt	88.16±0.18	84.42±0.70	83.01±0.88	80.82±1.19	74.09±1.43	76.13±1.15	66.16±1.71
	DivideMix	77.96±0.15	77.35±0.20	74.40±2.69	72.67±3.39	72.83±0.30	68.66±0.51	68.04±1.38
	Ours	**93.82±0.41**	**94.01±0.22**	**94.15±0.18**	**92.93±0.25**	**92.50±0.28**	**91.52±0.50**	**90.59±0.70**

FL and CL represent the federated settings and centralized setting (for reference only). The highest accuracy for each noise level is boldfaced.

Table 9.3 Average (five trials) and standard deviation of the best test accuracies on CIFAR-100 with IID setting.

Method	Best test accuracy (%) $\pm$ Standard Deviation(%)			
	$\rho = 0.0$ $\tau = 0.0$	$\rho = 0.4$ $\tau = 0.5$	$\rho = 0.6$ $\tau = 0.5$	$\rho = 0.8$ $\tau = 0.5$
JointOpt (CL)	72.94±0.43	65.87±1.50	60.55±0.64	59.79±2.45
DivideMix (CL)	75.58±0.14	75.43±0.34	72.26±0.58	71.02±0.65
FedAvg	72.41±0.18	64.41±1.79	53.51±2.85	44.45±2.86
FedProx	71.93±0.13	65.09±1.46	57.51±2.01	51.24±1.60
RoFL	67.89±0.65	59.42±2.69	46.24±3.59	36.65±3.36
ARFL	72.05±0.28	51.53±4.38	33.03±1.81	27.47±1.08
JointOpt	67.49±0.36	58.43±1.88	44.54±2.87	35.25±3.02
DivideMix	45.91±0.27	43.25±1.01	40.72±1.41	38.91±1.25
Ours	**72.56±2.07**	**74.43±0.72**	**66.78±4.65**	**59.10±5.12**

The bold values mean the highest accuracies for each noise level.

Table 9.4 Best test accuracies on Clothing1M with IID setting.

Settings	FedAvg	FedProx	RoFL	ARFL	JointOpt	Dividemix	Ours
FL	70.49	71.35	70.39	70.91	71.78	68.83	**72.55**
CL	–	–	–	–	72.23	74.76	–

CL results are the accuracies reported in corresponding papers.
The bold values mean the highest accuracies for each noise level.

Table 9.5 Average (five trials) and standard deviation of the best test accuracies of different methods on CIFAR-10 with different non-IID setting.

Method\\(ρ, α_{Dir})	$(0.7, 10)$	$(0.7, 1)$	$(0.3, 10)$
FedAvg	78.88±2.34	75.98±2.92	67.75±4.38
FedProx	83.32±0.98	80.40±0.94	73.86±2.41
RoFL	79.56±1.39	72.75±2.21	60.72±3.23
ARFL	60.19±3.33	55.86±3.30	45.78±2.84
JointOpt	72.19±1.59	66.92±1.89	58.08±2.18
DivideMix	65.70±0.35	61.68±0.56	56.67±1.73
Ours	**90.52±0.89**	**88.03±1.08**	**81.57±3.68**

The noise level is $(\rho, \tau) = (0.6, 0.5)$.

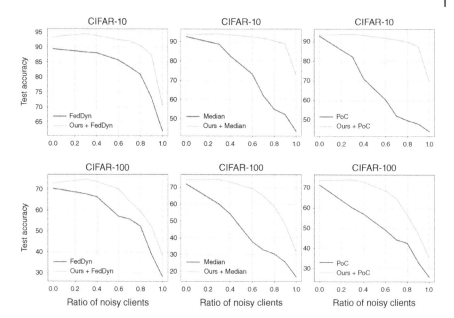

Figure 9.4 Best test accuracies of three FL methods combined with `FedCorr` on CIFAR-10/100 with multiple ρ and fixed $\tau = 0.5$.

9.4.2.3 Combination with Other FL Methods

The performance of three state-of-the-art methods are also investigated, when the first two stages of `FedCorr` are incorporated. As shown in Figure 9.4, significant accuracy improvements are consistently obtained on CIFAR-10/100 for various ratios of noisy clients.

9.4.2.4 Comparison of Communication Efficiency

The communication efficiency of different methods is also discussed. Here, given any implementation of an FL method, and any desired target accuracy ζ, its *targeted communication cost for ζ test accuracy* is defined to be the lowest total communication cost required (in the experiments) to reach the target ζ test accuracy. Informally, the lower the targeted communication cost, the higher the communication efficiency.

Table 9.6 shows the comparison of the communication efficiency on CIFAR-10, in terms of the targeted communication cost at test accuracies $\zeta = 80\%$. As the results show, `FedCorr` achieves improvements in communication efficiency, by a factor of at least 1.9 on CIFAR-10.

Table 9.6 A comparison of communication efficiency for different methods on CIFAR-10 with IID data partition, in terms of the targeted communication cost at $\zeta = 80\%$ test accuracy.

Method	$\rho = 0.0$ $\tau = 0.0$	$\rho = 0.4$ $\tau = 0.0$	$\rho = 0.4$ $\tau = 0.5$	$\rho = 0.6$ $\tau = 0.0$	$\rho = 0.6$ $\tau = 0.5$	$\rho = 0.8$ $\tau = 0.0$	$\rho = 0.8$ $\tau = 0.5$
Ours	150	210	230	230	330	360	510
FedAvg	370(2.6×)	450(2.1×)	470(2.0×)	550(2.4×)	930(2.8×)	810(2.3×)	—
FedProx	690(4.9×)	1050(5.0×)	1190(5.2×)	1230(5.3×)	1600(4.8×)	1730(4.8×)	4640(9.1×)
RoFL	990(7.1×)	1390(6.6×)	1580(6.9×)	1900(8.3×)	4200(12.7×)	2080(5.8×)	—
ARFL	290(2.1×)	740(3.5×)	1180(5.1×)	—	—	—	—
JointOpt	330(2.4×)	420(2.0×)	760(3.3×)	550(2.4×)	—	—	—
DivideMix	—	—	—	—	—	—	—

Values in brackets represent the ratios of the targeted communication costs as compared to our method FedCorr. Note that the test accuracies are evaluated after each communication round. In the case of methods and noise settings for which the target test accuracy ζ is not reached, we indicate '—'.

9.4.3 Ablation Study

Table 9.7 gives an overview of the effects of the components in FedCorr. Below, we consolidate some insights into what makes FedCorr successful:

- All components help to improve accuracy.
- Fraction scheduling has the largest effect. The small fraction used in the preprocessing stage helps to capture local data characteristics, as it avoids information loss brought by aggregation over multiple models.
- The highest accuracy among different noise levels is primarily achieved at a low noise level (e.g., $\rho = 0.4$) and not at the zero noise level, since additional label noise could be introduced during label correction.

9.5 Further Remarks

FedCorr is a general FL framework that jointly tackles the discrepancies in both local label quality and data statistics and that performs privacy-preserving label correction for identified noisy clients. Experiments have demonstrated the robustness and outperformance of FedCorr at multiple noise levels and diverse data settings. However, in its current formulation, FedCorr does not consider dynamic participation in FL, whereby clients can join or leave training at any time. New clients joining much later would always have relatively lower cumulative LID scores, which means new noisy clients could be categorized

Table 9.7 Ablation study results (average and standard deviation of five trials) on CIFAR-10.

| | ρ = 0.0 | ρ = 0.4 | | ρ = 0.6 | | ρ = 0.8 | |
| | Best test accuracy (%) ± Standard Deviation (%) | | | | | | |
Method	τ = 0.0	τ = 0.0	τ = 0.5	τ = 0.0	τ = 0.5	τ = 0.0	τ = 0.5
Ours	**93.82±0.41**	**94.01±0.22**	**94.15±0.18**	**92.93±0.25**	**92.50±0.28**	**91.52±0.50**	**90.59±0.70**
w/o correction	92.85±0.66	93.71±0.20	93.60±0.21	92.15±0.29	91.77±0.65	90.48±0.56	88.77±1.10
w/o frac. scheduling	86.05±1.47	85.59±1.10	78.44±7.90	80.29±2.62	77.96±3.65	76.67±3.48	72.71±5.03
w/o local proximal	93.37±0.05	93.64±0.15	93.46±0.17	92.34±0.14	91.74±0.47	90.45±0.94	88.74±1.72
w/o fine-tuning	92.71±0.18	93.06±0.15	92.62±0.28	91.41±0.14	89.31±0.90	89.62±0.40	83.81±2.59
w/o usual training	93.11±0.10	93.53±0.17	93.46±0.14	92.16±0.24	91.50±0.51	90.62±0.59	88.97±1.37
w/o mixup	90.63±0.70	88.83±1.88	91.34±0.39	87.79±0.89	87.50±1.33	87.86±0.53	83.29±1.78

The bold values mean the highest accuracies for each noise level.

incorrectly as clean clients. Thus, further work is required to handle dynamic participation.

Bibliography

Durmus Alp Emre Acar, Yue Zhao, Ramon Matas, Matthew Mattina, Paul Whatmough, and Venkatesh Saligrama. Federated learning based on dynamic regularization. In *International Conference on Learning Representations*, 2020.

Laurent Amsaleg, Oussama Chelly, Teddy Furon, Stéphane Girard, Michael E. Houle, Ken-ichi Kawarabayashi, and Michael Nett. Estimating local intrinsic dimensionality. In *Proceedings of the 21th ACM SIGKDD International Conference on Knowledge Discovery and Data Mining*, pages 29–38, 2015.

Eric Arazo, Diego Ortego, Paul Albert, Noel O'Connor, and Kevin McGuinness. Unsupervised label noise modeling and loss correction. In *International Conference on Machine Learning*, pages 312–321. PMLR, 2019.

Yiqiang Chen, Xiaodong Yang, Xin Qin, Han Yu, Biao Chen, and Zhiqi Shen. Focus: Dealing with label quality disparity in federated learning. In *International Workshop on Federated Learning for User Privacy and Data Confidentiality in Conjunction with IJCAI(FL-IJCAI'20)*, 2020. URL arXivpreprintarXiv:2001.11359.

Zihan Chen, Kai Fong Ernest Chong, and Tony QS Quek. Dynamic attention-based communication-efficient federated learning. In *International Workshop on Federated and Transfer Learning for Data Sparsity and Confidentiality in Conjunction with IJCAI (FTL-IJCAI'2021)*, 2021. URL arXivpreprintarXiv:2108.05765.

Yae Jee Cho, Jianyu Wang, and Gauri Joshi. Client selection in federated learning: Convergence analysis and power-of-choice selection strategies. arXiv preprint arXiv:2010.01243, 2020.

Shuhao Fu, Chulin Xie, Bo Li, and Qifeng Chen. Attack-resistant federated learning with residual-based reweighting. In *AAAI Workshop Towards Robust, Secure and Efficient Machine Learning*, 2021. URL arXivpreprintarXiv:1912.11464.

Bo Han, Quanming Yao, Xingrui Yu, Gang Niu, Miao Xu, Weihua Hu, Ivor W. Tsang, and Masashi Sugiyama. Co-teaching: Robust training of deep neural networks with extremely noisy labels. In *NeurIPS*, 2018.

Bo Han, Quanming Yao, Tongliang Liu, Gang Niu, Ivor W. Tsang, James T. Kwok, and Masashi Sugiyama. A survey of label-noise representation learning: Past, present and future. arXiv preprint arXiv:2011.04406, 2020.

Michael E. Houle. Dimensionality, discriminability, density and distance distributions. In *2013 IEEE 13th International Conference on Data Mining Workshops*, pages 468–473. IEEE, 2013.

Michael E. Houle. Local intrinsic dimensionality I: An extreme-value-theoretic foundation for similarity applications. In *International Conference on Similarity Search and Applications*, pages 64–79. Springer, 2017.

Tzu-Ming Harry Hsu, Hang Qi, and Matthew Brown. Measuring the effects of non-identical data distribution for federated visual classification. arXiv preprint arXiv:1909.06335, 2019.

Sohei Itahara, Takayuki Nishio, Yusuke Koda, Masahiro Morikura, and Koji Yamamoto. Distillation-based semi-supervised federated learning for communication-efficient collaborative training with non-IID private data. *IEEE Transactions on Mobile Computing*, 22(1):191–205, 2021.

Sai Praneeth Karimireddy, Satyen Kale, Mehryar Mohri, Sashank Reddi, Sebastian Stich, and Ananda Theertha Suresh. Scaffold: Stochastic controlled averaging for federated learning. In *International Conference on Machine Learning*, pages 5132–5143. PMLR, 2020.

Alex Krizhevsky and Geoffrey Hinton. Learning multiple layers of features from tiny images, 2009.

Xiang Li, Kaixuan Huang, Wenhao Yang, Shusen Wang, and Zhihua Zhang. On the convergence of FedAvg on non-IID data. arXiv preprint arXiv:1907.02189, 2019a.

Xiang Li, Kaixuan Huang, Wenhao Yang, Shusen Wang, and Zhihua Zhang. On the convergence of FedAvg on non-IID data. In *International Conference on Learning Representations*, 2019b.

Junnan Li, Richard Socher, and Steven C. H. Hoi. DivideMix: Learning with noisy labels as semi-supervised learning. In *International Conference on Learning Representations*, 2020a. URL https://openreview.net/forum?id=HJgExaVtwr.

Tian Li, Anit Kumar Sahu, Manzil Zaheer, Maziar Sanjabi, Ameet Talwalkar, and Virginia Smith. Federated optimization in heterogeneous networks. In *Proceedings of Machine Learning and Systems*, volume 2, pages 429–450, 2020b.

Tian Li, Shengyuan Hu, Ahmad Beirami, and Virginia Smith. Ditto: Fair and robust federated learning through personalization. In *International Conference on Machine Learning*, pages 6357–6368. PMLR, 2021.

Xingjun Ma, Bo Li, Yisen Wang, Sarah M. Erfani, Sudanthi Wijewickrema, Grant Schoenebeck, Dawn Song, Michael E. Houle, and James Bailey. Characterizing adversarial subspaces using local intrinsic dimensionality. In *International Conference on Learning Representations*, 2018a.

Xingjun Ma, Yisen Wang, Michael E. Houle, Shuo Zhou, Sarah Erfani, Shutao Xia, Sudanthi Wijewickrema, and James Bailey. Dimensionality-driven learning with noisy labels. In *International Conference on Machine Learning*, pages 3355–3364. PMLR, 2018b.

Brendan McMahan, Eider Moore, Daniel Ramage, Seth Hampson, and Blaise Aguera y Arcas. Communication-efficient learning of deep networks from decentralized data. In *Artificial Intelligence and Statistics*, pages 1273–1282. PMLR, 2017.

Sashank J. Reddi, Zachary Charles, Manzil Zaheer, Zachary Garrett, Keith Rush, Jakub Konečný, Sanjiv Kumar, and Hugh Brendan McMahan. Adaptive federated optimization. In *International Conference on Learning Representations*, 2021. URL https://openreview.net/forum?id=LkFG3lB13U5.

Sebastian U. Stich. Local SGD converges fast and communicates little. arXiv preprint arXiv:1805.09767, 2018.

Daiki Tanaka, Daiki Ikami, Toshihiko Yamasaki, and Kiyoharu Aizawa. Joint optimization framework for learning with noisy labels. In *Proceedings of the IEEE Conference on Computer Vision and Pattern Recognition*, pages 5552–5560, 2018.

Ching Pui Wan and Qifeng Chen. Robust federated learning with attack-adaptive aggregation. In *International Workshop on Federated and Transfer Learning for Data Sparsity and Confidentiality in Conjunction with IJCAI (FTL-IJCAI'2021)*, 2021. URL arXivpreprintarXiv:2102.05257.

Jianyu Wang, Qinghua Liu, Hao Liang, Gauri Joshi, and H. Vincent Poor. Tackling the objective inconsistency problem in heterogeneous federated optimization. In *Advances in Neural Information Processing Systems*, 2020.

Xiaobo Xia, Tongliang Liu, Bo Han, Nannan Wang, Mingming Gong, Haifeng Liu, Gang Niu, Dacheng Tao, and Masashi Sugiyama. Part-dependent label noise: Towards instance-dependent label noise. In *NeurIPS*, 33, 2020.

Tong Xiao, Tian Xia, Yi Yang, Chang Huang, and Xiaogang Wang. Learning from massive noisy labeled data for image classification. In *Proceedings of the IEEE Conference on Computer Vision and Pattern Recognition*, pages 2691–2699, 2015.

Xinyi Xu and Lingjuan Lyu. A reputation mechanism is all you need: Collaborative fairness and adversarial robustness in federated learning. In *International Workshop on Federated Learning for User Privacy and Data Confidentiality in Conjunction with ICML(FL-ICML'21)*, 2021.

Jingyi Xu, Tony Q. S. Quek, and Kai Fong Ernest Chong. Training classifiers that are universally robust to all label noise levels. In *2021 International Joint Conference on Neural Networks (IJCNN)*, pages 1–8. IEEE, 2021.

Jingyi Xu, Zihan Chen, Tony Q. S. Quek, and Kai Fong Ernest Chong. FedCorr: Multi-stage federated learning for label noise correction. In *IEEE Conference on Computer Vision and Pattern Recognition (CVPR)*, 2022.

Seunghan Yang, Hyoungseob Park, Junyoung Byun, and Changick Kim. Robust federated learning with noisy labels. *IEEE Intelligent Systems*, 37(2):35–43, 2022.

Dong Yin, Yudong Chen, Ramchandran Kannan, and Peter Bartlett. Byzantine-robust distributed learning: Towards optimal statistical rates. In Jennifer Dy and Andreas Krause, editors, *Proceedings of the 35th International Conference on Machine*

Learning, volume 80 of Proceedings of Machine Learning Research, pages 5650–5659. PMLR, 2018. URL https://proceedings.mlr.press/v80/yin18a.html.

Xingrui Yu, Bo Han, Jiangchao Yao, Gang Niu, Ivor Tsang, and Masashi Sugiyama. How does disagreement help generalization against label corruption? In *ICML*, pages 7164–7173. PMLR, 2019.

Hongyi Zhang, Moustapha Cisse, Yann N. Dauphin, and David Lopez-Paz. mixup: Beyond empirical risk minimization. In *International Conference on Learning Representations*, 2018. URL https://openreview.net/forum?id=r1Ddp1-Rb.

10

Analog Over-the-Air Federated Learning: Design and Analysis

Howard H. Yang[1], Zihan Chen[2], and Tony Q. S. Quek[2]

[1] *ZJU-UIUC Institute, Zhejiang University, Zhejiang Province, Haining, China*
[2] *Information System Technology and Design Pillar, Singapore University of Technology and Design, Singapore*

10.1 Introduction

Distributed optimizations in wireless networks have garnered considerable attention in recent years, especially with the rise of federated learning (FL) (McMahan et al. [2017], Li et al. [2020], Park et al. [2019]). The typical system is constituted of an edge server and a number of agents. The goal is to collaboratively optimize an objective function via the orchestration among the network elements. Particularly, each agent conducts on-device training based on its local dataset and uploads the intermediate result, e.g., the gradient, to the server for model improvement. Then, they download the new model for another round of local computing. This procedure repeats multiple rounds until the training converges. Upon each global iteration, the transmissions of model parameters need to go over the spectrum, which is resource limited and unreliable. Recognizing the conventional schemes that hinged on the separated communication-and-computation principle can encounter difficulty in accommodating massive access and stringent latency requirements, a recent line of studies (Zhu et al. [2020]) proposed utilizing the over-the-air computing to enable efficient model aggregation and hence achieve faster machine learning over many devices.

The essence of over-the-air computing is to exploit the waveform superposition property of multiaccess channels, where agents modulate the gradient on the waveform and use the air as an auto aggregator. In the presence of channel fading, it is suggested to invert the channel via power control at the end-user devices where the nodes that encounter deep fades suspend their transmissions (Zhu et al. [2019], Yang et al. [2020]). And the server shall adopt better scheduling methods in each communication round to rev up the model training process. To reduce

Federated Learning for Future Intelligent Wireless Networks, First Edition.
Edited by Yao Sun, Chaoqun You, Gang Feng, and Lei Zhang.

communication overheads, the devices can compress the gradient vectors by sending out a sparse (Amiri and Gündüz [2020]), or even a one-bit quantized (Zhu et al. [2020]), version, followed by quadrature amplitude modulation (QAM) modulation. At the edge server side, it can expand the antenna array to further mitigate the effects of channel fading, where the fading vanishes as the spatial dimension approaches infinity (Amiria et al. [Early Access]). Furthermore, Sery et al. [2020] devise a precoding scheme that gradually amplifies the model updates as the training progresses to handle the performance degradation incurred by the additive noise. With the help of feedback, Guo et al. [2021] optimize the transceiver parameters by jointly accounting for the data and channel states to cope with the nonstationarity of the gradient updates.

Inspired by the fact that machine learning algorithms need not operate under impeccably precise parameters, the authors of Sery and Cohen [2020] suggest the agents directly transmit the analog gradient signals without any power control or beamforming to invert the channel while the server updates the global model based on the noisy gradient. They also show that convergence is guaranteed. This approach substantially simplifies the system design while achieving virtually zero access latency Cai and Lau [2018]. What is more appealing, the data privacy is in fact enhanced by implicitly harnessing the randomness of wireless medium and the training procedure can be accelerated by adopting an analog alternating direction methods of multipliers (ADMM)-type algorithm Elgabli et al. [2021]. Despite the wealth of work in this area, a significant restriction in almost all the previous results lies at the presumption that the interference follows a normal distribution. While convenient, this assumption hardly holds in practice as the constructive property of the electromagnetic waves often results in heavy tails in the distribution of interference (Middleton [1977], Haenggi and Ganti [2009], Win et al. [2009]). In consequence, there is a nonnegligible chance that the magnitude of interference sheers to a humungous value in some communication rounds, which wreaks havoc on the global model. Understanding the impact of such a phenomenon on the performance of the learning algorithm is the focus of this chapter.

The present chapter leverages the theoretical model proposed in Sery and Cohen [2020] but differs from it by considering the heavy-tailed nature of the electromagnetic interference (Middleton [1977]). Specifically, we adopt the symmetric α-stable distribution – a widely used model in wireless networks (Haenggi and Ganti [2009], Win et al. [2009]) – to model the statistics of interference. The parameter α is commonly known as the *tail index* where smaller the α means heavier the tail in the distribution. Under such a setting, the aggregated global gradient admits diverging variance, and the conventional approaches that heavily rely on the existence of second moments for convergence analysis fail to function. In that respect, we take a new route toward the convergence analysis and verify that even though the intermediate gradients are severely distorted by channel fading and interference, GD-based methods can ultimately reach the optimal solution. We also provide simulations to validate the theoretical findings.

10.2 System Model

Let us consider an edge learning system consisting of one server and N agents. Each agent n holds a local dataset $D_n = \{(x_i \in \mathbb{R}^d, y_i \in \mathbb{R})\}_{i=1}^{m_n}$ with size $|D_n| = m_n$, and we assume the local datasets are statistically independent across the clients. The server is endowed with an access point that connects to the agents via spectrum. The goal of all the entities in this system is to jointly learn a statistical model constituted from all the data samples of the clients. More precisely, they need to find a vector $w \in \mathbb{R}^d$ that minimizes a global loss function, given as

$$f(w) = \frac{1}{N}\sum_{n=1}^{N} f_n(w), \tag{10.1}$$

where $f_n(w)$ is the local empirical risk of agent n, given by

$$f_n(w) = \frac{1}{m_n}\sum_{i=1}^{m_n} \ell(w; x_i, y_i). \tag{10.2}$$

The solution is commonly known as the empirical risk minimizer, denoted by

$$w^* = \arg\min f(w). \tag{10.3}$$

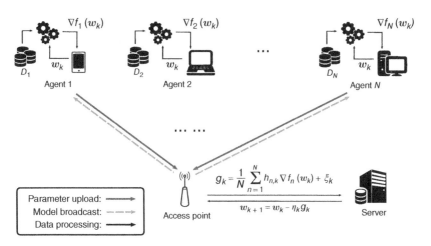

Figure 10.1 An illustration of the model training process. In each round, agents evaluate local gradients based on their own dataset and upload to the server via analog transmissions. The server receives the aggregated noisy gradient to update the global model. Then the new model is sent back to the agents, and the process is repeated.

Due to privacy concerns, the agents do not disclose their local dataset. Hence, the minimization of (10.1) needs to be accomplished by means of distributed learning. Particularly, the agents minimize their local loss and upload the intermediate gradients to the server, with which the server performs a global aggregation to improve the model and feeds it back to the agents for the next round of local training, as shown in Figure 10.1. Such interactions between the server and agents repeat until the model converges. As the communications among the server and the agents are taken place over the spectrum, which is by nature resource limited and unreliable. By virtue of its efficacy in spectral utilization, we adopt the analog over-the-air computing (Nazer and Gastpar [2007]) for the training of the statistical model. We term such a framework analog over-the-air federated learning (AirFL) and detail its implementation in the sequel.

10.3 Analog Over-the-Air Model Training

Let us consider the global training has progressed to the kth round and denote by $\boldsymbol{w}_k$ the global model broadcasted by the server. Because of the access point's high transmit power, we assume the global model is successfully received by all the agents. Then, each client n updates its gradient $\nabla f_n(\boldsymbol{w}_k)$ based on $\boldsymbol{w}_k$ and constructs the following analog signal:

$$x_n(t) = \langle \mathbf{s}(t), \nabla f_n(\boldsymbol{w}_k) \rangle, \tag{10.4}$$

where $\langle \boldsymbol{u}, \boldsymbol{v} \rangle$ represents the inner product between two vectors $\boldsymbol{u}$ and $\boldsymbol{v}$; and $\mathbf{s}(t) = (s_1(t), s_2(t), \dots, s_d(t))$, $0 < t < T$, is a set of orthonormal baseband waveforms that satisfies:

$$\int_0^T s_i^2(t)\, dt = 1, \quad i = 1, 2, \dots, d, \tag{10.5}$$

$$\int_0^T s_i(t) s_j(t) = 0, \quad \text{if } i \neq j. \tag{10.6}$$

According to (10.4), the signal $x_n(t)$ is a superposition of the analog waveforms whereas the magnitude of $s_i(t)$ equals to the ith element of $\nabla f_n(\boldsymbol{w}_k)$.

The agents transmit the analog waveforms $\{x_n(t)\}_{n=1}^N$ concurrently into the air once they have been assembled. And the signal received at the access point can be expressed as follows:

$$y(t) = \sum_{n=1}^N h_{n,t} P_n x_n(t) + \xi(t), \tag{10.7}$$

where $h_{n,t}$ is the channel fading experienced by agent n, P_n stands for the corresponding transmit power, and $\xi(t)$ represents the interference. Without loss of generality, we assume the channel fading is independent and identically distributed (i.i.d.) across the agents and communication rounds, with mean μ and variance σ^2. We further assume the transmit power is set to compensate for the large-scale path loss. In order to characterize the heavy-tailed nature of wireless interference, we consider $\xi(t)$ follows a symmetric α-stable distribution.

The edge server passes the received signal $y(t)$ to a set of matched filters, where filter i is tuned in accordance to $s_i(t)$. At the output, it obtains the following vector:

$$\boldsymbol{g}_k = \frac{1}{N}\sum_{n=1}^{N}h_{n,k}\nabla f_n(\boldsymbol{w}_k) + \boldsymbol{\xi}_k, \qquad (10.8)$$

where $\boldsymbol{\xi}_k$ is a d-dimensional random vector with each entry being i.i.d. and following an α-stable distribution. The server then updates global parameter as follows:

$$\boldsymbol{w}_{k+1} = \boldsymbol{w}_k - \eta_k\boldsymbol{g}_k, \qquad (10.9)$$

where η_k is the learning rate. The updated global parameter is sent back to the agents for the next round of local computing.

10.3.1 Salient Features

It is worthwhile to highlight the following properties of AirFL:

- *High spectral efficiency*: Compared with the conventional digital communication-based FL that needs to select a portion of the agents for parameter updating in each communication round, AirFL significantly improves the spectral utilization by allowing all the agents to simultaneously access the spectrum and upload their parameters to the edge server, regardless of the total number of agents in the network. Consequently, the bandwidth no longer constrains the number of participants, expediting the large-scale deployment of edge learning systems.
- *Low communication cost*: AirFL can be realized via elementary communication techniques such as amplitude modulation. Besides, the agents are also disburdened with estimating the instantaneous channel state information before each global transmission. Moreover, as opposed to digital communication-based FL, an increase in the number of agents in the network can improve the system's energy efficiency, hence allowing the agents to reduce their transmit power (Sery and Cohen [2020]). These salient features make it possible to establish an intelligent edge system over massively distributed agents with low-cost communication modules for cognitive network management.
- *Improved privacy protection*: The channel fading and interference noise induced by analog over-the-air computations impose a random perturbation to each agent's uploaded parameter. As a result, even if there are eavesdroppers in

the network, no original information can be recovered from the aggregated gradients. Such an enhancement to privacy protection in network management is particularly relevant in privacy-aware edge learning systems.

- *Reduced access latency*: By virtue of analog over-the-air computing, the local updates would be automatically aggregated at the edge server's output. This inherent integration of communication and computation significantly reduces the access, as well as processing latency, since the system no longer needs to go through the encoding (resp. decoding) and modulation (resp. demodulation) processes to obtain the individual gradients of each agent before adding them up.

10.3.2 Heavy-Tailed Interference

The spectrum is by nature a shared medium. Therefore, signals sent over the wireless channels inevitably suffer interference from the other concurrent transmitters. And it has been amply demonstrated from both theoretical (Middleton [1977]) and empirical (Clavier et al. [2021]) perspectives that electromagnetic interference generally obeys a heavy-tailed distribution. In that respect, we adopt the symmetric α-stable distribution to model the statistics of interference $\xi(t)$.

Definition 10.1 The random variable $\xi(t)$ is said to follow a symmetric α-stable distribution if its characteristic function takes the following form:

$$\mathbb{E}\left[e^{j\omega\xi(t)}\right] = \exp(-\delta^\alpha|\omega|^\alpha), \tag{10.10}$$

where $\delta > 0$ and $\alpha \in (0, 2)$. The parameters δ and α are commonly known as the scale parameter and tail index, respectively.

It is important to note that generally, α-stable distributions do not possess an explicit form of the probability density function, aside from two special cases, i.e., if $\alpha = 1$, the distribution reduces to Cauchy and when $\alpha = 2$, it reduces to Gaussian. The tail index α determines the heaviness of tail in the probability density function of $\xi(t)$. Particularly, as depicted in Figure 10.2, smaller the α, thicker the tail in the distribution, which implies the random variable $\xi(t)$ has a higher chance to attain a very large value. More theoretically, at any time instance t, the probability that the magnitude of the analog signal exceeds a large value is given by $\mathbb{P}(|\sin(t) + \xi(t)| > \lambda) \sim \lambda^{-\alpha}$, where $\lambda \gg 1$ (Samorodnitsky and Taqqu [2017]). This implies that a smaller tail index of $\xi(t)$ results in a higher probability of observing severe fluctuations in the analog signal magnitude.

Actually, α-stable random variables $\xi(t)$ have finite moments only up to the order α, i.e., $\mathbb{E}[|\xi(t)|^p] < \infty$ only in the range of $0 \leq p \leq \alpha$. In the context of wireless communications, we consider $1 < \alpha \leq 2$, i.e., the interference $\xi(t)$ has zero mean but (potentially) infinite variance. This, together with the random channel fading,

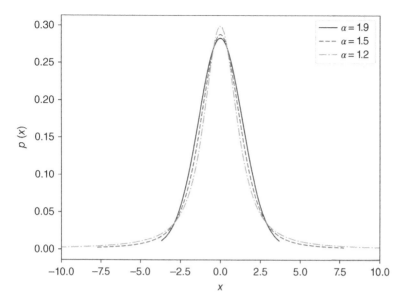

Figure 10.2 The examples of the α-stable random variables. (a) Plots the probability density function under different values of the tail index. (b) Illustrates the effects of heavy-tailed noise imposed on three different sinusoid signals.

makes the aggregated gradient $\boldsymbol{g}_k$ severely distorted. A natural question then arises as: *Does analog over-the-air machine learning algorithms converge?*

We give an affirmative answer in Section 10.4.

10.4 Convergence Analysis

This section constitutes the main technical part of our chapter, in which we derive analytical expressions for the convergence rate of the model training algorithm.

10.4.1 Preliminaries

Because the second moment of the aggregated gradient may not exist, in lieu of the conventially used $L-2$ norm (a.k.a. the Euclidean norm), we opt for the α moment as an alternative distance metric. Correspondingly, we need the concepts of *signed power* and α-*positive definite matrix* (Wang et al. [2021]), respectively, for further mathematical manipulations.

Definition 10.2 For a vector $\boldsymbol{w} = \left(w_1, \ldots, w_d\right)^\top \in \mathbb{R}^d$, we define its signed power as follows:

$$\boldsymbol{w}^{\langle \alpha \rangle} = \left(\operatorname{sgn}\left(w_1\right) |w_1|^\alpha, \ldots, \operatorname{sgn}\left(w_d\right) |w_d|^\alpha\right)^\top, \tag{10.11}$$

where $\operatorname{sgn}(x) \in \{-1, +1\}$ takes the sign of the variable x.

Definition 10.3 A symmetric matrix Q is said to be α-positive definite if $\langle v, Qv^{\langle\alpha-1\rangle}\rangle > 0$ for all $v \in \mathbb{R}^d$ with $\|v\|_\alpha > 1$.

To facilitate the analysis, we make the following assumptions.

Assumption 10.1 *The objective function $f : \mathbb{R}^d \to \mathbb{R}$ is γ-strongly convex, i.e., for any $w, v \in \mathbb{R}^d$ it is satisfied:*

$$f(w) \geq f(v) + \langle \nabla f(v), w - v\rangle + \frac{\gamma}{2}\|w - v\|^2. \tag{10.12}$$

Assumption 10.2 *The objective function $f : \mathbb{R}^d \to \mathbb{R}$ is λ-smooth, i.e., for any $w, v \in \mathbb{R}^d$ it is satisfied:*

$$f(w) \leq f(v) + \langle \nabla f(v), w - v\rangle + \frac{\lambda}{2}\|w - v\|^2. \tag{10.13}$$

Assumption 10.3 *For any given vector $w \in \mathbb{R}^d$, the Hessian matrix of $f(w)$, i.e., $\nabla^2 f(w)$, is α-positive definite.*

Since the magnitude of the transmitted waveforms cannot be arbitrarily large, we further assume the gradients of each agent is bounded, i.e., $\|\nabla f_n(w)\| \leq G, \forall n \in \{1, \ldots, N\}$, for some constant G. Furthermore, because each element of ξ_k has a finite α moment, we consider the α moment of ξ_k is upper bounded by a constant C, i.e., $\mathbb{E}\left[\|\xi_k\|_\alpha^\alpha\right] \leq C$.

10.4.2 Convergence Rate of AirFL

Armed with the notion of signed power, we can establish a Taylor expansion-type inequality for the α norm vectors.

Lemma 10.1 *Given $1 < \alpha \leq 2$, for any $w, v \in \mathbb{R}^d$, the following holds:*

$$\|w + v\|_\alpha^\alpha \leq \|w\|_\alpha^\alpha + \alpha\langle w^{\langle\alpha-1\rangle}, v\rangle + 4\|v\|_\alpha^\alpha. \tag{10.14}$$

Proof: Please refer to Krasulina [1969]. □

Furthermore, we need the following two results for technical operations.

Lemma 10.2 *Let Q be an α-positive definite matrix, for $\alpha \in [1, 2]$, there exists $\kappa, L > 0$, such that*

$$\|I - kQ\|_\alpha^\alpha \leq 1 - Lk, \qquad \forall k \in [0, \kappa]. \tag{10.15}$$

Proof: Please see Theorem 10 of Wang et al. [2021]. □

Lemma 10.3 *For a sequence of real numbers* $\{b_k\}$, $k \geq 1$, *that satisfies:*

$$b_{k+1} \leq \left(1 - \frac{c}{k}\right) b_k + \frac{c_1}{k^{p+1}}, \tag{10.16}$$

where $c > p > 0$, $c_1 > 0$. *The following relationship holds*

$$b_k \leq \frac{c_1}{c-p} \cdot \frac{1}{k^p} + o\left(\frac{1}{k^{p+1}} + \frac{1}{k^c}\right), \tag{10.17}$$

where $o\,(\cdot)$ *is the "little o" notation, meaning that if* $f\,(k) = o\,(g\,(k))$ *then* $\forall B > 0$, *there exists* k_0 *such that* $f\,(k) \leq Bg\,(k)$ *for all* $k \geq k_0$.

Proof: Please see Lemma 10.1 of Chung [1954]. □

We are now in position to present the main theoretical finding of this chapter.

Theorem 10.1   Under the employed edge learning system, if the learning rate is set as $\eta_k = \theta/k$ where $\theta > \frac{\alpha-1}{\mu L}$, then the algorithm converges as

$$\mathbb{E}\left[\|\mathbf{w}_k - \mathbf{w}^*\|_\alpha^\alpha\right] \leq \frac{4\theta^\alpha \left(C + \frac{\sigma^\alpha G^\alpha d^{1-\frac{1}{\alpha}}}{N^{\alpha/2}}\right)}{\mu\theta L - \alpha + 1} \cdot \frac{1}{k^{\alpha-1}}. \tag{10.18}$$

Proof: For ease of exposition, let us denote by $\Delta_k = \mathbf{w}_k - \mathbf{w}^*$. Then, in a specific communication round k, we can leverage (10.4) and (10.9) to write the update of global parameter $\mathbf{w}_k$ as follows:

$$
\begin{aligned}
\|\Delta_{k+1}\|_\alpha^\alpha &= \left\| \mathbf{w}_k - \mathbf{w}^* - \eta_k \left(\frac{1}{N}\sum_{n=1}^N h_{n,k}\nabla f_n\left(\mathbf{w}_k\right) + \xi_k\right) \right\|_\alpha^\alpha \\
&\overset{(a)}{\leq} \left\| \Delta_k - \frac{\eta_k}{N}\sum_{n=1}^N h_{n,k}\nabla f_n\left(\mathbf{w}_k\right) \right\|_\alpha^\alpha + 4\eta_k^\alpha \|\xi_k\|_\alpha^\alpha \\
&\quad + \alpha \left\langle \left(\Delta_k - \frac{\eta_k}{N}\sum_{n=1}^N h_{n,k}\nabla f_n\left(\mathbf{w}_k\right)\right)^{\langle\alpha-1\rangle}, \eta_k\xi_k\right\rangle,
\end{aligned}
\tag{10.19}
$$

where (a) follows from Lemma 10.1.

By noticing that each entry of ξ_t is independent and has a zero mean, we can take an expectation on both sides of (10.19) and arrive at the following:

$$
\begin{aligned}
\mathbb{E}\left[\|\Delta_{k+1}\|_\alpha^\alpha\right] &\leq \mathbb{E}\left[\left\| \Delta_k - \frac{\eta_k}{N}\sum_{n=1}^N h_{n,k}\nabla f_n\left(\mathbf{w}_k\right) \right\|_\alpha^\alpha\right] + 4\eta_k^\alpha \mathbb{E}\left[\|\xi_k\|_\alpha^\alpha\right] \\
&\leq \mathbb{E}\left[\left\| \Delta_k - \frac{\eta_k}{N}\sum_{n=1}^N h_{n,k}\nabla f_n\left(\mathbf{w}_k\right) \right\|_\alpha^\alpha\right] + 4C\eta_k^\alpha.
\end{aligned}
\tag{10.20}
$$

The first term on the right-hand side of (10.20) can be further bounded as follows:

$$
\mathbb{E}\left[\left\|\Delta_k - \frac{\eta_k}{N}\sum_{n=1}^{N}h_{n,k}\nabla f_n\left(\mathbf{w}_k\right)\right\|_{\alpha}^{\alpha}\right]
$$

$$
= \mathbb{E}\left[\left\|\Delta_k - \mu\eta_k\nabla f\left(\mathbf{w}_k\right) - \frac{\eta_k}{N}\sum_{n=1}^{N}\left(h_{n,k}-\mu\right)\nabla f_n\left(\mathbf{w}_k\right)\right\|_{\alpha}^{\alpha}\right]
$$

$$
\leq \mathbb{E}\left[\left\|\Delta_k - \mu\eta_k\nabla f\left(\mathbf{w}_k\right)\right\|_{\alpha}^{\alpha}\right] + \frac{4\eta_k^{\alpha}}{N^{\alpha}}\mathbb{E}\left[\left\|\sum_{n=1}^{N}\left(h_{n,k}-\mu\right)\nabla f_n\left(\mathbf{w}_k\right)\right\|_{\alpha}^{\alpha}\right]
$$

$$
+ \alpha\mathbb{E}\left[\left\langle\left(\Delta_k - \mu\eta_k\nabla f\left(\mathbf{w}_k\right)\right)^{\langle\alpha-1\rangle}, \frac{\eta_k}{N}\sum_{n=1}^{N}\left(h_{n,k}-\mu\right)\nabla f_n\left(\mathbf{w}_k\right)\right\rangle\right]
$$

$$
\overset{(a)}{=} \underbrace{\mathbb{E}\left[\left\|\Delta_k - \mu\eta_k\nabla f\left(\mathbf{w}_k\right)\right\|_{\alpha}^{\alpha}\right]}_{Q_1} + \underbrace{\frac{4\eta_k^{\alpha}}{N^{\alpha}}\mathbb{E}\left[\left\|\sum_{n=1}^{N}\left(h_{n,k}-\mu\right)\nabla f_n\left(\mathbf{w}_k\right)\right\|_{\alpha}^{\alpha}\right]}_{Q_2}, \quad (10.21)
$$

where (a) follows from the fact that $\{h_{n,k}\}_{n=1}^{N}$ are i.i.d. and satisfy $\mathbb{E}\left[h_{n,k}\right] = \mu$, $n = 1, \ldots, N$.

For the empirical risk minimizer $\mathbf{w}^*$, we have $\nabla f\left(\mathbf{w}^*\right) = 0$. Hence, Q_1 can be expanded as follows:

$$
Q_1 = \mathbb{E}\left[\left\|\Delta_k - \mu\eta_k\left[\nabla f\left(\mathbf{w}_k\right) - \nabla f\left(\mathbf{w}^*\right)\right]\right\|_{\alpha}^{\alpha}\right]
$$

$$
\overset{(a)}{=} \mathbb{E}\left[\left\|\Delta_k - \mu\eta_k\nabla^2 f\left(\mathbf{w}_k^{\sharp}\right)\Delta_k\right\|_{\alpha}^{\alpha}\right]
$$

$$
\leq \left\|\mathbf{I}_d - \mu\eta_k\nabla^2 f\left(\mathbf{w}_k^{\sharp}\right)\right\|_{\alpha}^{\alpha} \times \mathbb{E}\left[\left\|\Delta_k\right\|_{\alpha}^{\alpha}\right], \quad (10.22)
$$

where (a) follows from the mid value theorem, and $\mathbf{I}_d$ denotes a $d \times d$ identity matrix. Furthermore, by the property, we have

$$
\left\|\mathbf{I}_d - \mu\eta_k\nabla^2 f\left(\mathbf{w}_k^{\sharp}\right)\right\|_{\alpha}^{\alpha} \leq \left(1 - \eta_k\mu L\right). \quad (10.23)
$$

Therefore, Q_1 can be bounded as

$$
Q_1 \leq \left(1 - \eta_k\mu L\right)\mathbb{E}\left[\left\|\Delta_k\right\|_{\alpha}^{\alpha}\right]. \quad (10.24)
$$

On the other hand, Q_2 can be bounded via the following:

$$
Q_2 = \mathbb{E}\left[\left\|\sum_{n=1}^{N}\left(h_{n,k}-\mu\right)\nabla f_n\left(\mathbf{w}_k\right)\right\|_{\alpha}^{\alpha}\right]
$$

$$
\overset{(a)}{\leq} d^{1-\frac{1}{\alpha}} \cdot \mathbb{E}\left[\left(\left\|\sum_{n=1}^{N}\left(h_{n,k}-\mu\right)\nabla f_n\left(\mathbf{w}_k\right)\right\|_{2}^{2}\right)^{\frac{\alpha}{2}}\right]
$$

$$
\overset{(b)}{\leq} d^{1-\frac{1}{\alpha}} \cdot \left(\mathbb{E} \left[\left\| \sum_{n=1}^{N} \left(h_{n,k} - \mu \right) \nabla f_n \left(\boldsymbol{w}_k \right) \right\|_2^2 \right] \right)^{\frac{\alpha}{2}}
$$

$$
= d^{1-\frac{1}{\alpha}} \cdot \left(\mathbb{E} \left[\sum_{n,m=1}^{N} \left\langle \left(h_{n,k} - \mu \right) \nabla f_n \left(\boldsymbol{w}_k \right), \left(h_{m,k} - \mu \right) \nabla f_m \left(\boldsymbol{w}_k \right) \right\rangle \right] \right)^{\frac{\alpha}{2}}
$$

$$
= d^{1-\frac{1}{\alpha}} \cdot \left(\sum_{n=1}^{N} \mathbb{E} \left[\left(h_{n,k} - \mu \right)^2 \cdot \left\| \nabla f_n \left(\boldsymbol{w}_k \right) \right\|_2^2 \right] \right)^{\frac{\alpha}{2}}
$$

$$
\leq d^{1-\frac{1}{\alpha}} \cdot \left(N\sigma^2 G^2 \right)^{\frac{\alpha}{2}} = d^{1-\frac{1}{\alpha}} \cdot N^{\frac{\alpha}{2}} \sigma^\alpha G^\alpha, \tag{10.25}
$$

where (a) and (b) follow from the Holder's inequality and Jensen's inequality, respectively. To this end, by substituting (10.24) and (10.25) into (10.20), we have

$$
\mathbb{E} \left[\left\| \Delta_{k+1} \right\|_\alpha^\alpha \right] \leq \left(1 - \eta_k \mu L \right) \mathbb{E} \left[\left\| \Delta_k \right\|_\alpha^\alpha \right] + 4 \left(C + \frac{\sigma^\alpha G^\alpha d^{1-\frac{1}{\alpha}}}{N^{\alpha/2}} \right) \eta_k^\alpha, \tag{10.26}
$$

and the proof is completed by invoking Lemma 10.2 and removing the higher order terms as they become infinitesmall as k goes large. $\square$

Notably, the learning rate is also amenable, and the convergence rate can be characterized accordingly.

Corollary 10.1 *Under the employed edge learning system, if the learning rate is set as $\eta_k \asymp k^{-\rho}$ with $\rho \in (0, 1)$, the training algorithm converges as*

$$
\mathbb{E} \left[\left\| \boldsymbol{w}_k - \boldsymbol{w}^* \right\|_\alpha^\alpha \right] = \mathcal{O} \left(k^{-\rho(\alpha-1)} \right). \tag{10.27}
$$

Proof: In order to prove this corollary, let us first recall the following lemma.

(Lemma 10.4, Fabian [1967]) *For a sequence of real numbers $\{b_k\}$, $k \geq 1$, that satisfies the following recursion:*

$$
b_{k+1} = \left(1 - \frac{c}{k^p} \right) b_k + \frac{c_1}{k^{p+q}}, \tag{10.28}
$$

where $0 < p < 1$ and $c, c_1, q > 0$. We have $b_k = \mathcal{O}(k^{-q})$.

Then, if we assign $\eta_k = k^{-\rho}$ with $\rho \in (0, 1)$ in (10.26), it yields

$$
\mathbb{E} \left[\left\| \Delta_{k+1} \right\|_\alpha^\alpha \right] \leq \left(1 - \frac{\mu L}{k^\rho} \right) \mathbb{E} \left[\left\| \Delta_k \right\|_\alpha^\alpha \right] + 4 \left(C + \frac{\sigma^\alpha G^\alpha d^{1-\frac{1}{\alpha}}}{N^{\alpha/2}} \right) \frac{1}{k^{\alpha\rho}}, \tag{10.29}
$$

and the result follows by invoking Lemma 10.4. $\square$

From Corollary 10.1, we can see that the learning rate has a direct impact on the convergence property of the algorithm. Specifically, reducing the learning rate, as we decreases ρ, leads to a slowdown in convergence of the algorithm. In fact, the analog over-the-air GD converges even for very slowly decaying learning rate with ρ being close to 0.

10.4.3 Key Observations

Following the derivations above, let us enlist a few essential observations in the following.

Observation 1: *Although the aggregated gradient is corrupted by random channel fading and heavy-tailed interference with infinite variance, GD-based model training with a diminishing learning rate can converge to the global optimum without any modification, neither to the loss function nor the algorithm itself. In other words, GD-based algorithms are resilient to parameter distortions and hence are particularly suitable for edge learning systems.*

Observation 2: *The convergence rate is the order of $\mathcal{O}\left(\frac{1}{k^{\alpha-1}}\right)$, which is dominated by the tail index α. Specifically, a small α leads to a heavy tail in the distribution of interference, and that results in a slow convergence of the learning algorithm.*

Observation 3: *The tail index, α, also has an influence on the multiplication term in the convergence rate. Particularly, when $\mu L + 1 > \alpha$, we can set $\theta = 1$ and because N is usually a large number, it is safe to assume $\sigma G/N < 1$. Then, from (10.18), we can see that a decrease in α increases the multiplier term, which results in a slow convergence rate.*

Observation 4: *If the variance of channel fading, σ, goes up, the channel will have a higher chance to encounter deep fade which inflicts the model training process. The effect is reflected in the multiplier of the convergence rate.*

Observation 5: *An increase in the number of agents, N, can mitigate the impact of channel fading and accelerate the convergence rate. Therefore, scaling up the system can be beneficial to the federated learning. This is in line with conclusions made in Sery and Cohen [2020]. Nonetheless, even if the channel fadings vanished, the convergence rate is still determined by the interference.*

10.5 Numerical Results

In this section, we conduct experimental evaluation of the wireless machine learning algorithm. Particularly, we examine the performance of the analog over-the-air GD for training a multilayer perceptron (MLP) on the MNIST dataset, which

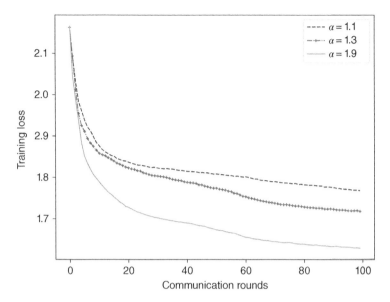

Figure 10.3 Simulation results of the training loss of MLP on the MNIST data set, under different tail index α.

contains the hand-written digits (LeCun et al. [1998]). The MLP is consisted of two hidden layers, each has 64 units and adopts the ReLu activations. We extract 60,000 data points from the MNIST dataset for training, where each agent is assigned with an independent portion that contains 600 data samples. Furthermore, we adopt the Rayleigh fading to model the channel gain. Unless otherwise stated, the following parameters will be used: Learning rate exponent $\rho = 1$, tail index $\alpha = 1.5$, number of agents $N = 100$, and average channel gain $\mu = 1$. The experiments are implemented with Pytorch on Tesla P100 GPU and averaged over 4 trials.

In Figure 10.3, we plot the training loss as a function of the communication rounds under different values of the tail index. We can see that the training loss declines steadily along with the communication rounds, regardless of the heaviness of the tail in the interference's distribution. Nonetheless, the tail index α still plays a vital role in the rate of convergence. Particularly, with an increase in the tail index, the convergence rate goes up accordingly, whereas the improvement is nonlinear with respect to α. These trends corroborate the statement of Observation 1.

Next, we evaluate the effects of the number of participating agents on the model training procedure. It can be seen from Figure 10.4 that the convergence rate of AirFL increases with N, revealing a positive influence from the enlarged number of agents in the system. The reason attributes to two aspects: (i) as we fix the size of dataset per agent, an increase in N boosts up the utilization of data information

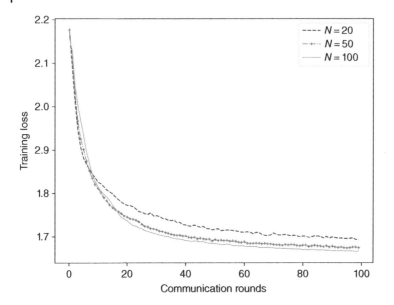

Figure 10.4 Simulation results of the training loss of MLP on the MNIST data set, under different number of agents N.

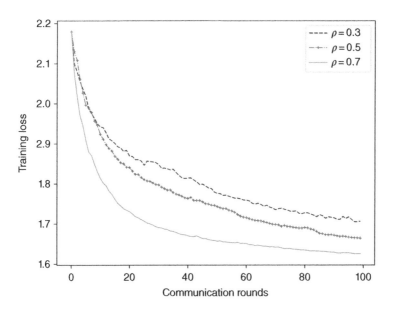

Figure 10.5 Simulation results for training on the MNIST data set, under different ρ.

because all the agents can concurrently access the radio channel and participate in each round of global iteration and (ii) allowing more agents to partake in the analog transmission can reduce the impact of channel fading. Nevertheless, we also observe that such an effect is less significant compared to the tail index because it only influences the multiplier in the convergence rate.

Figure 10.5 illustrates the impact of learning rate, which is controlled by the parameter ρ, on the convergence rate of the algorithm. The figure shows that a decrease in ρ, which slows down the learning rate, can impede the convergence of the model training. Note that compared to the number of agents, N, the variants in the learning rate have a more pronounced impact on the training algorithm as it affects the exponential parameter in the convergence rate.

10.6 Conclusion

We conducted a theoretical study of the analog over-the-air model training that solves a FL problem in a wireless network. Specifically, the agents concurrently transmit an analog signal that is comprised of the local gradients. The edge server receives a superposition of the gradients, which are, however, distorted by the channel fading and interference, and uses this noisy gradient to update the global model. Due to the heavy-tailed intrinsic of interference, the aggregated gradient admits an infinite variance that hinders the use of conventional techniques for convergence analysis that relies on the existence of second moments. To that end, we took a new route to establish the analyses of convergence rate of the training algorithm. Our analyses unveiled that while heavy-tailed noise slows down the convergence rate, the system can support more participating agents, which on the other hand improves the convergence performance. These new results advance the understanding of wireless machine learning, and the techniques demonstrated in this chapter also provided an entryway toward the analysis of FL algorithms in the context of heavy-tailed noise.

Bibliography

Mohammad Mohammadi Amiri and Deniz Gündüz. Machine learning at the wireless edge: Distributed stochastic gradient descent over-the-air. *IEEE Transactions on Signal Processing*, 68:2155–2169, Mar. 2020.

Mohammad Mohammadi Amiria, Tolga M. Dumanb, Deniz Gündüzc, Sanjeev R. Kulkarni, and H. Vincent Poor. Collaborative Machine Learning at the Wireless Edge with Blind Transmitters. *IEEE Global Conference on Signal and Information*

Processing (GlobalSIP), Pages. 1–5, Ottawa, ON, Canada, 2019. doi: 10.1109/GlobalSIP45357.2019.8969185.

Songfu Cai and Vincent K. N. Lau. Modulation-free M2M communications for mission-critical applications. *IEEE Transactions on Signal and Information Processing over Networks*, 4(2):248–263, Jun. 2018.

Kai Lai Chung. On a stochastic approximation method. *Annals of Mathematical Statistics*, 25(3):463–483, 1954.

Laurent Clavier, Troels Pedersen, Ignacio Rodriguez, Mads Lauridsen, and Malcolm Egan. Experimental evidence for heavy tailed interference in the IoT. *IEEE Communications Letters*, 25(3):692–695, Mar. 2021.

Anis Elgabli, Jihong Park, Chaouki Ben Issaid, and Mehdi Bennis. Harnessing wireless channels for scalable and privacy-preserving federated learning. *IEEE Transactions on Communications*, 69(8):5194–5208, 2021.

Vaclav Fabian. Stochastic approximation of minima with improved asymptotic speed. *Annals of Mathematical Statistics*, 38(1):191–200, 1967.

Huayan Guo, An Liu, and Vincent K. N. Lau. Analog gradient aggregation for federated learning over wireless networks: Customized design and convergence analysis. *IEEE Internet of Things Journal*, 8(1):197–210, Jan. 2021.

Martin Haenggi and Radha Krishna Ganti. *Interference In Large Wireless Networks*. Now Publishers Inc., 2009.

Tatiana Pavlovna Krasulina. On stochastic approximation processes with infinite variance. *Theory of Probability and Its Applications*, 14(3):522–526, 1969.

Yann LeCun, Léon Bottou, Yoshua Bengio, and Patrick Haffner. Gradient-based learning applied to document recognition. *Proceedings of the IEEE*, 86(11):2278–2324, Nov. 1998.

Tian Li, Anit Kumar Sahu, Ameet Talwalkar, and Virginia Smith. Federated learning: Challenges, methods, and future directions. *IEEE Signal Processing Magazine*, 37(3):50–60, May 2020.

Brendan McMahan, Eider Moore, Daniel Ramage, Seth Hampson, and Blaise Aguera y Arcas. Communication-efficient learning of deep networks from decentralized data. In *Proceedings of the International Conference on Artificial Intelligence and Statistics (AISTATS)*, pages 1273–1282, Fort Lauderdale, USA, Apr. 2017.

David Middleton. Statistical-physical models of electromagnetic interference. *IEEE Transactions on Electromagnetic Compatibility*, EMC-19(3):106–127, Aug. 1977.

Bobak Nazer and Michael Gastpar. Computation over multiple-access channels. *IEEE Transactions on Information Theory*, 53(10):3498–3516, Oct. 2007.

Jihong Park, Sumudu Samarakoon, Mehdi Bennis, and Mérouane Debbah. Wireless network intelligence at the edge. *Proceedings of the IEEE*, 107(11):2204–2239, Oct. 2019.

Gennady Samorodnitsky and Murad S. Taqqu. *Stable Non-Gaussian Random Processes: Stochastic Models with Infinite Variance: Stochastic Modeling.* Routledge, 2017.

Tomer Sery and Kobi Cohen. On analog gradient descent learning over multiple access fading channels. *IEEE Transactions on Signal Processing*, 68:2897–2911, Apr. 2020.

Tomer Sery, Nir Shlezinger, Kobi Cohen, and Yonina C. Eldar. Over-the-air federated learning from heterogeneous data. Available as ArXiv:2009.12787, 2020.

Hongjian Wang, Mert Gürbüzbalaban, Lingjiong Zhu, Umut Şimşekli, and Murat A. Erdogdu. Convergence rates of stochastic gradient descent under infinite noise variance. Available as ArXiv:2102.10346, 2021.

Moe Z. Win, Pedro C. Pinto, and Lawrence A. Shepp. A mathematical theory of network interference and its applications. *Proceedings of the IEEE*, 97(2):205–230, Feb. 2009.

Kai Yang, Tao Jiang, Yuanming Shi, and Zhi Ding. Federated learning via over-the-air computation. *IEEE Transactions on Wireless Communications*, 19(3):2022–2035, Mar. 2020.

Guangxu Zhu, Yong Wang, and Kaibin Huang. Broadband analog aggregation for low-latency federated edge learning. *IEEE Transactions on Wireless Communications*, 19(1):491–506, Oct. 2019.

Guangxu Zhu, Yuqing Du, Deniz Gündüz, and Kaibin Huang. One-bit over-the-air aggregation for communication-efficient federated edge learning: Design and convergence analysis. *IEEE Transactions on Wireless Communications*, 20(3):2120–2135, Mar. 2020. doi: 10.1109/GlobalSIP45357.2019.8969185.

11

Federated Edge Learning for Massive MIMO CSI Feedback

Shi Jin, Yiming Cui, and Jiajia Guo

School of Information Science and Engineering, National Mobile Communications Research Laboratory, Southeast University, Nanjing, China

11.1 Introduction

Massive multiple-input multiple-output (MIMO) is recognized as a key enabler for fifth generation and future communication systems. A massive MIMO system typically employs numerous antennas at the base station (BS), reaping the advantages of multiplexing gains and spatial diversity (Lu et al. [2014]). To harvest the advantages of massive MIMO, accurate downlink channel state information (CSI) needs to be obtained at the transmitter (BS). In time division duplex mode, the BS can infer the downlink CSI from the uplink CSI on the basis of bi-directional reciprocity. However, in frequency division duplex (FDD) mode, the reciprocity no longer holds. The downlink CSI should be sent to the BS through feedback links, leading to excessive communication overhead due to the large size of CSI (Liang et al. [2018]).

Developing an efficient downlink CSI feedback mechanism has long been an essential but challenging problem in massive MIMO systems. Related works can be roughly divided into three categories, including codebook-based (Love et al. [2008]), compressive sensing (CS)-based (Cheng and Chen [2014]), and deep learning (DL)-based feedback (Wen et al. [2018]). Codebook-based CSI feedback has been adopted in long-term evolution and fifth-generation new-radio systems. The feedback overhead of the codebook-based method increases with more antennas. The codebook-based feedback is only applicable for regular antenna arrays, and the codebook design becomes more complicated as the antenna number becomes larger. Besides, CS theory is also introduced to reduce feedback overhead. The main drawback of CS-based feedback is that the performance is heavily dependent on the sparsity of CSI matrices, which is not always valid in practical scenarios. In addition, the reconstruction of CSI in CS-based feedback

Federated Learning for Future Intelligent Wireless Networks, First Edition.
Edited by Yao Sun, Chaoqun You, Gang Feng, and Lei Zhang.

usually employs iterative algorithms, which has high computational complexity and is infeasible to practical communication systems.

Motivated by the wide success of DL, it has been introduced to CSI feedback to resolve the aforementioned problems. In DL-based CSI feedback, the CSI matrices are regarded as images and compressed and reconstructed by an autoencoder network. DL-based feedback method exhibits outstanding feedback accuracy and low computational complexity (Guo et al. [2022]). The related works can be divided into four categories, including introducing novel autoencoder architectures (Cai et al. [2019], Yu et al. [2020]), leveraging different types of correlation (Wang et al. [2019], Mashhadi et al. [2021]), jointly designing feedback and other modules (Guo et al. [2021], Mao et al. [2021]), and resolving practical deployment problems (Guo et al. [2020], Guo et al. [2020]).

Conventionally, centralized learning (CL) is adopted to train autoencoder networks for DL-based CSI feedback. In CL, all user equipments (UEs) transmit the local CSI datasets to a center server. The center server mixes the collected CSI datasets and trains an autoencoder network for all UEs. However, CL incurs potential privacy disclosure problems. CSI datasets are usually privacy sensitive. The location of UEs can be inferred from CSI samples (Wu et al. [2013]), and the user habits can also be inferred. To settle the aforementioned problems, a distributed learning method called federated learning (FL) is proposed in McMahan et al. [2017]. Instead of CL where all datasets are collected together at the center server, the datasets are kept local to protect data privacy. Notably, different from the applications such as next-word prediction (Hard et al. [2018]) and recommendation systems (Yang et al. [2020]), where the center server is located at the cloud, the center server for physical layer applications is usually an edge server placed near the BS because the UEs participating in FL are in the same cells. To emphasize the interaction between the BS and UEs, FL is referred to as federated edge learning (FEEL) in these physical layer applications. FEEL has also been applied to channel estimation (Elbir and Coleri [2022]), detection (Yang et al. [2021]), etc.

In this work, a communication-efficient personalized FEEL-based training framework is proposed for DL-based CSI feedback, in which UEs collaborate to train a shared global autoencoder for CSI feedback under the coordination of the BS. Specifically, each UE trains the autoencoder with local CSI datasets and uploads the gradient information to the BS. The BS aggregates the gradients to update the global autoencoder. The BS and UEs only exchange the model parameters or gradients rather than CSI datasets, and therefore the privacy is better protected. A parameter quantization method is then introduced to FEEL to further improve the communication efficiency in FEEL, which quantizes the gradient information transmitted from UEs to BS into lower bits. Finally, the proposed FEEL-based training framework is evaluated with simulated CSI datasets generated by QuaDRiGa channel generation software. The contributions of this work are listed as follows:

- A FEEL-based training framework is proposed for DL-based CSI feedback. Different from centralized training strategies, the FEEL-based training framework can effectively prohibit privacy disclosure problems.
- Considering the model transmission is still a bottleneck of FEEL, a parameter quantization strategy is introduced to further reduce the communication overhead in FEEL-based training framework.

11.2 System Model

11.2.1 Channel Model and Signal Model

Considering a single-cell FDD massive MIMO orthogonal frequency division multiplexing, the BS is equipped with a uniform linear array with $N_t \gg 1$ antennas. The UE is equipped with a single antenna. The number of subcarriers is N_c. The received signal $y_n \in \mathbb{C}$ at the nth subcarrier is formulated as follows:

$$y_n = \tilde{\mathbf{h}}_n^H \mathbf{v}_n x_n + z_n, \tag{11.1}$$

where $\tilde{\mathbf{h}}_n \in \mathbb{C}^{N_t \times 1}$, $\mathbf{v}_n \in \mathbb{C}^{N_t \times 1}$, $x_n \in \mathbb{C}$, and $z_n \in \mathbb{C}$ are the channel vector, precoding vector, data-bearing symbol, and additive noise of the nth subcarrier, respectively. The channel vector of N_c subcarriers is stacked together as $\tilde{\mathbf{H}} = [\tilde{\mathbf{h}}_1, \ldots, \tilde{\mathbf{h}}_{\tilde{N}_c}] \in \mathbb{C}^{N_t \times N_c}$ to form the CSI. To achieve better sparsity, $\tilde{\mathbf{H}}$ is transformed into the angular-delay domain as follows:

$$\mathbf{H} = \mathbf{F}_a \tilde{\mathbf{H}} \mathbf{F}_d, \tag{11.2}$$

where $\mathbf{F}_a \in \mathbb{C}^{N_t \times N_t}$ and $\mathbf{F}_d \in \mathbb{C}^{N_c \times N_c}$ are discrete Fourier transformation (DFT) matrices.

11.2.2 DL-Based CSI Feedback

In DL-based CSI feedback, an autoencoder neural network composed of an encoder network and a decoder network is employed for CSI compression and reconstruction. The CSI is first compressed into a low-dimensional codeword $\mathbf{s} \in \mathbb{C}^{N_s \times 1}$ by an encoder network, which can be formulated as follows:

$$\mathbf{s} = f_{en}(\mathbf{H}; \mathbf{\Theta}_{en}), \tag{11.3}$$

where $f_{en}(\cdot)$ denotes the encoder network, and Θ_{en} represents the parameters of the encoder network. The codeword is then transmitted to the BS. Perfect channel is assumed for feedback links. Once the BS receives the codeword, the CSI is reconstructed by a decoder network, which can be written as follows:

$$\hat{H} = f_{de}(s; \Theta_{de}), \tag{11.4}$$

where $f_{de}(\cdot)$ represents the decoder network, and Θ_{de} denotes the parameters of the decoder network. The BS then performs inverse DFT to obtain the CSI in the spatial-frequency domain.

To train the autoencoder networks, mean square error (MSE) loss function is most frequently adopted as follows:

$$l_{MSE}(H, \hat{H}) = \|\hat{H} - H\|_2^2, \tag{11.5}$$

where $\| \cdot \|_2$ is the Euclidean norm.

11.3 FEEL for DL-Based CSI Feedback

In this section, a FEEL-based training framework is first introduced to DL-based CSI feedback. Then, neural network parameter quantization is proposed to reduce communication overhead of the FEEL-based training framework.

11.3.1 Basic Autoencoder Architecture

In this work, an autoencoder similar to CsiNet in Wen et al. [2018] is employed for DL-based CSI feedback. The architecture of the autoencoder network is shown in Figure 11.1. The real and imaginary parts of the CSI are separated into two channels and normalized between 0 and 1 for inputs. Then, the encoder network consists of a convolutional layer with batch normalization and LeakyReLU activation for feature extraction and a dense layer for compression. The CSI is compressed into a N_s-dimensional real-valued vector referred to as a codeword. The compression ratio can be defined as follows:

$$\gamma = \frac{N_s}{2N_t N_c}. \tag{11.6}$$

The codeword is then reconstructed by the decoder at the BS. A dense layer is first employed to recover the dimension of the CSI. Then, two cascade RefineNets are followed to refine the initially reconstructed CSI for higher accuracy. The RefineNet is composed of three convolutional layers. Residual structure is adopted to avoid gradient vanishing and explosion (He et al. [2016]). To further accelerate the convergence, a ReZero module is added before skip connection,

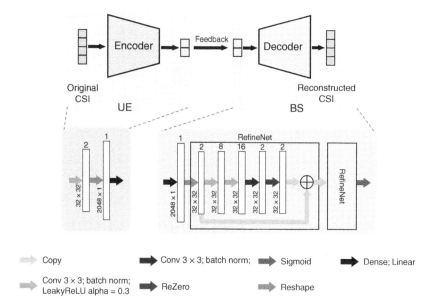

Figure 11.1 Illustration of DL-based CSI feedback and the detailed architecture of the proposed autoencoder network.

which includes a trainable scalar initialized as zero (Bachlechner et al. [2021]). After the two RefineNets, a Sigmoid activation is used to ensure the output values still in (0,1).

11.3.2 FEEL-Based Training Framework

11.3.2.1 Motivation
The BS can collect CSI samples from different UEs in different areas to form a mixed datasets for centralized training. However, location information of UEs can also be partially or completely inferred on the basis of collected CSI, that is, the uploading of CSI samples increases the risk of privacy disclosure. Therefore, FEEL is introduced to DL-based CSI feedback to protect the privacy of UEs, where all UEs perform local training and avoid uploading local CSI datasets to the BS.

11.3.2.2 Training Framework
Different from conventional centralized training, the FEEL-based training framework realizes collaborative model training by model parameter (or gradient) exchange between the UEs and BS. The UEs are not required to transmit the CSI samples to the BS, thereby protecting the UEs' privacy. Considering K UEs indexed from 1 to K participate in FEEL to collaboratively train an autoencoder for DL-based CSI feedback, the kth UE has a local datasets $\mathcal{D}_k$ and the number of

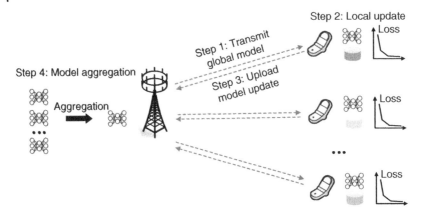

Figure 11.2 Illustration of FEEL-based training framework for DL-based CSI feedback.

samples in dataset D_k is denoted as n_k where $\sum_{k=1}^{K} n_k = N$. Figure 11.2 shows the main steps of FEEL-based CSI feedback as follows:

- *Step 1: Transmitting global model*: The BS schedules a certain group of available UEs for participation and transmits the current global model to these UEs for initialization.
- *Step 2: Local update*: Each scheduled UEs trains the received model with local CSI datasets for one or more epochs.
- *Step 3: Uploading model update*: After local training, the scheduled UEs report the model update (e.g., gradient information) to the BS.
- *Step 4: Model aggregation*: When the BS received all updates from the scheduled UEs, the updates are aggregated to generate a new global model by averaging weighted by the sample number in each local dataset. The performance of the global model is evaluated to determine whether to launch the next round of FEEL from Step 1.

The algorithmic description of FEEL is shown in Algorithm 11.1. If each UE only conducts gradient descent for one-epoch, that is, $E = 1$ in Algorithm 11.1, the optimization method is usually referred to as FederatedSGD. To further accelerate the convergence, the local update can be repeated for several epochs before uploading to the BS ($E > 1$). The approach is termed as FederatedAveraging (FedAvg) by McMahan et al. [2017]. In the current study, the FedAvg algorithm is adopted to improve the communication efficiency. The optimization objective of FEEL can be formulated as follows:

$$\min_{\omega \in \mathbb{R}^d} f(\omega) \quad \text{s.t.} \quad f(\omega) = \frac{n_k}{n} \sum_{k=1}^{K} \sum_{i=1}^{n_k} l(H_i, \hat{H}_i), \tag{11.7}$$

where d is the number of network parameters.

Algorithm 11.1 A FedAvg-based FEEL Algorithm for DL-based CSI feedback

Require: K UEs in FEEL indexed $1, \ldots, K$ with local CSI datasets $D_1, \ldots, D_K$, where $D_1 \cup \cdots \cup D_K = D$. M UEs are scheduled in each round. The local epoch number, batch size, local learning rate, and loss function are E, B, η, and $l(\cdot)$ respectively.

1: Initialize the global model parameters as ω_0
2: **while** $acc < acc_{\text{target}}$ **do**
3:     Schedule M UEs $S_t = \{s_1, \ldots, s_M\}$
4:     **for** $s \in S_t$ in parallel **do**
5:        Divide D_s into batches $B = \{b_1, \ldots, b_N\}$ with batch size B
6:        **for** i from 1 to E **do**
7:           **for** $b \in B$ **do**
8:              $\omega_t^s \leftarrow \omega_t^s - \eta \nabla l(\omega_t^s; b)$
9:           **end for**
10:       **end for**
11:     **end for**
12:     $\omega_{t+1} \leftarrow \frac{n_k}{N} \sum_{i=1}^{M} \omega_t^{s_i}$
13:     $acc \leftarrow \text{Evaluate}(\omega_{t+1})$
14: **end while**

11.3.3 Parameter Quantization in the FEEL-Based Training Framework

11.3.3.1 Key Idea

The frequent model parameter transmission between the UEs and BS is one of the major challenges of FEEL. In each round, UEs upload the gradient information to the BS through uplink and download the global model from the BS through downlink. The communication overhead is large in the proposed FEEL-based CSI feedback training framework because of the large parameter number of an autoencoder network and frequent parameter exchange. The parameters of an autoencoder network are usually stored as 32-bit float numbers, which are redundant for representation. In this work, parameter quantization is introduced to FEEL-based training framework to reduce the communication overhead. The quantization process can be formulated as follows:

$$y = x_{\min} + \text{round}\left(\frac{x - x_{\min}}{\text{interval}}\right) \times \text{interval}, \text{ s.t. interval} = \frac{x_{\max} - x_{\min}}{2^B - 1},$$
$$(11.8)$$

where x is the parameter to be quantized, y is the quantized parameter, $x_{\min}$ and $x_{\max}$ are the maximum and minimum of a set of parameters, and B is the number of quantization bits.

11.3.3.2 Details

The parameters of a CsiNet-like autoencoder network are mainly composed of the parameters of dense layers. For example, when the compression ratio is 1/64, the parameters of dense layers account for 97% of autoencoder parameters. The proportion becomes even higher when the compression ratio is larger, which means the quantization of dense layer parameters is essential to communication overhead reduction. Furthermore, the bias of dense layers has much fewer parameters and is more sensitive to quantization errors compared with the weights of dense layers. Therefore, only the weights of dense layers in the autoencoder networks are considered.

Besides, the uplink gradient information and downlink global model have different sensitivity to quantization error. For uplink gradient information, the update aggregation can mitigate the quantization error by averaging to a certain extent. Therefore, the quantization in uplink is less sensitive to the quantization error. However, the downlink global model is directly used as an initialization for local update and is more sensitive to quantization error than uplink gradient information.

11.4 Simulation Results

First, the channel generation settings and the training strategies are provided. Then, the performance of the FEEL-based training framework and the quantization method are evaluated.

11.4.1 Simulation Settings

11.4.1.1 Channel Generation

QuaDRiGa channel generation software is employed for CSI datasets generation (Jaeckel et al. [2014]). The channel generation settings are described in Table 11.1. The center frequency is set the same as the operating band "n7" as specified in 3GPP TS 38.101-1 (3GPP TS 38.101-1 [2021]). Other channel parameters are set according to 3GPP TR 38.901 3GPP TR 38.901 [2017]. In total, 5000 CSI samples are generated for each UE. Normalization is performed to each CSI sample. The CSI datasets are divided into training, validation, test sets by the proportion $8:1:1$.

11.4.1.2 Training Settings

All algorithm related to DL is realized with TensorFlow 2.4.0.[1] The Adam optimizer is used for FEEL. The batch size is set as 64. The learning rate is initially

---

1 https://github.com/tensorflow/tensorflow/releases/tag/v2.4.0.

Table 11.1 Channel generation settings.

Scenarios	3GPP_38.901_UMi_NLOS
	3GPP_38.901_UMi_NLOS_O2I
Antenna settings	BS: 32 omnidirectional antennas, uniform linear array (ULA)
	UE: a single omnidirectional antenna
Operating system	FDD-orthogonal frequency division multiplexing (OFDM) system
Subcarrier number	32
Center frequency	2.655 GHz
Bandwidth	70 MHz
Cell range	200 m
BS height	10 m
UE height	$3(n_{fl} - 1) + 1.5$ m
	O2I: $n_{fl} \sim U(1, N_{fl})$ $N_{fl} \sim U(4,8)$
	Outdoor: $n_{fl} = 1$
Indoor UE proportion	80%
Number of clusters	UMi NLOS: 19 UMi NLOS O2I: 12
Subpath number in a cluster	20

set as 10^{-3} and then reduced to 10^{-4}. The normalized mean square error (NMSE) is used to evaluate the accuracy of CSI feedback as follows:

$$NMSE = E\left\{ \frac{\|\mathbf{H} - \hat{\mathbf{H}}\|_2^2}{\|\mathbf{H}\|_2^2} \right\}. \tag{11.9}$$

Three training frameworks are tested for comparison including individual learning (IL), CL, and FEEL. For IL-based training framework, each UEs train an autoencoder network independently with local datasets. For CL-based training framework, all UEs upload the local datasets to the BS to train a shared autoencoder network. For FEEL-based training framework, UEs collaborate to train an autoencoder network via Algorithm 11.1.

11.4.1.3 Performance of FEEL-Based Training Framework

As shown in Figure 11.3, 100 UEs indexed from 1 to 100 are randomly generated. The first 50 UEs are prepared for training, and the last 50 UEs are used to test the performance of autoencoder networks. The positions of the UEs are uniformly distributed in the cell. Each UE is supposed to walk randomly in a circular area whose radius is 5 m. The BS is located at the center of the cell.

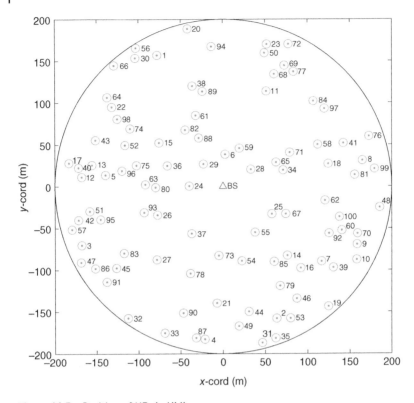

Figure 11.3 Position of UEs in UMi.

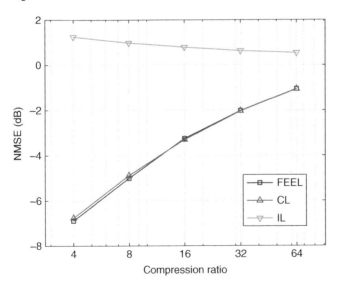

Figure 11.4 Performance of IL-, CL-, and FEEL-based training frameworks for DL-based CSI feedback under different compression ratio. The number of UE is set as 50, and each UE has 1000 training samples.

The performance of IL-, CL-, and FEEL-based training frameworks is first compared. As shown in Figure 11.4, FEEL-based training framework shows comparable performance with CL under different compression ratio, which demonstrates the effectiveness of FEEL in DL-based CSI feedback. Meanwhile, the accuracy of CL- and FEEL-based training frameworks significantly outperforms IL, emphasizing the significance of collaboration for DL-based CSI feedback.

The feedback accuracy is then evaluated under different UE number and training samples in Figure 11.5. With more UEs participating in FEEL and more training samples per UE, the performance of DL-based CSI feedback is improved. The improvement is attributed to more diverse channel observations included in training. Furthermore, the influence of UE number becomes more evident when the number of training sample per UE is small. When the number of training samples is relatively sufficient, the performance is more stable to UE number. Considering the limited storage of UEs, each UE cannot store considerable samples for FEEL. Therefore, more UEs should be encouraged to participant in FEEL to achieve higher accuracy.

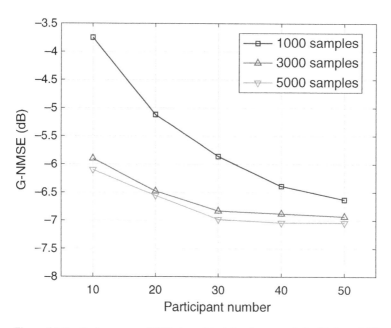

Figure 11.5 Performance of FEEL-based training framework for DL-based CSI feedback under different UE number and sample number per UE. The compression ratio is set as 1/4.

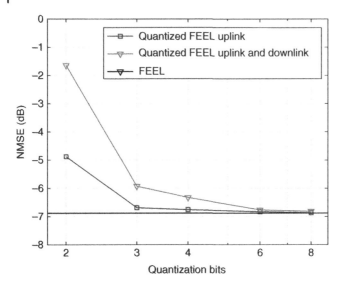

Figure 11.6 Performance of FEEL-based training framework for DL-based CSI feedback under different quantization bits. The compression ratio, the number of UEs, and the sample number per UE are set as 1/4, 50, and 1000, respectively.

Finally, the influence of quantization is tested under different quantization bits and quantization strategies. The parameters of autoencoder network are originally stored as 32-bit float number. As shown in Figure 11.6, when the uplink gradient information and the downlink global model are both quantized, 6 bits are required to achieve comparable performance with the nonquantized cases approximately. However, when only the uplink gradient information is quantized, the performance of the quantized FEEL-based training framework can approach the nonquantized one with only 3 bits. The result demonstrates that parameter quantization can significantly reduce the communication overhead with negligible accuracy loss.

11.5 Conclusion

In this work, a FEEL-based training framework is proposed for DL-based CSI feedback. The key idea of the study is that UEs collaborate by exchanging model parameters instead of directly uploading the CSI datasets. Therefore, the privacy of UEs is better protected compared with CL. Parameter quantization is then introduced to reduce the communication overhead of FEEL. Simulation results show that the FEEL-based training framework can achieve comparable performance with CL for DL-based CSI feedback.

Bibliography

3GPP TR 38.901. Study on channel model for frequencies from 0.5 to 100 GHz, Mar 2017. URL https://portal.3gpp.org/desktopmodules/Specifications/ SpecificationDetails.aspx?specificationId=3173.

3GPP TS 38.101-1. NR; user equipment (UE) radio transmission and reception; Part 1: Range 1 Standalone (Release 17), Mar 2021. URL https://portal.3gpp.org/ desktopmodules/Specifications/SpecificationDetails.aspx?specificationId=3283.

Thomas Bachlechner, Bodhisattwa Prasad Majumder, Henry Mao, Gary Cottrell, and Julian McAuley. ReZero is all you need: Fast convergence at large depth. In *Proceedings of the Thirty-Seventh Conference on Uncertainty in Artificial Intelligence*, volume 161, pages 1352–1361. PMLR, 27–30 Jul 2021.

Qiuyu Cai, Chao Dong, and Kai Niu. Attention model for massive MIMO CSI compression feedback and recovery. In *2019 IEEE Wireless Communications and Networking Conference (WCNC)*, pages 1–5, 2019. doi: 10.1109/WCNC.2019.8885897.

Peng Cheng and Zhuo Chen. Multidimensional compressive sensing based analog CSI feedback for massive MIMO-OFDM systems. In *2014 IEEE 80th Vehicular Technology Conference (VTC2014-Fall)*, pages 1–6, 2014. doi: 10.1109/VTCFall.2014.6966062.

Ahmet M. Elbir and Sinem Coleri. Federated learning for channel estimation in conventional and RIS-assisted massive MIMO. *IEEE Transactions on Wireless Communications*, 21(6):4255–4268, 2022. doi: 10.1109/TWC.2021.3128392.

Jiajia Guo, Jinghe Wang, Chao-Kai Wen, Shi Jin, and Geoffrey Ye Li. Compression and acceleration of neural networks for communications. *IEEE Wireless Communications*, 27(4):110–117, 2020. doi: 10.1109/MWC.001.1900473.

Jiajia Guo, Chao-Kai Wen, Shi Jin, and Geoffrey Ye Li. Convolutional neural network-based multiple-rate compressive sensing for massive MIMO CSI feedback: Design, simulation, and analysis. *IEEE Transactions on Wireless Communications*, 19(4):2827–2840, 2020. doi: 10.1109/TWC.2020.2968430.

Jiajia Guo, Chao-Kai Wen, and Shi Jin. Deep learning-based CSI feedback for beamforming in single- and multi-cell massive MIMO systems. *IEEE Journal on Selected Areas in Communications*, 39(7):1872–1884, 2021. doi: 10.1109/JSAC.2020.3041397.

Jiajia Guo, Chao-Kai Wen, Shi Jin, and Geoffrey Ye Li. Overview of deep learning-based CSI feedback in massive MIMO systems. *IEEE Transactions on Communications*, 70(12):8017–8045, 2022. doi: 10.1109/TCOMM.2022.3217777.

Andrew Hard, Kanishka Rao, Rajiv Mathews, Swaroop Ramaswamy, Françoise Beaufays, Sean Augenstein, Hubert Eichner, Chloé Kiddon, and Daniel Ramage. Federated Learning for Mobile Keyboard Prediction. *arXiv:1811.03604*, 2018, Available: http://arxiv.org/abs/1811.03604.

Kaiming He, Xiangyu Zhang, Shaoqing Ren, and Jian Sun. Deep residual learning for image recognition. In *Proceedings of the IEEE conference on computer vision and pattern recognition*, pages 770–778, 2016.

Stephan Jaeckel, Leszek Raschkowski, Kai Börner, and Lars Thiele. QuaDRiGa: A 3-D multi-cell channel model with time evolution for enabling virtual field trials. *IEEE Transactions on Antennas and Propagation*, 62(6):3242–3256, 2014. doi: 10.1109/TAP.2014.2310220.

Han-Wen Liang, Wei-Ho Chung, and Sy-Yen Kuo. FDD-RT: A simple CSI acquisition technique via channel reciprocity for FDD massive MIMO downlink. *IEEE Systems Journal*, 12(1):714–724, 2018. doi: 10.1109/JSYST.2016.2556222.

David J. Love, Robert W. Heath, Vincent K. N. Lau, David Gesbert, Bhaskar D. Rao, and Matthew Andrews. An overview of limited feedback in wireless communication systems. *IEEE Journal on Selected Areas in Communications*, 26(8):1341–1365, 2008. doi: 10.1109/JSAC.2008.081002.

Lu Lu, Geoffrey Ye Li, A. Lee Swindlehurst, Alexei Ashikhmin, and Rui Zhang. An overview of massive MIMO: Benefits and challenges. *IEEE Journal of Selected Topics in Signal Processing*, 8(5):742–758, 2014. doi: 10.1109/JSTSP.2014.2317671.

Yuanshang Mao, Xin Liang, and Xinyu Gu. DL-based Joint CSI feedback and user selection in FDD massive MIMO. In *2021 Workshop on Algorithm and Big Data*, WABD 2021, page 61–66, New York, NY, USA, 2021. Association for Computing Machinery. doi: 10.1145/3456389.3456399.

Mahdi Boloursaz Mashhadi, Qianqian Yang, and Deniz Gündüz. Distributed deep convolutional compression for massive MIMO CSI feedback. *IEEE Transactions on Wireless Communications*, 20(4):2621–2633, 2021. doi: 10.1109/TWC.2020 .3043502.

Brendan McMahan, Eider Moore, Daniel Ramage, Seth Hampson, and Blaise Aguera y Arcas. Communication-efficient learning of deep networks from decentralized data. In *Artificial Intelligence and Statistics*, pages 1273–1282. PMLR, 2017.

Tianqi Wang, Chao-Kai Wen, Shi Jin, and Geoffrey Ye Li. Deep learning-based CSI feedback approach for time-varying massive MIMO channels. *IEEE Wireless Communications Letters*, 8(2):416–419, 2019. doi: 10.1109/LWC.2018.2874264.

Chao-Kai Wen, Wan-Ting Shih, and Shi Jin. Deep learning for massive MIMO CSI feedback. *IEEE Wireless Communications Letters*, 7(5):748–751, 2018. doi: 10.1109/LWC.2018.2818160.

Kaishun Wu, Jiang Xiao, Youwen Yi, Dihu Chen, Xiaonan Luo, and Lionel M. Ni. CSI-based indoor localization. *IEEE Transactions on Parallel and Distributed Systems*, 24(7):1300–1309, 2013. doi: 10.1109/TPDS.2012.214.

Liu Yang, Ben Tan, Vincent W. Zheng, Kai Chen, and Qiang Yang. Federated recommendation systems. In *Federated Learning*, pages 225–239. Springer, 2020.

Yuwen Yang, Feifei Gao, Jiang Xue, Ting Zhou, and Zongben Xu. Federated dynamic neural network for deep MIMO detection. *arXiv:2111.12260*, 2021, Available: http://arxiv.org/abs/2111.12260.

Xiaotong Yu, Xiangyi Li, Huaming Wu, and Yang Bai. DS-NLCsiNet: Exploiting non-local neural networks for massive MIMO CSI feedback. *IEEE Communications Letters*, 24(12):2790–2794, 2020. doi: 10.1109/LCOMM.2020.3019653.

12

User-Centric Decentralized Federated Learning for Autoencoder-Based CSI Feedback

Shi Jin, Jiajia Guo, Yan Lv, and Yiming Cui

School of Information Science and Engineering, National Mobile Communications Research Laboratory, Southeast University, Nanjing, P. R. China

12.1 Autoencoder-Based CSI Feedback

12.1.1 CSI Feedback in Massive MIMO

In massive multiple-input multiple-output (MIMO) systems, the base station (BS) can utilize the accurate downlink channel state information (CSI) for beamforming or precoding, to improve link capacity and energy efficiency in communication systems (Larsson et al. [2014], Lu et al. [2014]). In time-division duplex (TDD) systems, the channel reciprocity holds, and the downlink CSI can be inferred from the uplink. However, the reciprocity is weak in frequency-division duplex (FDD) systems, where uplink and downlink operate on different frequency bands. Thus, the user equipment (UE) has to feed the estimated downlink CSI back to the BS via uplink. However, the huge number of antennas at the BS significantly increases the dimension of CSI matrices, which should be compressed to reduce the feedback overhead.

In the past few years, researchers have conducted numerous works to reduce the feedback overhead. The first commonly adopted feedback strategy is based on codebook (Love et al. [2008]). However, the accuracy of these codebook-based methods still cannot meet the requirement, and the feedback overhead increases with the CSI dimension. Moreover, the codeword search process is of high complexity. Therefore, the performance of feedback complexity and the overhead of the channel codebook needs to be improved. Compressive sensing (CS) is another widely used strategy to reduce feedback overhead (Qin et al. [2018]). The sparsity of massive MIMO channels in a certain domain can be exploited in CS to compress and reconstruct downlink CSI (Kuo et al. [2012]). However, conventional CS algorithms heavily rely on the CSI sparsity assumption and do

Federated Learning for Future Intelligent Wireless Networks, First Edition.
Edited by Yao Sun, Chaoqun You, Gang Feng, and Lei Zhang.

not make full use of channel structure characteristics. In addition, existing CS reconstruction algorithms are often iterative, resulting in low reconstruction speed. Therefore, a more efficient approach is needed in CSI feedback.

Deep learning (DL) has shown great success in wireless communications (Qin et al. [2019]), such as signal detection (Ye et al. [2018]) and channel estimation (Soltani et al. [2019]). Recently, DL has also been applied to the CSI feedback problem (Qin et al. [2019]). DL not only addresses the two above-mentioned problems but also introduces environment knowledge to CSI feedback. The first DL-based CSI feedback work is CsiNet (Wen et al. [2018]), which is based on an autoencoder (AE) architecture. The compression and reconstruction of the CSI are realized by the encoder and decoder, respectively. After that, the DL-based CSI feedback has been a popular research topic and Guo et al. [2022] detailedly summarizes the existing DL-based feedback works.

12.1.2 System Model

Considering a single-cell massive MIMO system, the BS is equipped with a uniform linear array (ULA) of $N_t \gg 1$ transmit elements, and the UE is equipped with a single receiver antenna. The system is operated in orthogonal frequency division multiplexing (OFDM) over N_c subcarriers. The received signal y_n over the nth subcarrier can be written as

$$y_n = \tilde{\mathbf{h}}_n^H \mathbf{v}_n x_n + z_n, \tag{12.1}$$

where $\tilde{\mathbf{h}}_n \in \mathbb{C}^{N_t \times 1}$, $\mathbf{v}_n \in \mathbb{C}^{N_t \times 1}$, x_n, and z_n represent the channel vector, precoding vector, data symbol, and complex additive random noise over the nth subcarrier, respectively. The entire downlink CSI matrix can be expressed as $\tilde{\mathbf{H}} = [\tilde{\mathbf{h}}_1, \tilde{\mathbf{h}}_2, \ldots, \tilde{\mathbf{h}}_{N_c}]^H \in \mathbb{C}^{N_c \times N_t}$. In FDD systems, the UE should feed $\tilde{\mathbf{H}}$ back to the BS. The total number of feedback parameters is $N = 2 \times N_t \times N_c$, which leads to a large overhead.

The simulation results in Wen et al. [2015] show that when the number of antennas at the BS tends to infinity, the CSI matrix shows sparsity in the angular-delay domain. In the practical system, although the number of antennas at the BS is a finite value, the CSI matrix is also approximately sparse in the angular-delay domain. Therefore, the two-dimensional discrete Fourier transform (2D-DFT) can be used to convert the spatial-frequency domain $\tilde{\mathbf{H}}$ into the angular-delay domain. The approximate sparse CSI matrix can be mathematically expressed as follows:

$$\mathbf{H} = \mathbf{F}_b \tilde{\mathbf{H}} \mathbf{F}_a, \tag{12.2}$$

where $\mathbf{F}_a \in \mathbb{C}^{N_t \times N_t}$ and $\mathbf{F}_b \in \mathbb{C}^{N_c \times N_c}$ are DFT matrices.

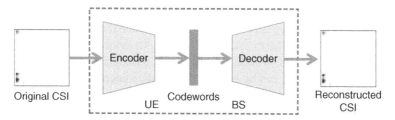

Figure 12.1 Illustration of the AE-based CSI feedback.

12.1.3 AE-Based Feedback Framework

Researchers have made many contributions to image compression using AEs. In general, since the DL framework cannot handle complex numbers, the CSI matrices are separated into two channels which represent the real part and the imaginary part, respectively. Then, the CSI matrices can be viewed as images that can be processed by AE networks. As shown in Figure 12.1, the encoder at the UE converts the original N-dimension downlink $\mathbf{H}$ into M-dimension codewords. The "CR" defined as M/N stands for compression ratio. Then, the UE sends the compressed codewords back to the BS through the feedback link. The compression process can be expressed as

$$\mathbf{s} = f_{\mathrm{en}}\left(\mathbf{H}; \Theta_1\right), \tag{12.3}$$

where $f_{\mathrm{en}}(\cdot)$ and Θ_1 denote the compression operation and the neural network (NN) parameters of the encoder at the UE, respectively. Once receiving the compressed codewords, the BS reconstructs the original CSI matrices via a decoder. The reconstruction process can be mathematically expressed as follows:

$$\widehat{\mathbf{H}} = f_{\mathrm{de}}\left(\mathbf{s}; \Theta_2\right), \tag{12.4}$$

where $f_{\mathrm{de}}(\cdot)$ and Θ_2 represent the reconstruction operation and the NN parameters of the decoder, respectively.

In the training phase, an end-to-end training approach is adopted. The loss function optimized in training is a mean square error (MSE):

$$L_{\mathrm{MSE}} = \|\widehat{\mathbf{H}} - \mathbf{H}\|_2^2. \tag{12.5}$$

12.2 User-Centric Online Training for AE-Based CSI Feedback

12.2.1 Motivation

In existing CSI feedback studies, the CSI samples used for training and testing are generated from the entire cell, and the NNs trained by these datasets are expected to perform well in the entire cell. However, if the channel in this cell changes greatly, or the trained NNs are deployed to another cell, then the new environment does not match the knowledge extracted by the NNs. Therefore, the performance may drop dramatically, which indicates the weak generalization of DL-based algorithms. Online training can be adopted in this phenomenon to improve NN performance with some DL tricks, such as transfer and meta-learning (Yang et al. [2020]).

The above-mentioned studies assume that the UEs are randomly distributed in the cell. In fact, although numerous UEs are located randomly within the cell, the location of a certain UE is regular according to the BS. For example, the phone holder may only remain in the office for a long time daily, and the environment around the phone is relatively stable. This phenomenon has been studied in computer vision (CV) (Guo et al. [2020], He et al. [2020]). A general AE is trained using collected data in advance, and the encoder is finetuned using the data for transmission without changes to the decoder (He et al. [2020]). Compared with the trained encoder for all data, the NNs trained on the required data perform better on the simpler distribution.

Inspired by this, a general AE can be trained for the entire cell, and then a specific NN can be trained to improve the feedback performance of the considered stable cell. Thus, the mismatch between the training and the practical datasets can be solved by training different NNs for different areas. However, the training is expensive and infeasible for the BS because this station cannot train thousands of NNs only for the CSI feedback task. Therefore, a user-centric online training strategy that maintains the BS is required.

12.2.2 Key Idea of User-Centric Online Training

The main idea of the user-centric online training strategy is shown in Figure 12.2. The proposed online training strategy contains two steps as follows: first, a general AE for the UE is trained in advance; then, the NN parameters of the trained decoder are fixed, and the UE collects numerous CSI samples to finetune the

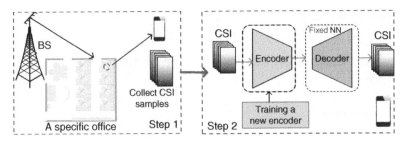

Figure 12.2 Illustration of the user-centric online training for AE-based CSI feedback.

Algorithm 12.1 User-centric online training framework

Input: Pretrained encoder: f_{en}, Pretrained decoder: f_{de}

Output: Novel encoder: f'_{en}

1: collect training CSI samples $(\mathbf{H}, \mathbf{H})$
2: fix the parameters of the decoder f_{de}
3: design a novel encoder f'_{en} using Eqs. 12.7, 12.8 or 12.9
4: **if** add an extra NN layer before/after the encoder f_{en} **then**
5:    fix parameters of the pre-trained encoder f_{en}
6: **end if**
7: **loop**
8:    randomly select a training batch $(\mathbf{H}_{selected}, \mathbf{H}_{selected})$
9:    update the parameters of the novel encoder f'_{en} with $(\mathbf{H}_{selected}, \mathbf{H}_{selected})$
10: **end loop**

---

encoder. Thus, the feedback performance of a specific UE can be improved. The optimization problem for training the new encoder $f'_{en}(\cdot)$ can be mathematically expressed as follows:

$$\hat{\Theta}'_1 = \text{argmin}_{\Theta'_1} \|\mathbf{H} - f_{de}(f'_{en}(\mathbf{H}; \Theta'_1); \Theta_2)\|^2_2, \tag{12.6}$$

where Θ'_1 represents the NN weights of the new encoder $f'_{en}(\cdot)$. The above training process is shown in Algorithm 12.1.

The purpose of user-centric online training is to change the latent representation of the CSI samples from a stable area to facilitate an accurate reconstruction of the CSI using the deployed decoder. Therefore, the compression operation by the novel encoder of the CSI samples is essential. In Section 12.3, three online training frameworks, which are different in considering the modification area of the original representation, are introduced.

12.2.3 Three Online Training Frameworks

12.2.3.1 Edit Before Encoder

In Talebi et al. [2021], the image reconstruction performance is considerably improved by only editing the images before compression. Inspired by this, the first framework also edits the original CSI samples before compression, that is, the novel encoder is the same as the trained one. Figure 12.3 shows the specific idea of this framework.

The convolutional layer first uses two 3×3 filters to edit the original CSI. The adopted activation function is Tanh. A residual learning framework is adopted to accelerate training. Then, the pretrained encoder and decoder, respectively,

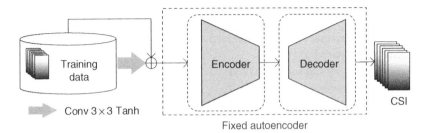

Figure 12.3 Illustration of the framework, which edits the CSI before compression.

compress the edited CSI and reconstruct the original CSI samples. Therefore, the reconstructed CSI at the BS can be expressed as

$$\hat{\mathbf{H}} = f_{\text{de}}\left(f_{\text{en}}\left(\mathbf{H}';\Theta_1\right);\Theta_2\right), \quad \text{s.t.,} \quad \mathbf{H}' = \mathbf{H} + f_{\text{conv}}\left(\mathbf{H};\Theta_{\text{conv}}\right), \tag{12.7}$$

where $f_{\text{conv}}(\cdot)$ and Θ_{conv} represent the extra convolution operation and its NN weights, respectively. The loss function for the online training is also an MSE.

12.2.3.2 Edit After Encoder

The first framework changes the output of the encoder by conducting some operations before the original CSI is inputted into the encoder. However, the direct way to change the compressed codewords is to refine the encoder, which is shown in Figure 12.4. Therefore, a fully connected layer with the Tanh activation function is followed by the pretrained encoder to edit the codewords. Similarly, residual learning is also adopted in training. The pretrained decoder then reconstructs the downlink CSI, which can be expressed as

$$\hat{\mathbf{H}} = f_{\text{de}}\left(s';\Theta_2\right), \quad \text{s.t.,} \quad s' = s + f_{\text{fc}}\left(\mathbf{s};\Theta_{\text{fc}}\right), \tag{12.8}$$

where $f_{\text{fc}}(\cdot)$ and Θ_{fc} represent the extra fully-connected (FC) layer and its NN weights, respectively.

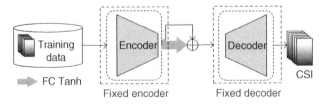

Figure 12.4 Illustration of the framework, which directly edits the codeword, that is, the output of the encoder.

Figure 12.5 Illustration of the framework, which maintains the original encoder but finetunes it using the collected CSI samples.

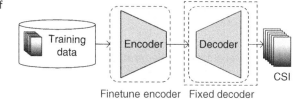

Finetune encoder   Fixed decoder

12.2.3.3 Finetuning Encoder

The two above-mentioned frameworks maintain the original encoder and add an extra NN layer before or after the encoder, thereby changing the codewords. Another framework, wherein no extra NN layers are added, is shown in Figure 12.5.

The new encoder has the same NN architecture as that of the original one and is finetuned from the original one using the newly collected CSI samples. The UE compresses the CSI using the finetuned encoder and the BS reconstructs the CSI, which can be expressed as

$$\hat{\mathbf{H}} = f_{de}\left(f_{en}'\left(\mathbf{H};\Theta_1'\right);\Theta_2\right) \tag{12.9}$$

12.3 Multiuser Online Training Using Decentralized Federated Learning

12.3.1 Motivation

The scenario considered in Section 12.2 can be regarded as a single-user case. In practical systems, UEs are often densely distributed in a stable area. Therefore, adjacent UEs may have similar channel characteristics. The feedback performance can be further enhanced by sharing channel knowledge among UEs, that is, making one UE realize the user-centric online training using the collected CSI samples of all UE. However, this approach may cause two problems. First, the CSI contains some sensitive information about the UE, and CSI sharing may lead to serious privacy concerns. Second, the NN training computational requirement is proportional to the training sample number. The online training will be computationally prohibitive if all CSI samples are centralized to a specific UE due to the limited computational capability of the UE. The above-centralized learning can be replaced by federated learning (FL) (Elbir et al. [2021]). However, each UE in the FL framework should frequently communicate with the BS (server), thereby occupying valuable uplink bandwidth. To solve this problem, user-centric online training should not need the participation of the BS, that is, without occupying uplink transmission resources. Therefore, a fully decentralized federated learning

(DFL) framework without BS participation is necessary to replace FL to improve feedback performance.

12.3.2 Decentralized Federated Learning Framework

12.3.2.1 Key Idea of Decentralized Federated Learning

DFL can improve the performance of NNs from distributed datasets (Hegedűs et al. [2019]). The computational center in the DFL framework is the UE instead of the BS. The basic idea of DFL is that the UEs exchange the trained local models and update their models based on the models of other UEs. Therefore, the communication between the UE and server can be replaced by the communication among the UEs, which does not need an uplink transmission. For example, the combination of DFL-based distributed machine learning and blockchain enables the decentralized training of machine learning models without central control and sharing of medical data (Warnat-Herresthal et al. [2021]).

The workflow of the DFL framework is shown in Algorithm 12.2. First, the UE initializes a local NN model. Assuming that the above method which finetunes the encoder is adopted in the online learning phase, thus, the new encoder can be initialized by finetuning the weights of the pretrained encoder. Then, the UE exchanges the trained NN model and the timestamp/age (t) of it to some randomly selected UE via device-to-device (D2D) communication. As is shown in Figure 12.6 which contains 8 UE, the 1st UE exchanges its NN weights Θ^1 with its neighbored UE, that is, 2nd and 8th UE, through D2D communications. Once receiving an NN model from another UE, the UE merges this model with the local one and then saves the merged NN as the local. The new merged model is continuously updated (trained) using the local datasets by stochastic gradient descent (SGD) as

$$\Theta \leftarrow \Theta - \eta \cdot \nabla\Theta, \tag{12.10}$$

where $\nabla\Theta$ and η represent the stochastic gradient of weight Θ and learning rate, respectively. This looping process stops once the NN training converges. The DFL

Algorithm 12.2 DFL framework

1: $(\mathbf{H}, \mathbf{H}) \leftarrow$ CSI sample for local training
2: currentModel $\leftarrow$ initialModel()
3: **loop**
4:     wait(Δ)
5:     $p \leftarrow$ select()
6:     send currentModel to p
7: **end loop**
8: **procedure** ONRECEIVEMODEL(m)
9:     m.updateModel(($\mathbf{H}, \mathbf{H}$))
10: **end procedure**

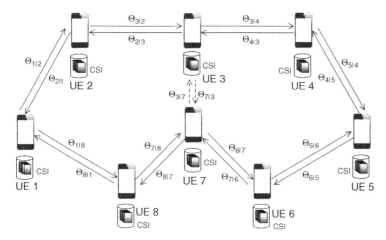

Figure 12.6 Illustration of the contact network among the neighbored UE.

performance is dependent on the D2D communication network topology, merge frequency, and merge function.

12.3.2.2 Merge Function

The merge technology is a key factor in learning performance. A commonly used technology is directly averaging the NN weights, which is simple and efficient. The detailed workflow is shown in Algorithm 12.3. Assuming that the ith UE is the local one and the jth UE sends its NN weight to the ith UE, the new NN weight is the weighted mean of the local and received NN weights, that is, Θ_i and $\Theta_{i|j}$, respectively, as well as the NN model age t (number of updates). Therefore, the updated NN weights can be expressed as follows:

$$\Theta \leftarrow (1-a)\Theta_i + a\Theta_{i|j}, \tag{12.11}$$

where $a = \frac{t_{i|j}}{t_i + t_{i|j}}$. If the t_i and $t_{i|j}$ are the same, then the new NN weight is the mean of the local and received ones.

Algorithm 12.3 Average merge function

1: **procedure** MERGEAVERAGE$((t_i, \Theta_i), (t_{i|j}, \Theta_{i|j}))$

2: $\quad a \leftarrow \frac{t_{i|j}}{t_i + t_{i|j}}$

3: $\quad t \leftarrow \max(t_i, t_{i|j})$

4: $\quad \Theta \leftarrow (1-a)\Theta_i + a\Theta_{i|j}$

5: $\quad$ **return** (t, Θ)

6: **end procedure**

12.3.2.3 Network Connectivity Architecture

The network topologies also play an important role in improving the performance of DFL. If each UE can exchange its local model with all other UE in the topology, then the DFL framework can be regarded as a special FL framework, in which each UE is equivalent to a central server. In fact, due to channel conditions limitations, each UE can only exchange its model with partial random UEs via D2D communication (Giaretta and Girdzijauskas [2019]). This part provides four different D2D network topologies, including line, ring, multiplex ring, and full mesh. The number of UEs contacted by D2D communications is different in these topologies. Figure 12.7a presents the line architecture, where the UE on the end only exchanges the NN weight with its single neighbor and other UEs contact with two neighbors. The ring architecture is shown in Figure 12.7b, in which each UE can exchange the NN weight with two UE. Figure 12.7c shows the multiplex ring architecture, where each UE contacts with four UE. The full-mesh architecture is presented in Figure 12.7d, in which each UE can exchanges NN weight Θ with all other UE.

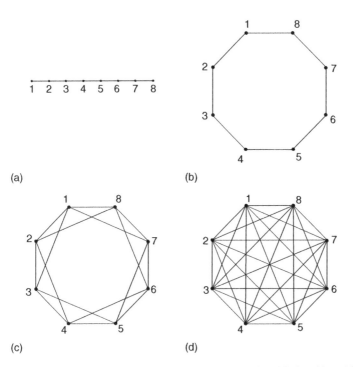

(a)   (b)

(c)   (d)

Figure 12.7 Four different network topologies. (a) Line, (b) ring, (c) multiplex ring, and (d) full mesh.

12.4 Numerical Results

12.4.1 Simulation Setting

12.4.1.1 Channel Generation

When evaluating the above frameworks, QuaDRiGa software (Jaeckel et al. [2014]) is adopted to generate channel datasets. The 3GPP urban macro (UMa) non-line-of-sight (NLOS) scenario, that is, 3GPP_38.901_UMa_NLOS channel model is considered. The carrier frequency is set as 2.655 GHz, and the considered subcarrier number is 64. The system bandwidth is 20 MHz. The number of antennas at the BS is set to $N_t = 32$, and the UE is equipped with an omnidirectional receive antenna. The cluster number N_c is set as 5, and each cluster has $N_s = 10$ subpaths. The considered location distribution of the UE when training the generalized AE is randomly located in a 40 m × 40 m square area. The location of the BS is fixed. The considered UE number is 10,000. The total datasets are divided into training and testing datasets, with 8000 and 2000 UE in each dataset.

12.4.1.2 Training Details

All experiments are conducted on TensorFlow 2.4.0. Both single-user and multi-user user-centric online training are sequentially based on a generalized AE, which is trained as a baseline. When training the generalized AE, the Adam algorithm is used to update the NN weights. The learning rate, batch size, and training epoch are set as 0.001, 256, and 300, respectively. The details of the training strategy used during online training are given later. The normalized mean square error (NMSE) is used to evaluate the performance of the above methods, which can be expressed as

$$NMSE = \mathbb{E} \left\{ \frac{\|\hat{\mathbf{H}} - \mathbf{H}\|_2^2}{\|\mathbf{H}\|_2^2} \right\}, \tag{12.12}$$

where $\| \cdot \|_2$ is the Euclidean norm.

12.4.2 Performance of User-Centric Online Training

The performance of user-centric online training is tested in an office scenario, where the UE moves in a 2 m × 2 m or 6 m × 6 m area for a long time. The training and testing CSI samples are both 800 for user-centric online training. The learning rate, batch size, and epoch are set as 0.01, 256, and 200, respectively.

The NMSE under different CRs is presented in Figure 12.8. For the two considered areas, editing CSI before the encoder only provides a slight gain for the pretrained AE. Besides, the two other online training frameworks show better CSI reconstruction performance obviously than the baseline AE. With the increase

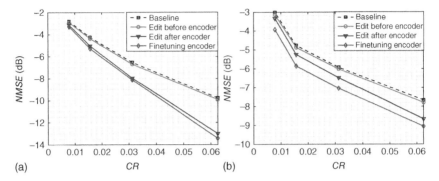

Figure 12.8 *NMSE* (dB) under different CRs when the UE moves in a 2 m × 2 m (a) or 6 m × 6 m (b) square area.

of *CR*, the CSI reconstruction gains achieved by these online training methods increase gradually. However, the NMSE gap between editing after encoder and finetuning encoder is different in areas of different sizes, as shown in Figure 12.8. When the UE moves in the large area in Figure 12.8b, the gap is larger than that in the stable area in Figure 12.8a. The reason is that the channel characteristic is complex in a large area. Therefore, the simple NNs cannot compress the original CSI efficiently. For the method editing after the encoder, only a simple FC layer is adopted to change the compressed codewords, which cannot extract all CSI features. However, the method of finetuning encoder can drastically change the compressed codewords, which makes the codewords more adaptive to the pretrained decoder. The above results show the performance benefits of the single-user user-centric online training methods.

12.4.3 Performance of Multiuser Online Training Using DFL

The feedback performance of the DFL-aided multiuser user-centric online training is evaluated in this section. Finetuning encoder is adopted to conduct the subsequent DFL-aided experiments. There are assumed to be 10 UE in the topologies, and they use the above online training methods to train their local models, respectively. During training, the epoch number is 100, and the learning rate is set as 0.01. As Figure 12.7 shows, the UE can randomly exchange the local training model with timestamps to selected UEs through D2D communication to realize the sharing of channel knowledge.

12.4.3.1 Performance Evaluation of DFL framework

Table 12.1 presents the NMSEs of the methods which use online learning. The gossip number in the table presents the D2D communication frequency. When the

Table 12.1 *NMSE* in dB of multiuser user-centric online training using DFL.

CR	Sample number	Gossip number	Original	Line	Ring	Multiplex ring	Full mesh
1/16	400	5	−13.5	−14.8	−15.0	−15.5	**−15.8**
		10		−15.4	−15.5	−15.8	**−16.0**
	800	5	−14.4	−15.8	−15.8	−16.2	**−16.4**
		10		−16.0	−16.1	−16.3	**−16.5**
1/32	400	5	−7.6	−8.8	−8.9	−9.1	**−9.3**
		10		−9.0	−9.0	−9.3	**−9.4**
	800	5	−8.2	−9.2	−9.3	−9.5	**−9.6**
		10		−9.3	−9.4	−9.6	**−9.7**

gossip number is 5, the UE contacts with selected UE and shares the networks five times during the entire training process. The "Original" represents the user-centric online training method. The NMSE of the baseline AE without online training when $CR = 1/16$ and $CR = 1/32$ are −13.9 and −7.7 dB, respectively. The simulation is conducted when the gossip number is 5 and 10, respectively. Besides, the considered CSI samples for training are 400 and 800, respectively.

Considering that the CSI samples and gossip number are 400 and 5, respectively. As the table shows, the user-centric online training method does not show better performance than the baseline AE which is without online training. For example, the NMSEs of the baseline AE and online training methods are −7.7 and −7.6 dB, respectively, when the CR is 1/32. That is because the limited datasets will make the NN easy to become overfitting. However, the DFL-aided methods which adopt arbitrary network topologies can improve the feedback accuracy dramatically. For example, the line network topology can reduce the NMSE by about 1 dB when CR is 1/32, which indicates the potential of DFL-aided methods in feedback. In addition, the feedback accuracy can be substantially improved if using more training datasets.

In addition, as Figure 12.7 shows, the full mesh network topology is a special FL framework where the UE can be regarded as a central server. Therefore, the performance of the full mesh will outperform the other three topologies, and the feedback performance increases with the increase of the network topology complexity. Table 12.1 compares the NMSEs of the four topologies. As the bold values indicate, the full mesh network topology shows the lowest NMSE. Moreover, the feedback accuracy increases with the increase of D2D communication links.

12.4.3.2 NMSE Under Different Gossip Numbers

The gossip number also has an important influence on feedback accuracy. To reduce the communication overhead, the gossip number should be as few as possible. Thus, the NMSEs of the DFL-aided methods are compared under different gossip numbers. The training epoch is set as 100, and two different gossip numbers are tested. As is shown in Table 12.1, the NMSEs with the gossip number 10 are less than that of 5. Therefore, the feedback accuracy can be improved by introducing the additional communication overhead.

12.4.3.3 NMSE Under Different Numbers of UE Participation

Table 12.2 compares the NMSE of the DFL-aided methods under different UE numbers. As the table shows, with the increase of the UE number, the CSI reconstruction accuracy increases. Besides, under large CRs, the gap between the methods with different UE numbers is obvious. However, the feedback accuracy has a slight gain when the training datasets are sufficient. The reason is that the DFL method can obtain sufficient knowledge from the training datasets.

12.4.3.4 NMSE Under Mismatched Channels

To further evaluate the generalization of the DFL-aided method, a new $6\,m \times 6\,m$ area, in which the CSI samples are not involved in the training phase, is considered. The UE number, training samples number, and Gossip times are set as 10, 800, and 10, respectively. The NMSE under different CRs is plotted in Figure 12.9. As expected, the DFL-aided methods can further improve the feedback performance remarkably. Besides, the simulation results also show that the NMSE gap between different topologies is not obvious under sufficient

Table 12.2 *NMSE* in dB of multiuser user-centric online training using DFL with different UE numbers.

CR	Sample number	UE number	Original	Line
1/16	400	8	−13.5	−15.2
		10		**−15.4**
	800	8	−14.4	−15.9
		10		**−16.0**
1/32	400	8	−7.6	−8.9
		10		**−8.9**
	800	8	−8.2	−9.3
		10		**−9.3**

Figure 12.9 *NMSE* (dB) against *CR* when the UE moves to a new 6 m × 6 m area.

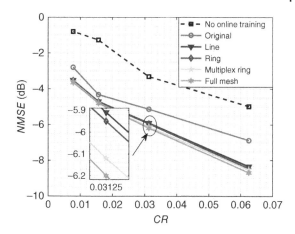

training datasets. This point indicates the importance of the number of training samples to this DFL-aided method.

12.5 Conclusion

This chapter introduces the DFL-aided user-centric online method for AE-based CSI feedback. First, the background and basic idea of AE-based CSI feedback are introduced. Then, three user-centric online learning frameworks for AE-based CSI feedback are introduced to improve the CSI reconstruction accuracy. These frameworks can train different AEs for different areas to enhance the performance of the general AE-based method. To further improve the performance achieved by the above method, the online learning method is generalized to a multiuser scenario. A DFL-aided user-centric online training strategy without BS participation is introduced. Compared with single-user online training, the feedback accuracy can be improved through multiuser cooperation and exploiting crowd intelligence.

Bibliography

Ahmet M. Elbir, Anastasios K. Papazafeiropoulos, and Symeon Chatzinotas. Federated learning for physical layer design. *IEEE Communications Magazine*, 59(11):81–87, 2021. doi: 10.1109/MCOM.101.2100138.

Lodovico Giaretta and Sarunas Girdzijauskas. Gossip learning: Off the beaten path. In *2019 IEEE International Conference on Big Data (Big Data)*, pages 1117–1124, 2019. doi: 10.1109/BigData47090.2019.9006216.

Tiansheng Guo, Jing Wang, Ze Cui, Yihui Feng, Yunying Ge, and Bo Bai. Variable rate image compression with content adaptive optimization. In *Proceedings of the IEEE/CVF Conference on Computer Vision and Pattern Recognition (CVPR) Workshops*, pages 533–537, 2020. doi: 10.1109/CVPRW50498.2020.00069.

Jiajia Guo, Chao-Kai Wen, Shi Jin, and Geoffrey Ye Li. Overview of deep learning-based CSI feedback in massive MIMO systems. *IEEE Transactions on Communications*, 70(12):8017–8045, 2022. doi: 10.1109/TCOMM.2022.3217777.

Gang He, Chang Wu, Lei Li, Jinjia Zhou, Xianglin Wang, Yunfei Zheng, Bing Yu, and Weiying Xie. A video compression framework using an overfitted restoration neural network. In *Proceedings of the IEEE/CVF Conference on Computer Vision and Pattern Recognition (CVPR) Workshops*, pages 593–597, 2020. doi: 10.1109/CVPRW50498.2020.00082.

István Hegedűs, Gábor Danner, and Márk Jelasity. Gossip learning as a decentralized alternative to federated learning. In *IFIP International Conference on Distributed Applications and Interoperable Systems*, pages 74–90, 2019. doi: 10.1007/978-3-030-22496-75·.

Stephan Jaeckel, Leszek Raschkowski, Kai Börner, and Lars Thiele. QuaDRiGa: A 3-D multi-cell channel model with time evolution for enabling virtual field trials. *IEEE Transactions on Antennas and Propagation*, 62(6):3242–3256, 2014. doi: 10.1109/TAP.2014.2310220.

Ping-Heng Kuo, H. T. Kung, and Pang-An Ting. Compressive sensing based channel feedback protocols for spatially-correlated massive antenna arrays. In *2012 IEEE Wireless Communications and Networking Conference (WCNC)*, pages 492–497, 2012. doi: 10.1109/WCNC.2012.6214417.

Erik G. Larsson, Ove Edfors, Fredrik Tufvesson, and Thomas L. Marzetta. Massive MIMO for next generation wireless systems. *IEEE Communications Magazine*, 52(2):186–195, 2014. doi: 10.1109/MCOM.2014.6736761.

David J. Love, Robert W. Heath, Vincent K. N. Lau, David Gesbert, Bhaskar D. Rao, and Matthew Andrews. An overview of limited feedback in wireless communication systems. *IEEE Journal on Selected Areas in Communications*, 26(8):1341–1365, 2008. doi: 10.1109/JSAC.2008.081002.

Lu Lu, Geoffrey Ye Li, A. Lee Swindlehurst, Alexei Ashikhmin, and Rui Zhang. An overview of massive MIMO: Benefits and challenges. *IEEE Journal of Selected Topics in Signal Processing*, 8(5):742–758, 2014. doi: 10.1109/JSTSP.2014.2317671.

Zhijin Qin, Jiancun Fan, Yuanwei Liu, Yue Gao, and Geoffrey Ye Li. Sparse representation for wireless communications: A compressive sensing approach. *IEEE Signal Processing Magazine*, 35(3):40–58, 2018. ISSN 1558-0792. doi: 10.1109/MSP.2018.2789521.

Zhijin Qin, Hao Ye, Geoffrey Ye Li, and Biing-Hwang Fred Juang. Deep learning in physical layer communications. *IEEE Wireless Communications*, 26(2):93–99, 2019. doi: 10.1109/MWC.2019.1800601.

Mehran Soltani, Vahid Pourahmadi, Ali Mirzaei, and Hamid Sheikhzadeh. Deep learning-based channel estimation. *IEEE Communications Letters*, 23(4):652–655, 2019. doi: 10.1109/LCOMM.2019.2898944.

Hossein Talebi, Damien Kelly, Xiyang Luo, Ignacio Garcia Dorado, Feng Yang, Peyman Milanfar, and Michael Elad. Better compression with deep pre-editing. *IEEE Transactions on Image Processing*, 30:6673–6685, 2021. doi: 10.1109/TIP.2021.3096085.

Stefanie Warnat-Herresthal, Hartmut Schultze, Krishnaprasad Lingadahalli Shastry, Sathyanarayanan Manamohan, Saikat Mukherjee, Vishesh Garg, Ravi Sarveswara, Kristian Händler, Peter Pickkers, N. Ahmad Aziz, et al. Swarm learning for decentralized and confidential clinical machine learning. *Nature*, 594:265–270, 2021. doi: 10.1038/s41586-021-03583-3.

Chao-Kai Wen, Shi Jin, Kai-Kit Wong, Jung-Chieh Chen, and Pangan Ting. Channel estimation for massive MIMO using Gaussian-mixture Bayesian learning. *IEEE Transactions on Wireless Communications*, 14(3):1356–1368, 2015. doi: 10.1109/TWC.2014.2365813.

Chao-Kai Wen, Wan-Ting Shih, and Shi Jin. Deep learning for massive MIMO CSI feedback. *IEEE Wireless Communications Letters*, 7(5):748–751, 2018. doi: 10.1109/LWC.2018.2818160.

Yuwen Yang, Feifei Gao, Zhimeng Zhong, Bo Ai, and Ahmed Alkhateeb. Deep transfer learning-based downlink channel prediction for FDD massive MIMO systems. *IEEE Transactions on Communications*, 68(12):7485–7497, 2020. doi: 10.1109/TCOMM.2020.3019077.

Hao Ye, Geoffrey Ye Li, and Biing-Hwang Juang. Power of deep learning for channel estimation and signal detection in OFDM systems. *IEEE Wireless Communications Letters*, 7(1):114–117, 2018. doi: 10.1109/LWC.2017.2757490.

Index

Federated Learning for Future Intelligent Wireless Networks, First Edition.
Edited by Yao Sun, Chaoqun You, Gang Feng, and Lei Zhang.
© 2024 The Institute of Electrical and Electronics Engineers, Inc. Published 2024 by John Wiley & Sons, Inc.

Printed and bound by CPI Group (UK) Ltd, Croydon, CR0 4YY

16/04/2025

14658573-0001